Manual of Water Supply Practices

M56

Nitrification Prevention and Control in Drinking Water

Third Edition

Errata incorporated April, 2026

American Water Works Association

Manual of Water Supply Practices—M56, Third Edition

Nitrification Prevention and Control in Drinking Water

Sr. Editorial Manager, Books: Suzanne Snyder
Manager – Manuals: Molly Beach
Manager – Publishing Operations: Gillian Wink
Specialist – Copyright and Permissions: Peggy Tyler
Director of Publishing: John Fedor
Technical Editor: Anne Lanctôt
Cover Image: Philipp Turr/Shutterstock.com
Production: Innodata

Library of Congress Cataloging-in-Publication Data
CIP data pending from the Library of Congress.

ISBN 978-1-64717-219-0 ISBN, electronic 978-1-61300-744-0
http://doi.org/10.12999/AWWA.M56ed3

American Water Works Association

American Water Works Association
6666 West Quincy Avenue
Denver, CO 80235-3098
awwa.org

Contents

This page intentionally blank.

Figures

Tables

Preface/How to Use This Manual of Practice

A group of volunteers from the Inorganic Contaminants Research and Distribution System Water Quality Committees of AWWA updated the third edition of this manual of practice. The need for a manual on nitrification prevention and control results from the increased use of chloramine as a residual disinfectant in drinking water distribution systems and the ubiquitous presence of nitrifying bacteria in the environment. Consequently, all chloraminating systems and systems with source water ammonia need to consider nitrification prevention and control.

The manual is organized into two main parts: Chapters 1–5 provide background information on the occurrence and microbiology of nitrification in various water environments. Chapters 7–9 are intended to provide current practical approaches to nitrification prevention and response. Each chapter can be read independently; therefore, there is some limited repetition between the chapters to provide necessary background of important concepts or to reference another chapter. Some distribution system maintenance techniques may also be used for multiple purposes. For example, breakpoint chlorination or flushing can be used both as nitrification prevention and response methods; therefore, they are discussed in multiple chapters. However, one subject is discussed in detail only in the most appropriate chapter, and typically shorter discussion is provided elsewhere. The index can be helpful in finding information on a subject of interest presented in different context throughout the manual.

Several nitrification prevention methods are commonly used for regular distribution system maintenance. For example, cleaning, flushing, and reduction of water age have been thoroughly discussed in other AWWA and Water Research Foundation resources. Interested readers should refer to those references. These distribution system maintenance practices are discussed in this manual specifically from the nitrification point of view.

The materials included herein provide a compendium of the state-of-the-art knowledge as of the writing of this manual. Several new advances are being made, such as

- development of molecular methods for identification of nitrifying bacteria,
- improvement of reservoir mixing techniques,
- better understanding of the effects of booster chlorination and booster chloramination, and
- investigation of chlorite as a nitrification-prevention measure.

These advances will have an impact on the application of chloramine and nitrification control. Therefore, the present manual will likely be updated in the future to reflect these advances and improved understanding of nitrification in drinking water distribution systems.

As this is the third edition of AWWA Manual M56, Nitrification Prevention and Control in Drinking Water, the Water Quality and Technology Division's Distribution System Water Quality Committee and AWWA welcome comments and suggestions for improving future editions of this manual. Please send an e-mail attachment to the AWWA Engineering and Technical Services Department at *etssupport@awwa.org* or hard copy correspondence to the AWWA ETS Department, 6666 West Quincy Avenue, Denver, CO 80235.

This page intentionally blank.

Acknowledgments

The AWWA Technical & Educational Council, the Water Quality & Technology Division, and the Distribution System Water Quality Committee gratefully acknowledge the contributions made by those volunteers who drafted, edited, and provided the significant and critical commentary essential to updating M56.

Co-Chairs

Kimberly Gupta, Portland Water Bureau, Portland, Ore.
Amlan Ghosh, Corona Environmental, Flower Mound, Tex.

Chapter Leads of the Third Edition

Katherine Alfredo, University of South Florida, Tampa, Fla.
Hélène Baribeau, California Water Resources Control Board, San Diego, Calif.
Helen Buse, US Environmental Protection Agency, Cincinnati, Ohio
Amlan Ghosh, Corona Environmental, Flower Mound, Tex.
Tarrah Henrie, California Water Service Company, San Jose, Calif.
Christopher Hill, AECOM, Tampa, Fla.
Alexander Mofidi, Confluence Engineering Group, Seattle, Wash.
Ashley Pifer, Garver, Fayetteville, Ark.
Susan Teefy, East Bay Municipal Utility District, Oakland, Calif.

Steering Committee Members

Amlan Ghosh, Corona Environmental, Flower Mound, Tex.
Kimberly Gupta, Portland Water Bureau, Portland, Ore.
Mary Jo Kirisits, University of Texas, Austin, Tex.
Quirien Muylwyk, AECOM, Toronto, Ont.
Kira Smith, US Environmental Protection Agency, Washington, D.C.
Josh Fortenbaugh, AWWA, Denver, Colo.

Contributors to the Third Edition

Danny Allison, City of Portland, Portland, Ore.
Melinda Friedman, Confluence Engineering Group, Seattle, Wash.
Richard Giani, CDM Smith, Tuscaloosa, Ala.
Robin Giguere, cleanwater1, Milpitas, Calif.
Kimberly Gupta, Portland Water Bureau, Portland, Ore.
Rob Harvey, Hazen and Sawyer, Baltimore, Md.
Stephen Hubbs, Retired, Louisville, Ky.
Mary Jo Kirisits, University of Texas, Austin, Tex.
Darren Lytle, US Environmental Protection Agency, Cincinnati, Ohio
Quirien Muylwyk, AECOM, Toronto, Ont.
Kimaya Rajwade, American Water Works Association, Denver, Colo.
John Regan, Pennsylvania State University, University Park, Pa.

Sarah Richardson, San Jose Water Company, San Jose, Calif.
Meg Roberts, Hazen and Sawyer, Greensboro, N.C.
Virpi Salo-Zieman, Confluence Engineering Group, Seattle, Wash.
Meric Selbes, Hazen and Sawyer, Fairfax, Va.
Rebecca Slabaugh, Arcadis, Indianapolis, Ind.
Kira Smith, US Environmental Protection Agency, Washington, D.C.
Kenneth Thompson, Retired, Greenwood Village, Colo.
Matt Umberg, US Environmental Protection Agency, Cincinnati, Ohio
David Wahman, US Environmental Protection Agency, Cincinnati, Ohio
Shannon Wedding, Ferguson Waterworks, Redmond, Ore.
Andrzej Wilczak, San Francisco Public Utilities Commission, San Francisco, Calif.
Xueqing (Eric) Zhu, Louisville Water Company, Louisville, Ky.

Chapter 1

Introduction to Nitrification and Potential Impacts on Regulatory Compliance

Katherine Alfredo, Steve Hubbs, Virpi Salo-Zieman, Melinda Friedman

INTRODUCTION

Nitrification is the microbial oxidation of reduced nitrogen compounds such as ammonia to nitrite and nitrate. Nitrification is an important part of the nitrogen cycle on the Earth but problematic when occurring in drinking water systems. Over the past few years, federal efforts to identify and prioritize distribution system water quality issues have included nitrification as a possible public health issue. The National Research Council's (NRC) assessment of distribution system risks determined that "the most important problem exacerbated by both nitrification and by long retention times is loss of disinfectant residuals" (NRC 2006). The NRC designated this to be a medium-priority concern due to indirect health impacts caused by the loss of residual, which may result in compromising biological integrity of the system and promoting microbial regrowth (NRC 2006). In the same report, the formation of nitrate and nitrite by nitrification was assigned as a low priority relative to other distribution system water quality issues. Finally, the NRC also noted that nitrification in low alkalinity waters can cause pH to decrease and affect lead and copper release. Water utilities across the United States have struggled with minimizing nitrification within their distribution system while still maintaining compliance with competing regulations. Nitrification may also occur in premise plumbing and is not as easily observed, monitored, or controlled as in the distribution system, yet can have significant impacts on the water quality at taps (Zhang et al. 2009b). The results from a recent study of water quality in a green building (Salehi et al. 2020) suggested seasonal nitrification occurred in the hot water plumbing supplied by a system that maintained a free chlorine residual but had seasonal free ammonia concentrations present.

This manual concentrates on nitrification that can happen in potable water systems using chloramination for secondary disinfection. The objectives of this manual are to

- summarize existing knowledge and current practices of water suppliers related to nitrification;
- provide water utilities with the latest information on nitrification in water distribution systems;
- provide resources, tools, and best management practices to help utilities maintain a chloramine residual and overall water quality in tap water; and
- provide resources, tools, and best management practices to help utilities detect, prevent, and mitigate nitrification episodes that may occur in their systems.

There are many sources that discuss the various aspects of nitrification. Several Water Research Foundation (WRF) reports discuss nitrification in conjunction with other topics, primarily drinking water disinfection and chloramination. Since the first edition of this manual, the field has evolved, and the second edition illustrated a better understanding of chloramination treatment, chloramine residual maintenance, and nitrification detection and control within distribution systems. This third edition captures further advancements in these fields in addition to describing the significant advances made in identifying the microbiota involved in the nitrification processes and nitrogen-transforming metabolic pathways.

Another improvement in the practice of nitrification control is the use and maintenance of a slightly higher chloramine residual level throughout the distribution system that allows for more effective nitrification detection and preventive monitoring and control to encompass water age management. This third edition also emphasizes the importance of monitoring chloramine chemistry so that it includes not only total chlorine but also free chlorine, monochloramine, free ammonia, nitrite, and nitrate in the distribution system. Furthermore, this edition contains significant updates related to the microbial indicators. Monitoring a combination of parameters is critical for understanding the risk of nitrification.

The Stage 2 Disinfectants and Disinfection Byproduct Rule necessitates that utilities comply with locational disinfection byproducts maximum contaminant levels (MCLs) (USEPA 2006). The use of a chloramine residual is seen by many utilities as a possible option for distribution system disinfection while controlling chlorinated disinfection byproducts and enabling systems to maintain detectable disinfectant residuals. Many states have promulgated minimum disinfectant residual levels, necessitating that systems improve their residual management or even switch from free chlorine to chloramine. Hence, there is a need for easy-to-access information related to nitrification occurrence, prevention, and control in drinking water distribution systems consolidated in a single location.

This manual combines the theories developed during foundational research with references to scientific papers and field practices applied by water utility personnel; it also includes examples of water utility programs and practices. Experimental approaches are also discussed in some sections and are clearly identified as nonpeer reviewed results or as experimental, anecdotal, or emerging issues. The final product is a manual that can be used by those utilities currently using chloramines, as well as those considering a change from free chlorine to chloramine. This manual is also useful to consulting engineers and researchers trying to control or mitigate nitrification through operational practices, engineering improvements that reduce or eliminate conditions that promote biological growth, and treatments that inactivate nitrifying microorganisms.

The following summarizes each chapter.

- Chapter 1, "Introduction to Nitrification and Potential Impacts on Regulatory Compliance," provides background information on disinfection practices, nitrification, and the impact of nitrification on regulatory compliance and other potential public health risks.
- Chapter 2, "Nitrification in Water Treatment," provides an overview of nitrification occurrences (intentional and unintentional) in drinking water facilities and resulting impacts on distribution system water quality. This overview includes biological treatment for ammonia removal (typically groundwater) and nitrification that may occur in biofilters downstream of ozonation (typically surface water).
- Chapter 3, "Nitrification in Chloraminated Drinking Water Distribution Systems—Occurrence," summarizes mechanisms and factors related to nitrification occurrence in drinking water distribution systems. This third edition offers updated content based on recent literature.
- Chapter 4, "Overview of Causes and Control of Nitrification in Chloraminated Drinking Water Distribution Systems," provides an overview of source water quality, water treatment, operations, and maintenance conditions that contribute to or cause nitrification in the distribution system. Control strategies for nitrification—including maintaining appropriate chloramine residual, monitoring, and distribution system operations—are also summarized in this chapter. This third edition offers updated and expanded content based on recent literature.
- Chapter 5, "Microbiology of Nitrifying Microorganisms," presents information on the microbiology fundamentals of nitrifying microorganisms. This third edition includes updated information on the taxonomy, morphology, and physiology of nitrifying microorganisms, as well as techniques for their isolation and enumeration, growth conditions, and inactivation by various disinfectants. Modifications and updates to this edition also include information on the complete ammonia-oxidizers group; abundance and behavior of ammonia-oxidizing microorganisms that include ammonia-oxidizing bacteria and archaea, and nitrite-oxidizing bacteria in drinking water systems; updated phylogenetic trees and Monod growth kinetics; and interactions between nitrifiers and other bacterial community members. This chapter also discusses next-generation sequencing and other molecular techniques to enhance our understanding of microbially mediated nitrification processes in the distribution system.
- Chapter 6, "Monitoring for Nitrification Prevention and Control," discusses water quality monitoring plans and programs to evaluate nitrification. This chapter also provides key monitoring parameters, monitoring locations, and monitoring frequencies for detecting and predicting nitrification.
- Chapter 7, "Operational and Treatment Practices to Prevent Nitrification," describes treatment and operational practices designed to prevent nitrification and makes recommendations for best practices. Strategies like booster chloramination, impact of high pH, Cl_2:N ratios, controlling water age, application of chlorite ions for nitrification control in the distribution system, and ultraviolet (UV) light to control nitrification in storage facilities are covered in this chapter with several utility examples.
- Chapter 8, "Assessment and Operational Responses to Nitrification Episodes," provides guidelines to assess whether nitrification is occurring, including water

quality indicators of nitrification. This chapter also provides recommendations on operational response to nitrification in the distribution system and finished water storage facilities and the development of a nitrification response plan.

- Chapter 9, "Engineering Changes and Capital Improvements for Nitrification Prevention," discusses nitrification prevention and control methods that are related to engineering practices and capital improvements. The methods discussed in this chapter focus on decreasing water age (e.g., improving storage tank and distribution system piping designs), homogenizing storage tank water (e.g., using active mixers), and directly adjusting water quality (e.g., using chloramine boosting systems). Such improvements require more planning, time, and financial resources than the monitoring and operational prevention methods discussed in earlier chapters.

Each chapter begins with an introduction and summary of key points, which are designed to help the reader in reviewing the contents of the chapter. Table 1-1 presents the summary of key points derived from Chapter 1.

Table 1-1 Key points from Chapter 1

Background information	• Free chlorine and chloramine are two disinfectants used in the distribution system; each has advantages and disadvantages. Free chlorine provides a strong disinfectant residual but reacts with organic matter to form disinfection byproducts. In general, chloramine is a weaker oxidant than free chlorine but provides a more stable residual and reduces the potential for trihalomethanes (THMs) and haloacetic acids (HAAs) formation. • The use of chloramine in the United States started in the early 1920s, but usage decreased due to ammonia shortages during WWII. Renewed interest in chloramination occurred after the introduction of total THM regulations by the US Environmental Protection Agency (USEPA) and the potential to reduce THM and HAA formation by using chloramines. • Because chloramines provide a longer-lasting residual, the desire to better control *Legionella* in building plumbing systems and more stringent disinfectant residual requirements have further increased the use of chloramination as a secondary disinfectant strategy.
Nitrification basics	• The nitrogen cycle, as it occurs in nature, consists of many biological reversible transformations of nitrogen among ammonia, nitrite, nitrate, cellular organic nitrogen, and nitrogen gases. • Ammonia nitrogen is converted to chloramine nitrogen at the point of chloramine formation. As the chloramine autodecomposes in the distribution system, the combined N converts to NH_3-N. • Nitrification is an oxidation process carried out by microorganisms in which reduced nitrogen compounds (primarily ammonia) are sequentially oxidized to nitrite and nitrate or directly to nitrate.
Nitrification and regulatory compliance and public health	• In the US Safe Drinking Water Act, primary maximum contaminant levels (MCLs) for nitrite and nitrate at the entry to the distribution system are 1 mg/L and 10 mg/L as N, respectively, and both compounds are regulated as acute contaminants with potential for immediate adverse health impacts. Currently there are no regulations or MCLs for nitrate or nitrite within the distribution system. If these MCLs were applied to locations in the distribution system, it is possible to exceed the nitrite MCL during severe nitrification episodes. • Nitrification may lead to violation of the USEPA Surface Water Treatment Rule (SWTR) for lack of detectable disinfectant residual or the Revised Total Coliform Rule due to increased microbiological activity and the possibility of coliform growth. Furthermore, many states have set a minimum disinfectant residual concentration above the detectable concentration as required by the SWTR. • Nitrification may impact USEPA Lead and Copper Rule compliance due to reduction in pH and alkalinity and subsequent impacts to lead and copper solubility.

DISTRIBUTION SYSTEM DISINFECTION PRACTICES

The practice of disinfecting drinking water and carrying a disinfectant residual throughout the distribution system began in the early 1900s to provide a barrier to waterborne diseases such as typhoid fever, cholera, and dysentery. Disinfectants used for distribution system residuals in the United States are primarily free chlorine and chloramines. Chlorine dioxide is also approved for use but is not widely used (USEPA 2009, AWWA Disinfection Systems Committee 2008a, 2017). Free chlorine and chloramines are effective at destroying some pathogenic microorganisms, controlling the growth of microorganisms in the distribution system; therefore, residual chlorine levels are used as an indicator of water quality integrity within the distribution system (Friedman et al. 2010).

The application of secondary disinfection (disinfectant applied with the objective of providing a residual throughout the drinking water distribution system) serves several objectives.

- To control the growth of microorganisms and opportunistic pathogens in the distribution system: Maintaining an adequate disinfectant residual in the distribution system, in addition to other measures, can help prevent or minimize the growth of microorganisms and opportunistic pathogens such as *Escherichia coli* in the distribution system or *Legionella* spp in premise plumbing systems.
- To protect against contamination occurring through external sources: Microbial contamination of the distribution system may occur through a cross-connection and during main breaks and main repairs. Contamination may also occur from intrusion due to pressure transients that may allow contaminants to enter the distribution system when the pressure within the distribution system is lower than the pressure external to the distribution system (backflow) and from potential microbial contamination of distribution system storage tanks and reservoirs.
- To limit biofilm growth: Disinfectant residuals help minimize biofilm growth on pipe and tank walls, especially in areas with high water age or experiencing water stagnation due to distance from the distribution system entry point, system dead-ends, or storage tank hydraulics, provided that the disinfectant residual can be maintained in these locations.
- To maintain oxidizing conditions: Disinfectant residuals, along with any remaining dissolved oxygen, provide oxidizing environments inside the distribution mains, which is critical for maintaining scale stability and controlling metals release from the pipeline deposits. Highly oxidizing conditions, such as those found in some systems using free chlorine as a secondary disinfectant, may also promote the formation of less soluble lead scales (i.e., Pb^{4+}), reducing the amount of lead at the customers' taps. Chloramine residuals are not strong enough oxidants to form Pb^{4+}.
- To control the development of tastes and odors: Drinking water or tap water is expected to have good aesthetic quality. By minimizing biological growth, the development of biologically formed tastes and odors is also controlled. However, chlorine and chloramine can impart tastes and odors to the water that are deemed objectionable by some customers. In chloramine systems, especially those practicing annual free chlorine conversions, the control of chloramine speciation (monochloramine, dichloramine, trichloramine) is particularly important in controlling objectionable disinfectant-related taste issues.
- To act as a water quality indicator for water quality deterioration and distribution system integrity: A sudden or unusual reduction in the level of disinfectant residual may indicate a contamination event within the distribution system. Similarly, a loss in residual over time is useful in identifying areas susceptible to long water ages and/or active corrosion.

HISTORY OF CHLORAMINATION IN THE UNITED STATES

As reported by Kirmeyer et al. (2004), in the early 1900s, chloramines were found to be effective in destroying pathogenic organisms and were easy to use and cost-effective. The Denver Union Water Company used chloramines as early as 1917 to prevent bacteriological growth problems, and in Canada, a water-treatment facility in Ottawa used them in 1918. Between 1920 and 1936, chloramines gained popularity in water treatment because they were more stable or longer lasting than free chlorine and caused fewer taste and odor problems compared to free chlorine. By the end of the 1930s, a survey of 2,541 water utilities in 36 states found that 16 percent used chloramines. However, following the discovery of breakpoint chlorination in 1939 and due to an ammonia shortage during World War II in the 1940s, the use of chloramines dropped (White 1999). A survey of 11,500 municipal water supplies in 1962 indicated use dropped from 16 percent to 2.6 percent (Kirmeyer 2004).

Following the discovery of THMs, a group of halogenated DBPs, and the promulgation of the Total Trihalomethane Rule in 1979 (USEPA 1979), more water utilities began using chloramines, which form substantially lower levels of chlorinated DBPs compared to free chlorine. USEPA classifies most of the regulated THMs as likely carcinogenic to humans; bromodichloromethane, bromoform, and chloroform are classified with the cancer descriptor "L," which means "likely to be carcinogenic to humans," and dibromochloromethane is classified as "S," which means there is "suggestive evidence of carcinogenic potential." (USEPA 2018). They are formed when free chlorine reacts with natural organic matter (NOM) in source water. In the distribution system, the ongoing reaction of free chlorine with residual NOM produces additional THMs.

By the 1990s, chloramines were being used in approximately 20 percent of the water treatment facilities in the United States, mainly for controlling DBP levels in the distribution system (Kirmeyer et al. 2004), with the application of free chlorine during water treatment for primary disinfection.

The use of chloramines has become more popular due to the USEPA Stage 1 and Stage 2 Disinfectants/Disinfection Byproducts (D/DBP) Rules (USEPA 2006; USEPA 2001). These rules regulate the levels of various DBPs in drinking water. Compliance of the chlorination-based DBPs—the sum of four THMs and the sum of five haloacetic acids (HAAs)—is based on locational annual averages rather than system-wide averages. The USEPA forecasted that as a result of D/DBP Stage 2 implementation alone, 8 percent of surface water and 3 percent of groundwater community water systems would convert to chloramination (USEPA 2005). A total of up to 57 percent of all surface and 7 percent of groundwater systems (USEPA 2005) were predicted to use chloramination for secondary disinfection.

Past surveys of disinfection practices have reported varying percentages of chloramine use over other disinfection methods. For instance, AWWA's 2017 (AWWA Disinfection Systems Committee 2017) survey of 375 water systems reported that 20 percent of the responding utilities use chloramine and 70 percent use free chlorine, while a survey completed in 2007 reported that 30 percent of respondents used chloramine to maintain distribution system residuals (AWWA Disinfection Systems Committee 2008b; Seidel et al. 2005). The 2007 survey suggested that 8–12 percent of the utilities were contemplating switching to chloramination, which matches closely to the 11 percent of systems that reported to have installed chloramination in the past 10 years in the 2017 survey. However, only 5 percent of the systems were contemplating switching to chloramines in the 2017 survey, potentially indicating a slowing trend in conversions. It is important to note that the same systems are not included in each survey because the utilities are not required to participate and that solicitating a diversified, representative sample based on utility characteristics was not a focus of the survey methodology.

There are signs that an increased number of smaller water systems have converted to chloramination in the wake of more stringent disinfectant residual requirements as chloramine provides less reactive, more stable disinfectant residual than free chlorine. For instance, Louisiana instituted a minimum residual of 0.5 mg/L measured as free or total chlorine (for systems that add ammonia) in 2013 to combat *Naegleria fowleri* (amoeba) growth in the distribution systems (Louisiana Department of Health 2013).

Aside from the AWWA disinfection surveys that capture less than 10 percent of the total number of community water systems, no current data were available to assess national trends in number of systems or populations served that are supplied by chloraminated water. However, Furatian (2019) estimated that one-third of the population in the United States and Canada is estimated to receive chloraminated water supplies.

CHLORAMINE FORMATION

Chloramines used for drinking water disinfection are formed by a chemical reaction between chlorine and ammonia, ideally at a weight ratio of approximately 5:1 of Cl_2:NH_3-N (chlorine-to-ammonia-N) to form the preferred monochloramine species. This is further discussed in Chapters 6 and 9 as part of the description of breakpoint chlorination. Chlorine is introduced to the water either as a gas, as a sodium hypochlorite solution (in bulk liquid or generated on site), or by dissolving calcium hypochlorite tablets. Ammonia is introduced as dry ammonium sulfate, liquid ammonium hydroxide or liquid ammonium sulfate, or by dissolving anhydrous ammonia gas into solution.

The following are generalized inorganic chloramine formation reactions (a molecular depiction of chloramine species is presented in Figure 1-1):

Monochloramine (NH_2Cl)

$$NH_3 + HOCl \rightarrow NH_2Cl + H_2O \quad \text{(Eq. 1-1)}$$

Dichloramine ($NHCl_2$)

$$NH_2Cl + HOCl \rightarrow NHCl_2 + H_2O \quad \text{(Eq. 1-2)}$$

Trichloramine (NCl_3)

$$NHCl_2 + HOCl \rightarrow NCl_3 + H_2O \quad \text{(Eq. 1-3)}$$

Monochloramine is the preferred (inorganic) chloramine species for drinking water treatment and disinfectant residual in the distribution system. Monochloramine does not normally cause significant taste and odor problems, while dichloramine and trichloramine are known to produce detectable chlorinous tastes and odors at relatively low concentrations (Kirmeyer et al. 2004). After entering the distribution system, the chloramine residual starts to decay (autodecompose) due to water age and physical conditions, such as temperature and pH. A chloramine demand can be exerted due to water quality conditions, such as presence of NOM, biofilm, microorganisms, corrosion byproducts, and nitrification byproducts. Chapter 4 contains a detailed discussion of chloramine decay and demand. As discussed in the following sections, during these processes, free ammonia is released into the water.

If the source water contains bromide, it may also react with the added free chlorine and form hypobromous acid (HOBr) and then further bromamines once ammonia is added. These reactions may produce mono-, di-, and tribromamines and bromochloramines. Bromamines are more reactive and degrade faster than chloramines (e.g., Zhang et al. 2012; Pope et al. 2006) and may also lead to rapid releases of ammonia into the water.

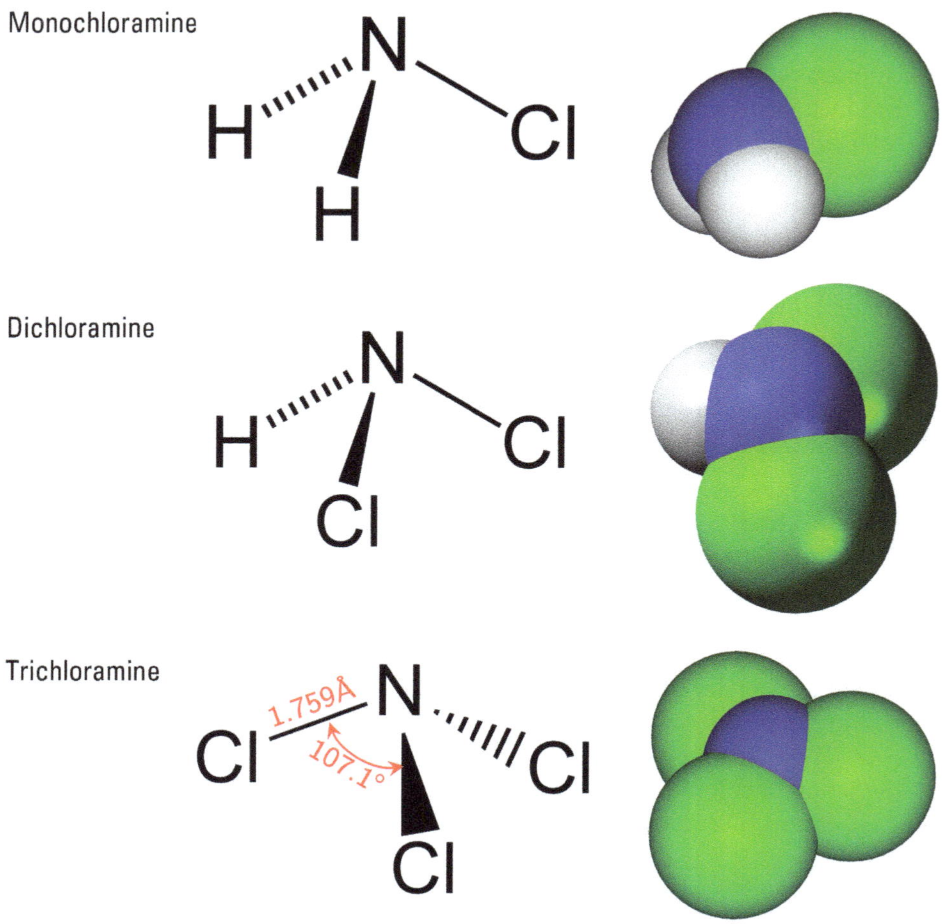

Source: Wikimedia Commons

Figure 1-1 Molecular diagrams for mono-, di-, and trichloramine

In addition to forming these inorganic chloramines, organic chloramines (also known as organochloramines) may also form when chlorine reacts with organic nitrogen-containing compounds such as amines, amides, and amino acids. Organic chloramines are likely present, but due to the complexity of the detection methods, their occurrence and levels in drinking water systems are largely unknown (How et al. 2017; Wahman and Pressman 2015). According to How et al. (2017), studies to date suggest that organic chloramines in the form of *N*-chloramines and *N*-chloramino acid formed from amines or amino acids would most likely be the species present in the drinking water system. According to the studies summarized by Wahman and Pressman (2015), organic chloramines are stable, have very little disinfecting power, continue to form in the distribution system, and will give positive signal in total chlorine measurements performed in the field.

NITRIFICATION BASICS

The Nitrogen Cycle

All biological growth processes require nitrogen for the synthesis of cellular proteins and nucleic acids. Microorganisms like ammonia-oxidizing microorganisms (AOM) that include bacteria (AOB) and archaea (AOA), nitrite-oxidizing bacteria (NOB), and complete ammonia-oxidizers (comammox) can use a range of nitrogen compounds under a variety of conditions; in some cases, the oxidation state of nitrogen is changed, while in others, it

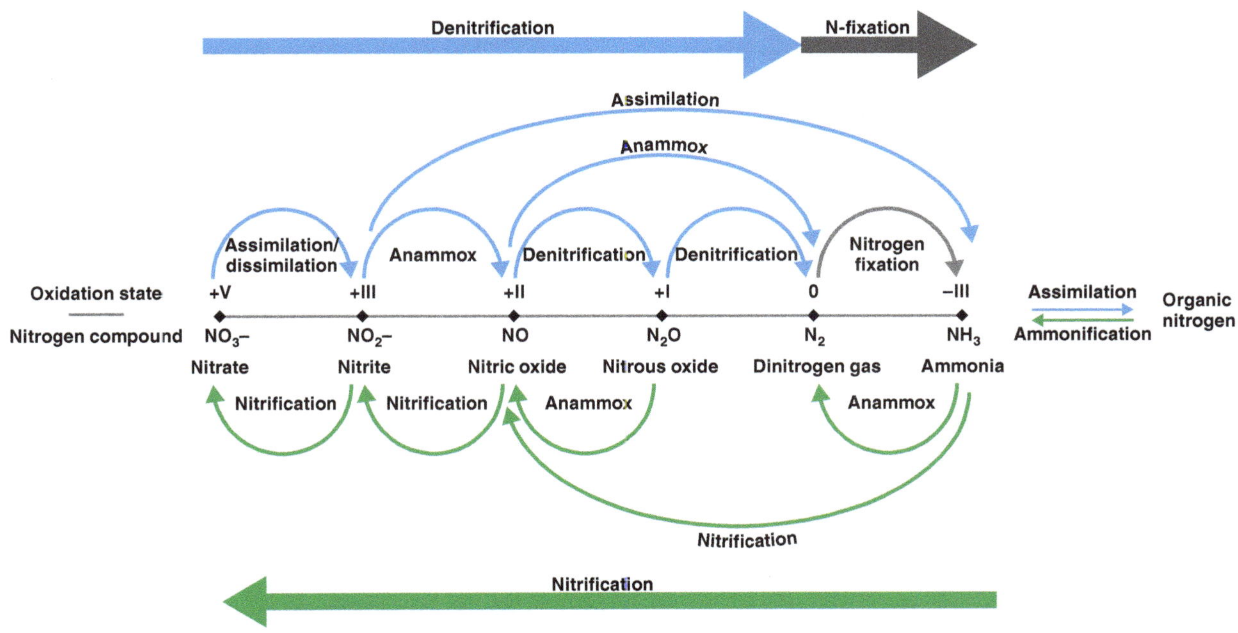

Figure 1-2 Main biological processes involving nitrogen transformation

is not. The main processes involving nitrogen transformations are shown in Figure 1-2. Together, these processes are often called the nitrogen cycle; although, as Kuypers et al. (2018) have pointed out, the transformation processes are distinct and have very different levels of nitrogen flux that do not result in one balanced nitrogen cycle.

Nitrogen fixation involves the synthesis of cellular nitrogen compounds from elemental nitrogen; such reactions commonly occur in soils, surface waters, and to a limited extent, in activated sludge wastewater processes. In assimilation, microorganisms use the oxidized nitrogen compounds such as nitrate for biosynthesis, thereby reducing the nitrogen compound. In dissimilative reactions, nitrogen compounds are used as electron acceptors for energy metabolism under anaerobic conditions and become reduced. Through various deamination reactions and due to cell lysis, organically bound nitrogen in the cells is released as ammonia (Painter 1970). Nitrification is the process of oxidizing ammonia to nitrite and further to nitrate under aerobic conditions. The microbial conversion of ammonium and nitrite to nitrogen gas (anaerobic ammonium oxidation or anammox conversion) and direct oxidation of ammonia to nitrate (comammox) are the most recent additions to our understanding of the biological nitrogen cycle (Koch et al. 2019; van Kessel et al. 2015; Kuenen et al. 2001; Strous et al. 1999). The latter discoveries have revealed novel metabolic pathways and interactions related to the nitrogen cycle, especially to the complexity of nitrification (Koch et al. 2019). All in all, nitrogen-transforming reactions and microorganisms carrying out those reactions are very diverse and occur under varying conditions. Many of these reactions may take place in drinking water distribution systems and affect the water quality.

Nitrification is carried out by a limited number of autotrophic bacteria or archaea; however, there exists a limited number of studies comparing AOB and AOA occurrence within distributions systems. One comparative study between two distribution systems found greater AOB concentrations than AOA, that there is no universal correlation between AOB and AOA concentrations within the systems, and that AOA did not possess the same potential for regrowth within the system (Scott et al. 2015). Some research suggests a selectivity can occur for specific AOB communities (Regan et al. 2002). Non-AOB nitrifying microorganisms in drinking water distribution systems is a topic of ongoing and needed research.

Free ammonia is metabolized by AOM, which are ubiquitous in the environment and chloraminated water distribution systems. These microorganisms metabolize the free ammonia and produce nitrite, which in turn is metabolized by NOB into nitrate. Nitrite also imparts a chlorine demand. The following are approximate equations for nitrification reactions:

Ammonia-oxidation reactions:

$$NH_3 + 3/2O_2 \rightarrow NO_2^- + H_2O + H^+ \qquad \text{(Eq. 1-4a)}$$

$$NH_4^+ + 3/2O_2 \rightarrow NO_2^- + H_2O + 2H^+ \qquad \text{(Eq. 1-4b)}$$

Nitrite-oxidation reaction:

$$NO_2^- + 1/2O_2 \rightarrow NO_3^- \qquad \text{(Eq. 1-5)}$$

More information on nitrification reactions by AOM and NOB, and the *Nitrospira* species that can directly oxidize ammonia to nitrate (commamox reaction), is provided in Chapter 5. The overall reaction for complete nitrification (Grady et al. 1999) is shown in Eq. 1-6.

$$\begin{aligned} &NH_4^+ + 1.856\ O_2 + 1.979\ HCO_3^- \rightarrow \\ &0.0205\ C_5H_7O_2N + 0.979\ NO_3^- + 1.041\ H_2O + 1.876\ H_2CO_3 \end{aligned} \qquad \text{(Eq. 1-6)}$$

As shown by the reactants and products of Eq. 1-6, complete nitrification results in the consumption of oxygen and alkalinity (HCO_3^-), reduction of pH (formation of carbonic acid, H_2CO_3), production of nitrate (NO_3^-), and increased biomass ($C_5H_7O_2N$). Both complete (to nitrate) and incomplete (to nitrite) nitrification have been frequently observed in chloraminated drinking water distribution systems and some activated-carbon point-of-use filters within those distribution systems (Alfredo et al. 2020).

Free ammonia, the electron donor (or food) for the nitrifying microorganisms, may enter the distribution system from the source or treatment plant due to an excess dose of ammonia compared to chlorine during chloramine creation (typical), incomplete reaction with free chlorine, or may be released due to chloramine decay, cell lysis, or deamination reactions. The growth of nitrifying microorganisms will result in eventual cell lysis and increased chloramine demand. The soluble microbial products produced by active bacteria also exert a chloramine demand. A decrease in relative concentrations of inorganic chloramines and an increase in organic chloramines have been reported in storage reservoirs subject to long detention times; whether cell lysis could contribute to this transformation is currently poorly understood.

Nitrification in Storage Tanks

While nitrification can occur in all types of distribution system tanks, the occurrence is often exacerbated in those that do not have complete water exchange on a routine cycle, allowing stagnation and stratification to occur within the tanks. For instance, often only about 20 percent of the storage volume is moved in the daily cycle for a standpipe storage tank, which can lead to increased water age and minimal "fresh" water input. At a certain point, the nitrification process becomes embedded in the tank and is essentially "feeding" on the increased chloramine. This has been dubbed "feeding the beast" and "the death spiral" (McGuire 2013) by operators frustrated by the process. Some utilities are replacing standpipes built in the 1980s with more traditional elevated tanks to minimize problems associated with water age and nitrification.

Nitrification in the Environment

Nitrification is a very important process in agriculture, making the ammonia in fertilizers available for utilization by plants. As mentioned above, microorganisms that can use reduced nitrogen species are ubiquitous in the environment. They are found in soils, compost piles, wastewater, fresh water, marine habitats, and in most other aerobic environments. Many environments with suboptimal conditions still support the growth of nitrifying bacteria. For example, nitrifying bacteria are strict aerobes, yet they can be isolated from wastewater aeration tanks that are extremely low in dissolved oxygen.

The highest concentrations of nitrifying bacteria are found in the upper 10-cm layer in soils, at the sediment–water interface in rivers and streams, and attached to the sides of the aeration tanks in wastewater treatment plants (Watson et al. 1981). Considerable amounts of nitrifying bacteria (as well as ammonia, nitrate, and other nitrogen-based compounds) are brought into rivers through the discharge of treated and untreated wastewater. This discharge essentially "seeds" the receiving water body with nitrifying bacteria. However, high levels of ammonia can remain present for a longer time in the water column of some lakes and deeper rivers (Brion and Billen 2000), indicating that nitrification is a sediment-based process in these environments. Sediment resuspension might transport the nitrifiers into the water column, and macrophytes may serve as surfaces for nitrifiers to proliferate in the water column.

Ammonia, nitrate, and nitrite concentrations are typically present in surface water sources due to natural processes. Surface water nitrogen concentrations can increase from anthropogenic inputs, such as agricultural runoff to include fertilization and livestock wastes, sewage discharges, and industrial discharges. Groundwater is also susceptible to ammonia and nitrogen contamination. In some areas, ammonia can naturally occur; however, runoff percolating into the groundwater aquifers from agricultural lands and improperly managed septic tanks can increase nitrogen concentrations. The concentration of nitrite nitrogen in surface water and groundwater is normally far below 0.1 mg/L (Sawyer and McCarty 1978). As described in Chapter 7, the level can vary rapidly and dramatically due to watershed events. Seasonal highs in surface water ammonia concentrations typically occur in winter when nitrification rates decline. Groundwater generally contains relatively low concentrations of ammonia because of the cation exchange capacity of soil, unless there have been anthropogenic inputs (Bouwer and Crowe 1988).

REGULATORY COMPLIANCE AND PUBLIC HEALTH

Nitrification and the Safe Drinking Water Act

Because nitrification can lead to chemical and biological degradation of water quality, it presents a potential risk to public health and compliance with the following drinking water rules:

- SWTR
- Revised Total Coliform Rule (RTCR)
- Lead and Copper Rule (LCR)
- Stage 1 and Stage 2 DBP Rules
- Any local regulations on maintenance of minimum disinfectant residual level

In addition to these rules, nitrification can also lead to other public health concerns such as elevated nitrate or nitrite levels at taps above the USEPA health advisory levels (USEPA 2018).

The Total Coliform Rule/Distribution System Advisory Committee recommended that a research and information collection partnership be formed to inform and support the drinking water community in developing future national risk management decisions pertaining to drinking water distribution systems. The advisory committee recommended seven priority areas for research and information collection, one of which was nitrification. The document titled *Priorities of the Distribution System Research and Information Collection Partnership* (USEPA 2010) states that adverse public health impacts are suspected to be associated with nitrification, although available information is more anecdotal in nature, and additional research and information collection are necessary to better define public health risks. Several projects identified in this document propose to further explore the occurrence and impacts of nitrification and associated public health risks (USEPA 2010).

Table 1-2 provides a summary of water quality and possible compliance issues as they relate to nitrification.

Surface Water Treatment Rule

Disinfectant depletion and heterotrophic bacteria increases are examples of water quality impacts associated with nitrification that are addressed under provisions of the SDWA. The loss of a disinfectant residual does not necessarily pose a direct public health threat; however, disinfectant loss can allow growth and/or survival of microbial organisms within the bulk water, in biofilms, or on accumulated sediments and leave the water unprotected against contamination.

Table 1-2 Water quality and compliance issues caused by nitrification

	Water Quality Issues	Concerns	Applicable Rules
Chemical	Disinfectant depletion	Microbial proliferation	Surface Water Treatment Rule, Revised Total Coliform Rule, local rules on minimum disinfectant levels
	Nitrite/nitrate formation	Elevated nitrite/nitrate levels	*
	Dissolved oxygen depletion	Increased corrosion	Lead and Copper Rule
	Reduction in pH and alkalinity	Increased corrosion	Lead and Copper Rule
	Reduction in oxidation-reduction potential	Increased metals solubility and risk of scale instability; aesthetic concerns (like discolored water)	Lead and Copper Rule, Secondary Standards
	DBP formation due to mitigation techniques	Increased public health risks due to exposure to DBPs	Stages I and II Disinfectants/ Disinfection Byproducts Rules
	Presence of organic nitrogen, formation of organochloramines	False positive signal for disinfectant residual measurement, inability to maintain detectable chlorine residual	Surface Water Treatment Rule, Revised Total Coliform Rule, local rules on minimum disinfectant levels
Biological	Heterotrophic plate count increase	Increased disinfectant demand	Surface Water Treatment Rule, Total Coliform Rule
	Coliform occurrences	Proliferation of opportunistic pathogens	Revised Total Coliform Rule
	Increase in nitrifying microorganisms	Increases in nitrate and nitrite; increased disinfectant demand	
Aesthetic	Taste and odor, color, turbidity	Potential customer complaints	Secondary standards

* Compliance with nitrate and nitrite maximum contaminant levels is required at the point of entry to the distribution system, not within the distribution system.

DBP = disinfection byproduct

SWTR establishes MCL goals for viruses, *Legionella,* heterotrophic plate count (HPC), and *Giardia duodenalis* (formerly known as *G. lamblia*). It also includes treatment technique requirements for filtered and unfiltered systems that are specifically designed to protect against the adverse health effects of exposure to these microbial pathogens. SWTR requires that a "detectable" disinfectant residual be maintained in at least 95 percent of samples collected throughout the distribution system monthly (or HPC measurements not exceeding 500 colony-forming units per mL.) The detectable residual is not defined in SWTR but may be considered the detection limit of the field test analysis employed, such as 0.01 mg/L for chlorine and chloramines (APHA et al. 2005; Harp 2002). A failure to comply with this requirement for any 2 consecutive months is a treatment technique violation under SWTR.

The minimum monitoring requirement for disinfectant residuals (or HPC levels) in the distribution system is at the same frequency and locations as total coliform measurements taken pursuant to the RTCR.

Minimum Disinfectant Residual Level

While SWTR requires systems utilizing surface water supplies to maintain a detectable disinfectant concentration, many systems relying on chloramines for secondary disinfection typically aim to maintain a distribution system residual greater than 1.0 mg/L as Cl_2 (USEPA 1999) due to the relatively weak disinfecting properties of chloramines and to prevent nitrification from developing. As discussed in Chapter 7, residual levels of >2.0 mg/L as Cl_2 leaving the treatment plant and >1.0 mg/L as Cl_2 throughout the distribution system are recommended for reducing risk of nitrification, although minimum levels are largely system specific. Thus, operating a chloraminated distribution system with only a detectable disinfectant residual level is not advisable.

Several states have also promulgated state-specific requirements that define detectable residual concentration and require higher minimum disinfectant residual levels. Wahman and Pressman (2015) summarized these state-specific requirements. At the time, nine states had assigned a numerical value that would be considered detectable disinfectant residual (range 0.02–0.5 mg/L as Cl_2 for total chlorine), and 22 states had specified minimum concentrations for the residuals. The minimum concentrations vary from detectable to 1.5 mg/L as Cl_2 of total chlorine, and the rules differ in allowable exceptions as well as how compliance is determined.

Revised Total Coliform Rule

Nitrification has been linked with total coliform positive samples and red water events (e.g., Ybanez et al. 2020; Do et al. 2017). Nitrification can have an indirect effect on RTCR compliance in three ways:

- disinfectant depletion caused by nitrification can allow for the increased growth of coliform bacteria;
- improper implementation of nitrification control practices, such as poorly managed flushing, can cause biofilm detachment from pipes and introduction to the bulk water (Hill et al. 2018); and
- the resulting growth of heterotrophic bacteria during nitrification events can interfere with the detection of coliform bacteria.

However, as described in several case studies presented in Friedman et al. (2009), the presence of disinfectant residual level alone is not an adequate indicator of microbial water quality or RTCR compliance status.

The *Revised Total Coliform Rule Assessments and Corrective Actions Guidance Manual* (USEPA 2014) recommends that systems collect and evaluate disinfectant residual levels as well as nitrite/nitrate levels at high-water age locations (for chloramine systems) during level 1 and 2 assessments. These extra tests assist in manually identifying management of water age as a potential corrective action for systems experiencing nitrification and coliform occurrences. Additionally, maintenance of a disinfectant residual, which can be affected by nitrification, is identified as a treatment technique. Although nitrification is not specifically identified as a cause of coliform treatment technique violations, systems with coliform problems will likely need to resolve nitrification should it be identified during the assessment process.

Lead and Copper Rule

Nitrification can have an indirect effect on level of corrosion and LCR compliance. Nitrification can reduce pH, alkalinity, dissolved oxygen, and, due to oxygen and chloramine depletion, oxidation-reduction potential. These can lead to lead and copper release from lead and/or copper-bearing materials in the distribution system (Zhang et al. 2010; Zhang et al. 2009a, 2009b; Boyd et al. 2008; USEPA 2007; NRC 2006). As reported by Zhang et al. (2009b), the magnitude of pH changes depends on the initial alkalinity, extent of nitrification, and acid production. Although reductions in pH and alkalinity may not pose a direct public health threat, such reductions could theoretically result in enough elevated lead or copper samples to contribute to an exceedance of the LCR action levels (USEPA 2007), as well as failure to maintain optimal water quality parameters. For instance, lead (Pb) tends to form Pb(IV) oxide scale under higher oxidizing conditions, but once oxidation-reduction potential drops, the thermodynamically stable Pb(IV) oxide scale changes to Pb(II) carbonate that has higher solubility (USEPA 2016). These kinds of changes in scale type may increase the soluble lead concentration in the water and release particulate lead while the scales are trying to reach a new equilibrium with the changed water quality conditions.

A set of pipe experiments exposed to different levels of orthophosphate P and high levels of free chlorine demonstrated that serious problems with lead leaching due to nitrification are not expected at the alkalinities and initial pHs encountered at many water utilities. However, utilities operating at pH values near 8.3 (the point of minimum buffering capacity for the carbonate system) and low alkalinities could be very susceptible to elevated soluble lead from nitrification (Zhang et al. 2009b).

Zhang et al. (2009b) also investigated nitrification by simulating homes with polyvinyl chloride (PVC) or copper plumbing and leaded brass faucets by placing a brass rod inside a PVC or copper pipe. Zhang et al. (2009b) observed complete nitrification in the PVC pipes after 2 months and not in the copper pipes, presumably due to copper toxicity. At an alkalinity of 35 mg/L as $CaCO_3$, nitrification resulted in a pH drop of 1.4 units (pH 8.2–6.8) and leached 45 percent more lead and 81 percent more zinc from leaded brass inserted into the PVC pipes compared to the leaded brass inserted into the copper pipes. The pH increased in the copper piping from 8.15–9.96, possibly due to corrosion.

The researchers pointed out that when nitrification has occurred and leaded brass faucets are in use, homes with PVC piping will demonstrate higher lead levels in first-draw samples than homes with copper piping. The researchers also determined that lead leaching was not significantly impacted by production of nitrite and nitrate or by reductions in inorganic carbon or dissolved oxygen via nitrification (Zhang et al. 2009b).

Copper can have both an inhibitory effect on nitrification or may promote nitrification possibly depending on the free copper concentration, copper complexes present, and the type of nitrifiers present (Zhang et al. 2009b and 2008). The same group's earlier research (Zhang et al. 2008) found that copper at 1–10 μg/L might stimulate nitrification, while levels greater than 100 μg/L might prohibit nitrification.

Nitrite and Nitrate Primary MCLs

Primary MCLs have been established for nitrite, nitrate, and the sum of nitrite plus nitrate applicable at the points of entry to the distribution system. The MCLs are 1 mg N/L for nitrite (equal to ~3.3 mg nitrite /L), 10 mg N/L for nitrate (~44 mg nitrate/L), and 10 mg N/L for total nitrate/nitrite. The USEPA Phase II Inorganic Contaminant regulations require water systems to sample for nitrite and nitrate at each entry point to the distribution system at least annually. Additional monitoring is required on a quarterly basis for at least 1 year following any one routine sample in which the measured concentration is greater than 50 percent of the MCL (USEPA 2001). In some states, if any drinking water sample is analyzed using certified methods and a regulated contaminant is found above its MCL, the result must be reported to the regulator/state, even though it might not be reported as part of the official regulatory compliance program.

Nitrite and nitrate are regulated as acute health risks and require 24-h public notice if a result above the MCL is confirmed. Both compounds can cause methemoglobinemia in infants, which can be fatal. In methemoglobinemia, the iron in the hemoglobin has been oxidized and cannot bind oxygen, leading to asphyxia. The health impacts language states, "[i]nfants below the age of six months who drink water containing nitrite or nitrate in excess of the MCL could become seriously ill and, if untreated, may die. Symptoms include shortness of breath and blue-baby syndrome" (USEPA 2006).

Ammonia can be released from chloramine through a series of complex reactions, as shown in Table 1-3. Reactions 2–6 describe five mechanisms of ammonia release presented by Woolschlager et al. (2001) and Vikesland et al. (1998). Three of the demand reactions (3, 4, and 6) essentially release one mole of ammonia for one mole of monochloramine. Comparatively, the autodecomposition reaction (major pathway) releases only one mole of ammonia for every three moles of monochloramine. This indicates that some demand reactions release significantly more ammonia than the major autodecomposition (decay) reaction.

Nitrite and nitrate are produced during nitrification through ammonia utilization by nitrifying bacteria. According to Table 1-3's reactions 1 and 2, for every mole of ammonia-N produced through chloramine decay, a one-mole equivalent of nitrite-N is produced by AOB. Subsequently, for every mole of nitrite-N produced by AOB, a one-mole equivalent of nitrate-N is produced by NOB. Figure 1-3 shows that by using reaction 3 in Table 1-3, which is the most conservative chloramine demand reaction in terms of quantity of

Table 1-3 Overview of nitrification and chloramine reactions

Reaction Description	Overall Reaction
1. Ammonia and nitrite utilization	$NH_3 + O_2 \rightarrow NO_2 + 3H^+ + 2e^-$ $NO_2 + H_2O \rightarrow NO_3 + 2H^+ + 2e^-$
2. Release of ammonia through chloramine decay (autodecomposition)	$3NH_2Cl \rightarrow N_2 + NH_3 + 3Cl^- + 3H^+$ (Major)† $4NH_2Cl + 3H_20 \rightarrow 4Cl^- + 3NH_3NO_3 + 5H^+$ (Minor)†
3. Release of ammonia through oxidation of organic matter by chloramine	$1/10C_5H_7O_2N + NH_2Cl + 9/10H_2O \rightarrow 4/10CO_2 +$ $1/10HCO_{3-} + 11/10NH_4^+ + Cl^-$
4. Release of ammonia through reaction of chloramine with corrosion products at pipe surfaces*	$1/2NH_2Cl + H^+ + Fe^2+ \rightarrow Fe^{3+} + 1/2NH_4^+ + 1/2Cl^-$
5. Release of ammonia through catalysis reactions of chloramine at pipe surfaces	$3NH_2Cl \rightarrow N_2 + NH_3 + 3Cl^- + 3H^+$
6. Release of ammonia through oxidation of nitrite by chloramine*	$NH_2Cl + NO_{2-} + H_2O \rightarrow NH_3 + NO_{3-} + HCl$

Adapted from Woolschlager et al. 2001

* These reactions may not be significant in distribution systems, especially in situations of biologically accelerated chloramine decay.

† *Source:* Vikesland et al. (1998).

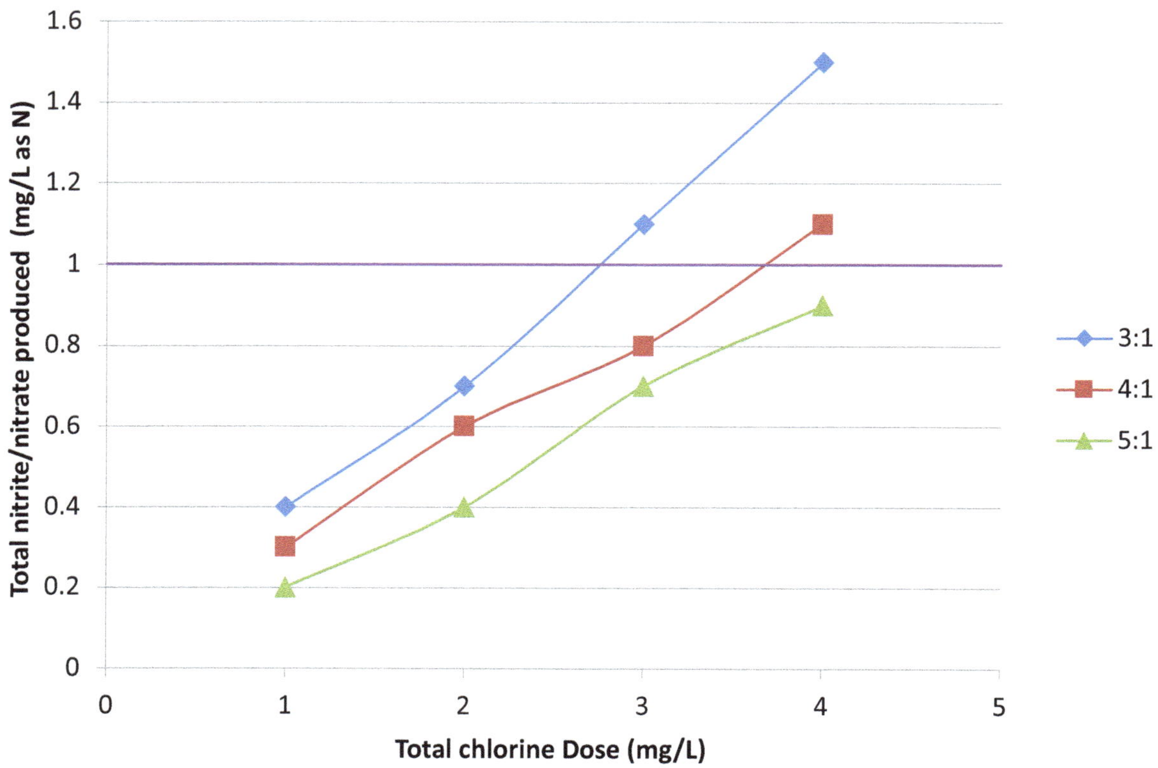

Figure 1-3 Theoretical nitrite/nitrate production based on chloramine decay stoichiometry as a function of chlorine-to-ammonia-N ratio. Assumes 100-percent decay according to reaction 3 in Table 1-3 and 100-percent conversion of ammonia to nitrite/nitrate-N.

ammonia produced per mole of monochloramine present, the nitrite-N MCL of 1 mg/L as N could theoretically be exceeded if the chloramine dose greater than approximately 3 mg/L (as total chlorine) and the Cl_2:NH_3-N weight ratio is less than 5:1. As the chloramine dose increases and the chlorine-to-ammonia-N weight ratio decreases, more ammonia becomes available for the nitrification process. In a pilot-scale distribution system, Harrington et al. (2002) measured nitrite concentrations equal to 1 mg/L as N with an average total chlorine concentration of 4.6 mg/L and a Cl_2:NH_3-N ratio of 3.9:1. The pilot train represented conventionally coagulated Lake Mendota, Madison, Wisc., water at a pH of 9.

Valentine et al. (1998) conducted a series of mass and redox balances on solutions of varying pH, NOM concentration, and initial chloramine concentration. For all conditions that were studied, the amount of nitrate formed as a percentage of monochloramine decay was less than 15 percent, and for all but three cases, the amount was less than 10 percent. The authors concluded that although nitrate is an important decomposition product of monochloramine decay, it is not the major nitrogen-containing species of decomposition.

Using data from a survey of 40 utilities that use chloramine as a disinfectant and a previous survey by Hack (1985), Wilczak et al. (1996) indicated that nitrite-N and nitrate-N levels may increase by 0.05–0.5 mg/L as N during nitrification events, although increases of greater than 1 mg/L are possible. Nitrite levels during nitrification episodes have been reported ranging from 0.005–0.5 mg/L as N, with levels more frequently ranging from 0.015–0.1 mg/L as N (Wolfe and Lieu 2001). As discussed previously, the amount of nitrite produced is a function of the amount of ammonia present.

High nitrite levels could be observed in systems with high source water ammonia concentrations, regardless of whether disinfection is practiced. Rezania (2010) reported ammonia levels ranging from 1.7 mg/L up to 7 mg/L in a study of 39 public water supply wells in Minnesota. In addition, it was reported that samples taken in the distribution system had nitrite levels ranging from less than the detection limit to 2.1 mg/L-N. Similarly, Lytle et al. (2007; 2009) reported that levels of nitrite greatly exceeded the maximum contaminant level (MCL) of 1.0 mg/L-N in the distribution systems of several communities in Illinois that had elevated ammonia in their source water.

Although the NRC's assessment of distribution system health risks determined that formation of nitrate and nitrite in the distribution system was a relatively low priority compared to other potential distribution system health risks (NRC 2006), nitrite and nitrate formation is a serious concern in systems with high source water ammonia concentrations. The additional nitrate and nitrite could cause levels higher than the respective MCLs.

Disinfection Byproducts

Nitrification in and of itself is not known to form DBPs; however, certain practices to prevent or respond to nitrification, such as breakpoint chlorination and chlorine burns, can form increased levels of regulated DBPs. While these are not likely to cause a compliance issue due to their short duration, the levels of DBPs formed during these events can be high. Nitrification has also been linked with presence and increase of nitrogenous DBPs, a group of DBPs that do not have federal regulatory limits or goals established yet (Zeng and Mitch 2016; Schreiber and Mitch 2006). Nitrogenous DBPs can form during chloramination of waters containing organic nitrogen precursors and are potentially carcinogenic (USEPA 2018). These precursors may be naturally present in some source waters but can also be released from the nitrifying biofilms or due to cell lysis in the distribution system (Zeng and Mitch 2016). The nitrogenous DBPs include nitrosamines like *N*-nitrosodimethylamine (NDMA), N-nitrosodiethylamine (NDEA), and N-nitrosodipropylamine (NDPA); haloacetonitriles; haloacetamides; and halonitromethanes. Out of these, the nitrosamines and more specifically, NDMA, are emerging as the most common concern for chloraminated water systems (Woods and Dickenson 2015).

Of the utilities monitoring NDMA as part of the second Unregulated Contaminant Monitoring Rule, 25 percent detected the presence of NDMA in at least one sample (Woods and Dickenson 2015), and two-thirds of those utilities detecting NDMA reported the use of chloramines (Russell et al. 2012). Other nitrosamines monitored as part of this effort appear to occur less frequently and in smaller quantities. USEPA has not established a regulatory limit for any of the nitrogenous DBPs but included five nitrosamines on the fourth Drinking Water Contaminant Candidate List and is likely to address the regulatory determination during the next DBP Rule review cycle. The World Health Organization (WHO) has established a health guideline of 100 ng/L for NDMA in drinking water. A WHO health guideline is the level of a compound in the water that the organization believes does not represent any significant risk to health over a lifetime of consumption. California has set a notification level at 10 ng/L for NDMA, NDEA, and NDPA nitrosamine species. Notification levels in California are health-based advisory levels, and if exceeded, certain requirements and recommendations will apply. California has also set a public health goal at 3 ng/L for all three species. Massachusetts has established a guideline of 10 ng/L for NDMA based on an acceptable health risk similar to the federal process of setting MCLs. Other states may also have standards, health advisory limits, or public health goals established, but a regulatory limit has yet to be established by USEPA.

Interestingly, THMs, byproducts of chlorination, may also be beneficial in limiting nitrification. Speitel et al. (2010, 2011) concluded that AOB can biodegrade THMs through cometabolism, an enzymatic process in which a specific microbial enzyme causes biodegradation of a chemical with no significant benefit to the microorganism. In these experiments, complete nitrification was never observed in the reactors containing THM concentrations, not even after 650 h.

More recent explorations into monochloramine cometabolism posit that it can more accurately account for monochloramine loss during distribution system nitrification events. In research conducted by Maestre et al. (2016), they found that cometabolism by a pure culture of AOB increased monochloramine demand with kinetics similar to those of ammonia metabolism. This cometabolism of monochloramine was reported to account for 30 percent of the monochloramine loss observed in their experimental systems conducted at a pH of 8.2–8.4. More research is needed to understand the full impacts of cometabolism on monochloramine loss and nitrification kinetics.

Distribution System Optimization—Partnership for Safe Water

The Partnership for Safe Drinking Water—an alliance of the American Water Works Association, USEPA, the Association of State Drinking Water Administration, the National Association of Water Companies, the Water Research Foundation, and the Association of Metropolitan Water Agencies—has developed an optimization program for distribution systems. The program identified water quality, hydraulic, and infrastructure integrity as the three key performance variables. Chlorine residual levels measured throughout the distribution system (beyond RTCR locations) serve as the indicator of water quality integrity. The measurable performance goals are to

- maintain a free chlorine residual ≥0.20 mg/L and ≤4.0 mg/L, total chlorine residual ≥0.50 mg/L and ≤4.0 mg/L, for at least 95% of the measurements taken;
- determine that no consecutive residual measurements are outside target concentrations at optimized routine sampling locations; and
- maintain DBPs within regulatory requirements.

The goal of the program is to provide utilities with tools to assess and improve their performance beyond regulatory compliance. Utilities can subscribe to the program and follow the guidelines by committing to the optimization goals, gathering data, assessing their own performance, developing action plans for improvements, and reporting performance on an annual basis to the partnership. Awards are given to the utilities that meet the performance goals.

KEY TAKEAWAYS

Nitrification is an oxidation process carried out by microorganisms in which reduced nitrogen compounds (primarily ammonia) are sequentially oxidized to nitrite and nitrate or directly to nitrate. The presence of ammonia (resulting from natural occurrence, residual in dosing for chloramine formation, or from chloramine decomposition) and nitrifying bacteria are necessary for nitrification in drinking water distribution systems. Because nitrifying bacteria are ubiquitous in the environment and the distribution system, nitrification will occur if the physical and chemical conditions that support the growth of these organisms exist and sufficient levels of free ammonia are present in the water.

There are several impacts of nitrification that can affect distribution system water quality. Of the water quality issues identified in the literature and summarized in Table 1-2, most impacts on regulatory compliance are indirect, such as loss of disinfectant residual, reduction in pH, increased corrosion, or risk of scale instability. Only the formation of nitrite and nitrate within the distribution system poses a potential direct public health threat, but nitrite and nitrate are not addressed within the distribution system through current provisions of SDWA. Although compliance with current regulations may not be impacted directly by nitrification, it is a good practice to minimize the adverse effects of nitrification in the distribution system, especially the loss of a disinfectant residual, the growth of HPC bacteria, the formation of nitrite and nitrate, and any decrease in pH and alkalinity.

The causes of nitrification, its prevention, and mitigation are discussed in detail in the following chapters. Prevention of nitrification should be considered an essential water quality maintenance practice because it helps to provide a more stable chloramine residual in the distribution system and water that is biologically stable in terms of limiting nitrifying bacterial growth, as well as the growth of heterotrophic bacteria prevalent during nitrification episodes. Maintenance of a disinfectant residual and the biological integrity of the distribution system are important from a customer point of view, as well as from a distribution system security point of view, because a disinfectant residual is commonly used as one indicator of distribution system integrity and overall water quality, and one aspect of how public health is protected in distribution systems.

REFERENCES AND BIBLIOGRAPHY

Alfredo, K., J. Lin, A. Islam, and Z.W. Wang. 2020. "Impact of Activated Carbon-Block Point-of-Use (POU) Filters on Chloraminated Water Quality." *Water Science*, 2:3:e1180.

American Public Health Association, American Water Works Association, and Water Environment Federation (APHA, AWWA, and WEF). 2005. *Standard Methods for the Examination of Water and Wastewater*. Washington, D.C.: APHA.

American Water Works Association and Economic and Engineering Services, Inc. (AWWA and EES). 2002. *Nitrification*. USEPA Distribution System White Papers. www.epa.gov/sites/default/files/2015-09/documents/nitrification_1.pdf (accessed February 26, 2025).

AWWA Disinfection Systems Committee. 2008a. "Committee Report: Disinfection Survey, Part 1—Recent Changes, Current Practices, and Water Quality." *Journal AWWA*, 100:10:76–90.

AWWA Disinfection Systems Committee. 2008b. "Committee Report: Disinfection Survey, Part 2—Alternatives, Experiences, and Future Plans." *Journal AWWA*, 100:11:110–124.

AWWA Disinfection Committee. 2017. Water Utility Disinfection Survey Report; April 2018, report prepared by Cornwell Engineering Group.

Blute, N., D.W. Pedersen, X. Wu, and E. Means. 2013. "Demonstration Testing of Nitrification Control Using Chlorite in Irvine, Calif." *Journal AWWA*, 105:4:E207–E216.

Bouwer, E.J., and P.B. Crowe. 1988. "Biological Processes in Drinking Water Treatment." *Journal AWWA*, 80:9:82–93.

Boyd, G.R., K.M. Dewis, G.V. Korshin, S.H. Reiber, M.R. Schock, A.M. Sandvig, and R. Giani. 2008. "Effects of Changing Disinfectants on Lead and Copper Release." *Journal AWWA*, 100:11:75–87.

Brion, N., and G. Billen. 2000. "Wastewater as a Source of Nitrifying Bacteria in River Systems: The Case of the River Seine Downstream from Paris." *Water Research*, 34:12:3213–3221.

Do, T.D., A.D. Pifer, Z.K. Chowdhury, D.G. Wahman, W. Zhang, and Fairey, J.L. 2017. Developing Fluorescence Sensor Systems for Early Detection of Nitrification in Chloraminated Drinking Water Distribution Systems, presentation, WQTC 2017, Portland, OR.

Friedman, M., A. Hanson, K. Dewis, G. Kirmeyer, M. LeChevallier, G. Gagnon, L. Truelstrup Hansen, C. Krentz, M. Mosher, S. J. Payne, J. Rosen, T. Hargy, J. Sobrinho, M.-C. Besner, and M. Prevost. 2009. *Strategies for Managing and Responding to Total Coliform and E. coli in Distribution Systems*. Denver: WRF.

Friedman, M. G. Kirmeyer, J. Lemieux, M. LeChevallier, S. Seidl, and J. Routt. 2010. *Criteria for Optimized Distribution Systems*. Denver: WRF.

Furatian, L. 2019. "Chloramination: An Evolving Tool of Public Health Protection." *Journal AWWA*, 111:7: 88–90.

Hack, D.J. 1985. "State Regulation of Chloramine." *Journal AWWA*, 77:1:46–49.

Harp, D., 2002. *Current Technology of Chlorine Analysis for Water and Wastewater. Technical Information Series – Booklet No. 17*. Loveland: Hatch Co.

Harrington, G.W., D.R. Noguera, A.I. Kandou, and D.J. Vanhoven. 2002. "Pilot-Scale Evaluation of Nitrification Control Strategies." *Journal AWWA*, 94:11:78–89.

Hill, A.S., M. Friedman, M. Hallett, V. Salo-Zieman, S. Booth, A. Hanson, K. Gupta, Y. Akagi, C. Kochiss, L. Koperski, P. Igoe, L. Kirby, and W. Harper. 2018. *Use of Flushing as a Corrective Action Under the Revised Total Coliform Rule*. Denver: WRF.

How, Z.T., I. Kristiana, F. Busetti, K.L. Linge, and C.A. Joll. 2017. "Organic Chloramines in Chlorine-based Disinfected Water Systems: A Critical Review." *Journal of Environmental Sciences*, 58:2–18

Grady, C.P.L., Jr., G.T. Daigger, and H.C. Lim. 1999. *Biological Wastewater Treatment* (2nd ed.). New York: Marcel Dekker.

Kihn, A., A. Andersson, P. Laurent, P. Servais, and M. Prévost. 2002. "Impact of Filtration Material on Nitrification in Biological Filters Used in Drinking Water." *Aqua*, 51:1:35–46.

Kirmeyer, G., L.H. Odell, J.G. Jacangelo, A. Wilczak, and R.L. Wolfe. 1995. *Nitrification Occurrence and Control in Chloraminated Water Systems*. Denver: WRF.

Kirmeyer, G., K. Martel, G. Thompson, L. Radder, W. Klement, M. LeChevallier, H. Baribeau, and A. Flores. 2004. *Optimizing Chloramine Treatment* (2nd ed.). Denver: WRF.

Koch, H., M.A.H.J. van Kessel, and S. Lücker. 2019. "Complete nitrification: insights into the ecophysiology of comammox *Nitrospira*." *Applied Microbiology and Biotechnology*, 103:1:177–189.

Kuenen, J.G., and M.S.M. Jetten. 2001. "Extraordinary Anaerobic Ammonium-oxidizing Bacteria." *ASM News*, 67:9:456–463 .

Kuypers, M.M.M., Marchant H.K., and Kartal B. 2018. "The Microbial Nitrogen-cycling Network." *Nature Reviews: Microbiology*, 16:5:263–276.

Louisiana Department of Health. 2013. "DHH Confirms Death of a Child Associated with Rare Amoeba Found in St. Bernard Parish Home." https://ldh.la.gov/news/2862 (accessed October 9, 2025).

Lytle, D.A., D. Williams, C. Muhlen, B. Almassalkhi, and C. White. 2009. "Consequences and reduction of elevated ammonia in Illinois groundwaters: U.S. EPA research efforts." *Illinois AWWA/IWEA Meeting*, March 16–19, 2009.

Lytle, D.A., C. Muhlen, and B. Almassalkhi. 2007. "Elevated Natural Source Water Ammonia and Nitrification in the Distribution Systems of Four Water Utilities." *AWWA Water Quality Technology Conference*. Charlotte, NC. Nov. 4–8, 2007.

Maestre, J.P., D.G. Wahman, and G.E. Speitel. 2016. "Monochloramine Cometabolism by Mixed-Culture Nitrifiers under Drinking Water Conditions." *Environmental Science & Technology*, 50:12:6240–6248.

McGuire, M.J. 2013. *The Chlorine Revolution: Water Disinfection and the Fight to Save Lives*. Denver: American Water Works Association.

Morel, F.M.M., and J.G. Hering. 1993. *Principles and Applications of Aquatic Chemistry*. New York: John Wiley & Sons.

National Research Council, Division on Earth and Life Studies, Committee on Public Water Supply Distribution Systems. 2006. *Drinking Water Distribution Systems: Assessing and Reducing Risks*. Washington, D.C.: National Academies Press.

Painter, H.A. 1970. "A Review of Literature on Inorganic Nitrogen Metabolism in Microorganisms." *Water Research*, 4:6:393–450.

Pauer, J.J., and M.T. Auer. 2000. "Nitrification in the Water Column and Sediment of a Hypereutorphic Lake and Adjoining River System." *Water Research*, 34:4:1247–1254.

Pope, P.G., G.E. Speitel, Jr., and M.R. Collins. 2006. "Kinetics of Dihaloacetic Acid Formation during Chloramination." *Journal AWWA*, 98:11:107–120

Regan, J.M., G.W. Harrington, and D.R. Noguera. 2002. "Ammonia- and Nitrite-oxidizing Bacterial Communities in a Pilot-scale Chloraminated Drinking Water Distribution System." *Applied and Environmental Microbiology*, 68:1:73 –81.

Rezania, L.W. 2010. "Minnesota Ammonia Study - Phase 1 Problem Assessment and Data Collection." AWWA Inorganic Contaminants Workshop, Denver. Feb. 28–Mar. 2, 2010.

Russell, C.G., N.K. Blute, S. Via, X. Wu, and Z. Chowdury. 2012. "Nationwide Assessment of Nitrosamine Occurrence and Trends." *Journal AWWA*, 104:3: E205-E217.

Salehi, M., T. Odimayomi, K. Ra, C. Ley, R. Julien, A.P. Nejadhashemi, J.S. Hernandez-Suarez, J. Mitchell, A.D. Shah, and A. Whelton. 2020. "An Investigation of Spatial and Temporal Drinking Water Quality Variation in Green Residential Plumbing." *Building and Environment*, 169:106566.

Sawyer, C.N., and P.L. McCarty. 1978. *Chemistry for Environmental Engineering* (3rd ed). New York: McGraw-Hill.

Seidel, C.J., M.J. McGuire, R.S. Summers, and S. Via. 2005. "Have Utilities Switched to Chloramines?" *Journal AWWA*, 97:10:87-97.

Schreiber, I.M. and W.A. Mitch. 2006. "Nitrosamine Formation Pathway Revisited: The Importance of Chloramine Speciation and Dissolved Oxygen." *Environmental Science & Technology*, 40:19:6007–6014.

Scott, D.B., M.I. Van Dyke, W.B. Anderson, and P.M. Huck. 2015. "Influence of Water Quality on Nitrifier Regrowth in Two Full-scale Drinking Water Distribution Systems." *Canadian Journal of Microbiology*, 61:12:965–976.

Speitel, G.E., Jr., B.M. Bayer, and R. Kannappan. 2010. *Significance of Trihalomethanes in Preventing Distribution System Nitrification in Chloraminated Waters*. Denver: WRF.

Speitel, G.E., Jr., R. Kannappan, and B. Bayer. 2011. "The Nitrification Index: A Unified Concept for Quantifying the Risk of Distribution System Nitrification." *Journal AWWA*, 103:1:69–80.

Strous, M., J.G. Kuenen, and M.S. Jetten. 1999. "Key Physiology of Anaerobic Ammonium Oxidation." *Applied and Environmental Microbiology*, 65:7:3248–3250.

USEPA (US Environmental Protection Agency). 1979. *National Interim Primary Drinking Regulations; Control of Trihalomethanes in Drinking Water*. Federal Register, 44:231:68624.

USEPA. 1999. *Alternative Disinfectants and Oxidants Guidance Manual*. EPA 815-R-99-014. Washington: Government Printing Office.

USEPA. 2001. *National Primary Drinking Water Regulations*. Federal Register, 19(141).

USEPA. 2005. *Economic Analysis for the Final Stage 2 Disinfectants and Disinfection By-Products Rule*. 815-R-010. Washington: USEPA.

USEPA. 2006. *National Primary Drinking Water Regulations*. Federal Register. 40 CFR Parts 9,141, and 142.

USEPA. 2007. *Simultaneous Compliance Guidance Manual for the Long Term 2 and Stage 2 DBP Rules*. 815-R-07-017. Washington: USEPA.

USEPA. 2009. *2006 Community Water System Survey*. 815-R-09-001. USEPA.

USEPA. 2014. *Revised Total Coliform Rule Assessments and Corrective Actions Guidance Manual*. Interim Final. 815-R-14-006. Washington: USEPA.

USEPA. 2010. *Priorities of the Distribution System Research and Information Collection Partnership*. Washington: USEPA.

USEPA. 2016. *Optimal Corrosion Control Treatment Evaluation Technical Recommendations for Primacy Agencies and Public Water Systems*. EPA 816-B-16-003. Washington: USEPA.

USEPA. 2018. *2018 Edition of the Drinking Water Standards and Health Advisories*. 822-F-18-001. Washington: USEPA.

Valentine, R.L., K. Ozekin, and P.J. Vikesland. 1998. *Chloramine Decomposition in Distribution System and Model Waters*. Denver: AWWA Research Foundation and AWWA.

van Kessel, M.A.H.J., D.R. Speth, M. Albertsen, P.H. Nielsen, H.J.M. Op den Camp, B. Kartal, M.S.M. Jetten, and S. Lücker. 2015. "Complete Nitrification by a Single Microorganism." *Nature*, 528: 7583:555–559.

Vikesland, P.J., K. Ozekin, and R.L. Valentine. 1998. "Effect of Natural Organic Matter on Monochloramine Decomposition: Pathway Elucidation through the Use of Mass and Redox Balances." *Environmental Science & Technology*, 32:10:1409–1416.

Wahman, D.G., and J.G. Pressman. 2015. "Distribution System Residuals—Is 'Detectable' Still Acceptable for Chloramines?" *Journal AWWA*, 107:8:53–63.

Watson, S.W., F.W. Valois, and J.B. Waterbury. 1981. "The Family Nitrobacteraceae." In *The Prokaryotes*. M.P. Starr, H. Stolp, H.G. Trüper, A. Balows, and H.G. Schlegel, editors. New York: Springer-Verlag.

White, G.C. 1999. *Handbook of Chlorination and Alternative Disinfectants*. (4th edition) New York: Wiley & Son.

Wilczak, A., J.G. Jacangelo, J.P. Marcinko, L.H. Odell, and G.J. Kirmeyer. 1996. "Occurrence of Nitrification in Chloraminated Water Distribution Systems." *Journal AWWA*, 88:7:74–85.

Wolfe, R.L., and N.I. Lieu. 2001. "Nitrifying Bacteria in Drinking Water." In *Encyclopedia of Environmental Microbiology*, Vol. 4. G. Bitton, editor. New York: John Wiley & Sons.

Woods, G.C., and E.R.V. Dickenson. 2015. "Evaluation of the Final USMR2 Database: Nationwide Trends in NDMA." *Journal AWWA*, 107:1:E58-E68

Woolschlager, J.E., B.E. Rittmann, P. Piriou, L. Kiene, and B. Schwartz. 2001. "Using a Comprehensive Model to Identify the Major Mechanisms of Chloramine Decay in Distribution Systems." *Water Supply*, 1:4:103–110.

Ybanez, C., G. Ramirez, W. Becker, and M. Santos. 2020. "When the Distribution System Needs a NAP: Tackling One City's Nitrification Action Plan." *Journal AWWA*, 112:1:40–51

Zeng, T., and W.A. Mitch. "Impact of Nitrification on the Formation of N-Nitrosamines and Halogenated Disinfection Byproducts within Distribution System Storage Facilities." 2016. *Environmental Science & Technology*, 50:6:2964–2973.

Zhang, Y., A. Griffin, and M. Edwards. 2008. "Nitrification in Premise Plumbing: Role of Orthophosphate, pH, and Pipe Corrosion." *Environmental Science & Technology*, 42:12:4280–4284.

Zhang, Y., N. Love, and M. Edwards. 2009a. "Nitrification in Drinking Water Systems". *Critical Reviews in Environmental Science & Technology*, 39:3:153–208.

Zhang, Y., A. Griffin, M. Rahman, A. Camper, H. Baribeau, and M. Edwards. 2009b. "Lead Contamination of Potable Water Due to Nitrification." *Critical Reviews* in *Environmental Science and Technology*, 43:6:1890–1895.

Zhang, Y., M. Edwards, A. Pinto, N. Love, A. Camper, M. Rahmen, and H. Baribeau. 2010. *Effect of Nitrification on Corrosion in the Distribution System*. Denver: WRF and USEPA.

Zhang, Y., T.J. Tseng, C. Andrews-Tate, R.C. Cheng, and K.L. Wattier. 2012. "Pilot-Scale Evaluation of Blending Desalinated Seawater into a Distribution System." *Journal AWWA*, 104:7:E416-E430.

Chapter 2

Nitrification in Water Treatment

Alex Mofidi, Darren Lytle, Andrzej Wilczak

INTRODUCTION

This chapter discusses available information on the occurrence of nitrification during water treatment and its potential impact on distribution system water quality. Nitrification (either intentionally encouraged or unknowingly occurring) as part of the water treatment process can occur whenever ammonia is present in or added to the source water and whenever water is not initially free chlorinated to achieve breakpoint, oxidize ammonia fully, and establish a free chlorine residual. Nitrification can be either controlled or uncontrolled. Controlled nitrification via biological or physiochemical treatment may be conducted, for example, when the concentrations of source water ammonia are high and the desire is to remove the ammonia partially or completely. Nitrification has been used and modeled extensively in wastewater treatment, and it is useful to understand the lessons learned there and draw certain comparisons and parallels that may improve the understanding of nitrification in water treatment and distribution systems. The key findings of this review are summarized in Table 2-1.

DRINKING WATER TREATMENT TO REMOVE AMMONIA

A key issue that causes nitrification is the amount of ammonia in water sources. Although there is no ammonia drinking water standard in the United States, the European Union (EU) has established a maximum limit of ammonium at 0.5 mg/L NH_4 and a guide level that calculates a result based on nitrate and nitrite levels for systems that use chloramines

Table 2-1 Key points from Chapter 2

Nitrification in water treatment	• Nitrification can occur whenever ammonia is present in source water (by addition or naturally), sufficient oxygen is present, and when water does not contain free chlorine and/or does not achieve ammonia breakpoint. • Ammonia removal can be effectively achieved using biological and physiochemical methods. Controlled and complete nitrification (to nitrate) can be accomplished biologically in various types of filter beds and media contactors. • Uncontrolled and incomplete nitrification (to nitrite) in filter beds is not desirable because it increases chlorine demand and can lead to nitrification in the distribution system. Furthermore, the primary drinking water standard for nitrite (1 mg/L NO_2-N) is much lower than the nitrate (10 mg/L NO_3-N) standard. • Temperature is one of the major factors impacting nitrification in a filter bed. • Low loading rates, long solids retention times, and continuous (instead of start/stop) operation are needed to achieve a high degree of nitrification in a filter bed. Nitrification can occur in very cold waters at 1°C (34°F) if long detention time and appropriate nutrient concentrations to establish microbial growth (such as phosphates) are available. • High filter media porosity is better for achieving nitrification in a filter bed, especially in cold water; in order of decreasing effectiveness: macroporous carbon > microporous carbon > anthracite > sand. • Preammoniation or intentionally passing chloramines through a granular activated carbon bed can lead to nitrifying bacteria colonization of the treatment train. • Postdisinfection of biological filters (with a residual disinfectant) is necessary to prevent passage of microorganisms and protect the distribution system.
Nitrification in wastewater treatment versus that occurring in drinking water treatment and distribution systems	• Compared to drinking water, nitrification in wastewater typically operates at much greater ammonia concentrations (>10 mg/L NH_3-N) and much greater bacterial biomass levels, allowing process retention times to be much shorter. • The growth rate of ammonia-oxidizing microorganisms (AOM) in drinking water systems may depend on free ammonia concentration, except that *Nitrosomonas oligotropha* (*N. oligotropha*) requires very little substrate and its growth rate may become substrate limited at very low NH_3 concentrations (0.1 mg/L NH_3-N or less). Nevertheless, keeping free ammonia levels low in the distribution system is considered key to nitrification control. • Based on experience gained from water and wastewater treatment practices, good nitrification control in drinking water storage facilities should occur in completely mixed, clean reservoirs with minimal deposits, having high water turnover to surpass AOM growth rate and cause their washout. A completely mixed tank offers the best chloramine concentration throughout the tank along with minimizing free ammonia levels.

as follows: [nitrate as NO_3]/50 + [nitrite as NO_2]/3 ≤1 (European Union [EU] 2018, 1998). Although there are no immediate indications that ammonia will become regulated within the United States, there are benefits for utilities to remove ammonia before the distribution system.

The removal of source water ammonia before it enters the distribution system has important benefits. The presence of ammonia in drinking water distribution systems has been correlated to increased biological activity (Wilczak et al. 1996; Servais et al. 1995), corrosion (Murphy et al. 1997), formation of nitrite and nitrates (Lytle et al. 2007a; Wilczak et al. 1996), and adverse impacts on taste and odor (Rittmann and Huck 1989; Bouwer and Crowe 1988). In addition, the presence of ammonia can interfere with the effectiveness of some water treatment processes. For example, the oxidation of arsenic (III) by free chlorine and subsequent removal from water may be impacted by the formation of chloramines that are less effective at oxidizing arsenic than is free residual chlorine (Chen et al. 2018;

Lytle et al. 2007b). When free chlorine is the desired residual disinfectant in the distribution system, the removal or complete oxidation of naturally occurring ammonia is required; otherwise, chloramines will be formed.

The below information summarizes nonbiological and biological treatment for ammonia removal from drinking water (typically from groundwater sources) and ozone-enhanced biological treatment (biofiltration) that has been observed to remove ammonia (typically during surface water treatment conditions).

Nonbiological Ammonia Treatment for Drinking Water

Both physicochemical and biochemical processes can be applied for removal of naturally occurring ammonia from drinking water. Several physicochemical processes may be effective—including reverse osmosis (RO), chemical oxidation (breakpoint chlorination), and ion exchange—but could be more costly when compared to biological treatment technologies. Some of these physicochemical processes are described here.

The option of adding chlorine to combine with ammonia and achieve a target monochloramine concentration can be very challenging. First, continuous monitoring of ammonia levels may be needed and can be difficult to achieve due to the current lack of readily available equipment to achieve highly accurate low-level ammonia (<0.5 mg/L NH_3-N) monitoring. Ammonia levels can change not only in static groundwater sources but also due to changing water sources, and mismatched chlorine-to-ammonia dosing increases the potential for free ammonia to remain and resulting nitrification to occur in the distribution system.

Adding excess chlorine to achieve breakpoint chlorination is an option that is increasingly being applied. For example, a system with 1 mg/L NH_3-N would require an approximate chlorine dose of 10 mg/L as Cl_2 to fully react with and oxidize available ammonia. During this chlorination process, it is important not only to complete breakpoint and oxidize all ammonia but also to allow for any additional reactions to complete before water enters the distribution system. For example, after breakpoint is achieved, it still may take several minutes (up to 30 min of contact time or more) for ratio of free residual chlorine-to-total residual chlorine (F:T) to surpass a ratio of 80% F:T. This 80% F:T level is a recommended goal derived from several groundwater treatment systems that apply the breakpoint process (Mofidi and Hill 2023).

Ion (cation) exchange softening is a relatively common drinking water treatment approach in some parts of the United States and can remove the ammonium ion (NH_4^+); however, there is no selective cation exchanger for the ammonium ion, and the role of pH on ammonia speciation must be considered (i.e., the dominant species shifts to unionized NH_3 as pH increases particularly above 9). Therefore, the ammonium ion capacity of the cation exchanger is impacted by important cations (e.g., calcium, magnesium) other than ammonium ion. Furthermore, ion exchange treatment systems typically bypass (to the treated water) a fraction of water (e.g., 20–30 percent), allowing for equivalent ammonia concentration fractions to remain; they can also produce a concentrated brine waste that must be considered. Perhaps more noteworthy is the tendency for exchanged ammonium ions to be eluted from the resin with operation cycle time by preferred cations (namely calcium) before regeneration (Lytle 2013). The resulting observation can be that ammonia is initially removed but reaches an exchange capacity earlier than calcium, and regeneration cycles are typically based on hardness ion removal. As a result, ammonia levels in the ion exchange effluent rapidly increase to influent levels. Continued operation can result in chromatographic peak effects where effluent ammonia levels can increase well above influent values. Considering multiple ion exchange vessels are often in operation, the final finished water ammonia levels can be variable and challenging to manage and may subsequently contribute to distribution system nitrification episodes (see Table 2-2).

Table 2-2 Summary of considerations for nonbiological treatment methods for ammonia

Treatment Method	Considerations
Chlorine addition to form monochloramine	• Challenge balancing ammonia and chlorine ratios, particularly with a changing water source • Potential for nitrification in the distribution system
Breakpoint chlorination (chemical oxidation)	• May require high chlorine doses • Taste and odor compound formation may require more advanced treatment, such as granular activated carbon adsorption to remove. • Disinfection byproduct formation
Ion exchange	• No selective cation exchanger so may be impacted by other cations present in water • Ammonium ion can be eluted by preferred cations before regeneration. • Multiple vessels can result in variable levels in finished water.
Reverse osmosis (RO)	• Removal of ammonia through RO may be poor at higher pH (>9). • In addition to ammonia, RO will remove other cations and anions from water. Permeate water will be corrosive and will need posttreatment for stabilization. • RO process will also generate a waste stream with high concentrations of ammonia in the concentrate. • RO process is costly to implement.

Biological Ammonia Treatment for Drinking Water

Biological nitrification can not only develop in holding or sedimentation basins but especially within filter beds if chlorination is absent/delayed until after physical treatment processes. Biological nitrification within the treatment plant can be an effective approach to reducing ammonia levels.

Biological nitrification consisting of water aeration and a subsequent (or simultaneous) biological ammonia oxidation step has been generally recommended in Europe as preferable to chemical methods (Janda and Rudovsky 1994; Rogalla et al. 1990) for reducing source water ammonia. Bouwer and Crowe (1988) documented the use of various biological ammonia removal methods throughout Great Britain, France, and Germany, including fluidized beds, rapid sand filters, biologically active granular activated carbon (GAC), and soil-aquifer treatment.

The use of biologically active filtration to oxidize ammonia as a full-scale drinking water treatment process has grown in acceptance in some regions of the United States. Some potential issues associated with biological water treatment include the potential release of excessive bacteria into finished waters, sensitivity of bacteria to changes in water chemistry and operating conditions, and a lack of long-term documentation of the effectiveness and reliability of biological water treatment processes.

The effectiveness of biological ammonia oxidation treatment to remove source water ammonia levels is dependent on a number of source water and engineering design factors, including temperature, ammonia concentration, presence of other contaminants (e.g., iron, manganese), dissolved oxygen concentration, oxygen delivery system, total organic carbon (TOC), phosphorus, pH, biomass quantity and microbiological community make-up, media type, and surface area, as well as support media characteristics, hydraulic loading rate and contact time (Lytle et al. 2015, 2013, 2007b; Andersson et al. 2001; Uhl and Gimbel 2000; Janda and Rudovsky 1994; Rittmann and Snoeyink 1984).

Canadian researchers (Kihn et al. 2002; Andersson et al. 2001) investigated the impact of GAC and temperature on biological nitrification using full and pilot-scale water treatment filters. Full-scale studies were conducted with Mille-Iles River (Quebec) water (dissolved organic carbon [DOC] of 4–10 mg C/L and alkalinity 20–50 mg $CaCO_3$/L) that contained

0.02–0.12 mg/L NH_3-N. Full-scale evaluations were performed using loading rates of 3.9–5 m/h (empty bed contact time [EBCT] of 20–30 min). At 15°C, an average of 94 percent ammonia oxidation was achieved, and at 5°C, ammonia oxidation was insignificant. Pilot-scale studies were conducted using Mille-Iles River water enriched with 0.4 or 1.2 mg/L NH_3-N of ammonia at a loading rate of 5 m/h and an EBCT of 3.6 min. Complete ammonia oxidation during pilot studies ranged from 40–90 percent at temperatures above 10°C. Ammonia removal dropped to 10–40 percent at moderate temperatures of 4–10°C. The results demonstrated the dependency of biological ammonia oxidation on temperature where oxidation improved as temperature increased.

Lytle et al. (2007b) monitored the performance of ammonia oxidation in a full-scale, 1.2 million gal per day (MGD) iron-removal filtration plant for more than a year. The treatment train consisted of a forced draft aerator followed by a basin located at the bottom of an aerator that provided a retention time of approximately 12 min, and three dual media (sand, anthracite) filters operated at a hydraulic loading rate of 2 gal/m/ft^2. Source water contained 1.2 mg/L NH_3-N and 2.3 mg/L iron with a pH of 7.48. The treatment plant consistently removed ammonia levels by greater than 95 percent, and nitrification was complete (nitrite was not detected in filtered water). Seasonal variations in ammonia oxidation effectiveness were not observed because yearly changes in water temperature and other water quality parameters were minimal in the groundwater source.

Filter media (sand, gravel, anthracite, or GAC) have been used as a support for the growth of nitrifying bacteria. Sand covered with manganese dioxide has also been reported to be an effective support for the attachment of nitrifying bacteria (Janda and Rudovsky 1994). Kimura et al. (2001) described operation of an experimental rotating biofilm membrane (polysulfone, 750,000 Dalton molecular weight cut-off) pilot reactor treating coagulated and settled river water for ammonia removal.

Nitrifying bacteria are recognized as slow growers, so colonization of filters will take time. During this time, ammonia will pass through the filter, and as the system transitions, finished nitrite levels will likely be elevated for a brief amount of time. Lytle et al. (2007b) conducted pilot tests using dual anthracite/sand filters to predict the amount of time required to achieve complete nitrification of 1.3 mg/L NH_3-N by three different seeding methods of new filters. Simply running aerated raw water through a virgin dual media filter (sand, anthracite) promoted bacterial growth and nitrification just as quickly as efforts to seed the new filters by either adding some biologically active full-scale media or by adding backwash solids collected during backwashing of full-scale nitrifying filters. The pilot filters achieved greater than 95 percent ammonia removal in less than 80 days. During microbial colonization of these filters, nitrite levels as high as 0.75 mg/L NO_2-N were measured. Pilot data were used to predict the time necessary to reach complete nitrification abilities of associated full-scale biologically active filters. Following a full-scale filter rebuild (sand, anthracite) to implement these results, complete nitrification of 1.3 mg/L NH_3-N was achieved by 150 days, slower than pilot predictions (Lytle et al. 2009). The difference was attributed to the fact that the full-scale system was only operated for approximately 12 h/day while the pilot operated 24 h/day.

Operations and media impacts on biological treatment for ammonia reduction. The key components to a biological treatment process to oxidize ammonia include the aeration/oxygenation delivery system, microbiological growth support system, filtration system (can also support microbiological growth), disinfection, pre- and posttreatment requirements, and handling of residuals. Rittmann and Snoeyink (1984) described four types of nitrification process units and conditions for filtration: biological filters (flooded or trickling), fluidized bed filters, rapid sand filters, and GAC beds. Low loading rates, long solids retention times, and steady-state operation are needed to achieve effluent ammonia concentration less than 0.1 mg/L NH_3-N. The effectiveness and reliability of nitrifying biological filters for ammonia removal are also dependent on other factors, including hydraulic loading rate, backwash frequency, duration and quality, the presence of iron, and other water quality variables.

Laurent et al. (1997) stated that temperature was one of the major factors impacting nitrification in a filter bed, with the optimum between 25 and 30°C. When temperature dropped below 7°C, nitrification in GAC/sand filters was more effective than in the sand filters. Filter media porosity, the filtration rate, and EBCT are the key parameters for the removal of biodegradable organic matter and ammonia at filtration rates varying from 5–16 m/h. Other studies showed that GAC filters provided better nitrification than sand or anthracite filters, especially when external conditions were unfavorable (low ammonia concentration, low temperature). On a volumetric basis, GAC has been shown to be three times more effective than sand (Kihn et al. 2002). It was found that a period of colonization is required following startup of biological filters before constant ammonia removal is reached. During this period, ammonia breakthrough and nitrite formation can alter the quality of filtered water. The GAC filters at 20°C showed that nitrification occurred two weeks after the filters were in service (Kihn et al. 2002).

More porous filter media were observed as better for achieving nitrification in the filter bed, especially in cold water, with the following order of efficacy: macroporous carbon > microporous carbon > anthracite > sand. Gravel filters treated the River Thames water at rates of 0.8–2.8 m/h with a raw water ammonia concentration of 2–3 mg/L NH_3-N. Despite lower efficiency, the biological filter maintained about 50 percent efficiency at 5°C and a loading rate of 2.4 m/h and maintained good nitrification, even at 3°C. In the fluidized filter reactor, nearly 100 percent removal was achieved at temperatures of 4–21°C, as long as the fluidized solids concentration was at least 30 percent by volume. About 50 percent removal of 1.3 mg/L NH_3-N occurred during rapid sand filtration of Illinois groundwater at 4.9 m/h. Activated carbon columns loaded at 4.84 m/h (9 min EBCT) removed 78 percent of ammonia across the bed. No information was given if nitrification was complete (to nitrate) or incomplete (to nitrite) in any of these cases (Rittmann and Snoeyink 1984).

Ozone-Enhanced Biofiltration for Ammonia Removal

Ammonia removal has been observed to occur during ozone-enhanced biological filtration (different from the above biological nitrification process treatment discussions), usually to remove ammonia that was added to control bromate formation (Wert 2011). Several studies, summarized below, have shown that nitrification does not occur until several months of steady-state operation.

Pilot studies demonstrated performance of a shallow (21 in. anthracite over 10 in. sand) and deep bed (72 in. anthracite over 8 in. sand) biologically active filter for nitrifying removal of ammonia removal (conversion to nitrate). The studies showed that lower filtration rates and higher filter EBCTs enhanced nitrification. However, it took nearly 100 days of operation to start converting ammonia to nitrite, and conversion of nitrite to nitrate was not achieved until 165 days (Neemann et al. 2004), with full nitrification at 170 days (Wert et al. 2008). The study showed that GAC performed with a higher removal efficiency than anthracite (Stoughtenger and Wert 2009).

Additional pilot studies conducted by Lauderdale and coworkers (2018) evaluated nutrient addition (adding phosphate with 0.025 mg/L-P phosphoric acid) for GAC biofilters to remove ammonia only (0.5– 1.5 mg/L NH_3-N), and ammonia from chloramine (0.5–1.5 mg/L total chlorine) downstream of ozonation. Like the above pilot, it took over 90 days to reach steady-state nitrification with both ammonia and chloramine conditions. Performance decreased with water temperature below 15°C, and it decreased when ammonia (and chloramine) levels decreased.

A full-scale 21-MGD facility with 0.2 mg/L NH_3-N ammonia added before ozonation was able to achieve ammonia reduction in biologically active anthracite/sand filters after many months of operation: specifically, 0.1 mg/L NH_3-N removal in 5 months and 0.15 mg/L NH_3-N removal in 9 months (Teefy et al. 2010, 2008).

A study evaluating both anthracite and GAC biofilters from ten utilities across the United States was conducted by Upadhyaya and coworkers (2017). This study showed challenges with biofiltration's ability to reduce ammonia in that although filters were all confirmed to be biologically active (TOC and DOC removal), normal operation of anthracite/sand filters through 18 weeks of observation showed no ammonia reduction. Ammonia reduction required filters to be operated in an "engineered" mode (pH adjustment from 7.2–7.8, addition of 0.02 mg/L-P phosphorus, and intermittent addition of 0.2 mg/L hydrogen peroxide to control excess microbial activity) with complete ammonia removal after 5 weeks of operation.

Cases of Incomplete Nitrification

Seasonal variations in temperature lead to periods of process instability and incomplete removal of ammonia or increased nitrite production, both of which are undesirable from a process control and water quality point of view. Other potential issues with the application of biological processes as part of the treatment process include biomass shedding and generation of microbial growth byproducts, which may lead to an increased disinfectant demand. Below are some examples of incomplete nitrification.

Andersson et al. (2001) reported 45–90 percent ammonia removal in full-scale filters at temperatures of 4–10°C, but this removal efficiency decreased to below 30 percent in cold water below 4°C. Uhl and Gimbel (2000) discussed the periodic inadequacy of the treatment process (ozonation, clarification, ozonation, dual-media filtration, GAC adsorption, groundwater infiltration) in Germany treating ammonia in river water. Ammonia concentration in the river water varied between 0.05 mg/L NH_3-N in summer and up to 4.2 mg/L in winter. During most of the year, effective ammonia removal by the treatment train was achieved. At temperatures above 20°C, up to 1 mg/L NH_3-N ammonia was already removed during the clarification step. When water temperature dropped below 5°C, ammonia concentrations in the raw water increased very quickly, and the removal capacity of nitrifying bacteria in the filters was not sufficient to treat these high concentrations.

Kors et al. (1998) discussed a case of incomplete nitrification in the Netherlands under extreme cold-water conditions below 4°C. Typically, the treatment plant (coagulation, sedimentation, holding reservoir with 100 days detention, rapid sand filtration, ozonation, softening, biological activated carbon (BAC) filtration, slow-sand filtration) can efficiently remove 1.5–3.0 mg/L NH_3-N raw water ammonia. The holding reservoir usually accounts for 50–90 percent of ammonia oxidation to nitrate, and the remaining ammonia (0.3–1.2 mg/L NH_3-N) is nitrified during passage through the rapid sand filters. Nitrite is generally not present in either reservoir or sand filter effluents. The temperature decreased during the winter of 1995–1996 to below 2.5°C for 100 days and was less than 1°C for 35 days. Water production decreased by 20 percent, backwash frequency decreased, and low-ammonia raw water was blended to help maintain nitrification. These steps improved the removal, but the rapid sand filter effluent ammonia concentration increased to 0.8 mg/L NH_3-N. A phosphate salt solution was dosed ahead of the rapid sand filters at 0.1–0.15 mg/L PO_4^{3-} and resulted in an improvement of nitrification efficiency after 7–10 days lag time.

Positive results in restoring nitrification by phosphate were known from the literature and other utilities' experiences. The authors indicated that a full restoration of nitrification was possible at 1°C with phosphate addition and other treatment changes previously discussed.

The oxygen demand of nitrification is also significant. For complete nitrification (complete conversion to nitrate), 4.57 mg O_2 is required per mg NH_4-N oxidized (USEPA

1975). If source water contains ammonia exceeding the stoichiometric amount of oxygen necessary to biologically oxidize it, the water will likely experience incomplete nitrification. Resulting finished water may contain elevated nitrite and ammonia.

Need for Treatment Following Nitrifying or Biological Filters

Very little information is available on the microbial ecology within a full-scale nitrifying biological water treatment system. White (2010) and White et al. (2009) reported on the microbial community within a biologically active, full-scale filtration water treatment plant in Ohio (details described in Lytle et al. 2007b) using molecular microbiological approaches. A culture-independent analysis of the filter media using 16S and *amoA* gene molecular markers revealed a complex microbial community. Specifically, the major ammonia oxidizers were identified as members of the genus *Nitrosomonas,* the majority being closely related to *Nitrosomonas oligotropha.* Other ammonia-oxidizing bacteria identified were *Nitrosomonas europaea* and *Nitrosospira multiformis.* The nitrite-oxidizing bacteria (NOB) guild ecology was more diverse. The majority (greater than 50 percent of clones) of identified nitrite oxidizers were of the genus *Nitrospira* of the phylum Nitrospirae. Although many clones were closely related to the order Rhizobiales, no sequences were identified as *Nitrobacter,* and no ammonia-oxidizing archaea were found in the filters. Aside from nitrifiers, members of *Sphingomonadales, Rhizobiales, Rhodobiales, Plantomycetes,* Bacterodietes, *Burkholderia,* and Methylophiales were identified. Additional research should be conducted to identify the bacterial colonies present in nitrification/denitrification processes such that better engineered process controls can be developed in the future.

As early as 1935, Feben observed that filters receiving ammoniated water were a near ideal environment for growth and multiplication of nitrifying bacteria and that the cells tended to clump into zoogleal masses. The tendency of sand grains to become coated with aluminum hydroxide floc allows ammonia-oxidizing microorganisms (AOM) growth to occur. The filter effluent contains these organisms, and if they survive postchlorination, they will enter the distribution system. Feben (1935) discussed nitrification in the filter beds and argued against preammoniation before the filters and chlorination after: "Sand filtration after ammoniation promotes their growth to a point where the chloramine process defeats its own purpose and becomes costly due to wasting of both ammonia to feed the bacteria and chlorine to oxidize their products." He also demonstrated the relative difficulty in disinfecting nitrifying bacteria with chlorine.

Bacterial aggregates from biological processes or bacteria attached to sand or GAC particles are more resistant to disinfection than individual bacterial cells. Thus, a higher chlorine dosage and a longer contact time may be necessary following a biological process to effectively disinfect the product water (Bouwer and Crowe 1988). Below are some documented examples and solutions that were employed.

Skadsen (1993) and Tokuno (1997) reported cases of detrimental impacts of GAC filtration on distribution system chloramine stability and nitrification. In both instances, raw water was chloraminated and high doses of free ammonia were applied to fresh GAC filters. Skadsen (1993) reported the first incident of chloramine loss and nitrification in the distribution system to coincide with the change from sand to GAC filter medium. Predisinfection with monochloramine and 1.3 mg/L NH_3-N excess ammonia applied to the GAC filters was thought to resultantly seed the system with nitrifying bacteria from the filter bed. Possible release of GAC fines was also implicated in increased heterotrophic plate counts (HPCs) in the system. Tokuno (1997) compared operation of the newly installed GAC filter to anthracite/sand filters fed with 4 mg/L total chlorine and >0.5 mg/L NH_3-N free ammonia. Nitrification in the GAC filter was evidenced within two months of installation with complete removal of total chlorine and ammonia, pH drop of 0.4 units, 3 mg/L NO_2-N nitrite, and loss of residual in the distribution system.

The above and other similar examples led to the conclusion that posttreatment is needed. Rittmann and Snoeyink (1984) stated that a single barrier (i.e., postdisinfection) between the consumer and a process in which microbiological growth is encouraged may be considered insufficient in the event undesirable organisms are produced by the process. A good alternative is the use of a biological process, such as a fluidized filter bed, as the first process in the treatment train and to follow biological process with a conventional treatment sequence. Wilczak et al. (2003) observed that a free chlorine contact time of 1 h or more was necessary after ozone/BAC filtration to avoid severe nitrification in the distribution system. Some modern applications of nitrifying anthracite filters in France are followed by membrane filtration, which ensures that neither undesirable organisms nor growth byproducts pass into the product water (Wilczak 2004). The impact of treatment processes on chloramine stability and subsequent nitrification is discussed in detail in Chapter 4.

Lytle et al. (2007b) showed that chlorination can be very effective in reducing bacteria counts. They performed biological analysis (HPCs and AOM) across a full-scale, nitrifying biological filtration plant treatment train. Raw water contained 4.1×10^{2} cfu/mL HPCs and only 7.9×10^{-1} MPN/mL AOM. Full-scale filter media contained a large amount of HPC bacteria (4.5×10^{4} cfu/mL) and AOM (2.9×10^{3} MPN/mL). Treatment plant effluent (after applying chlorine) bacteria levels were at nondetected levels.

Mofidi and Linden (2004) showed how ultraviolet (UV) disinfection can be effective in treating biological filter effluent water, although care needs to be maintained in monitoring disinfection effectiveness and long-term performance degradation issues. Testing with UV irradiation was compared alongside chlorine and chloramines contact at both the bench- and full-scale (3 MGD). Bench-scale results showed that UV was comparable to the chlorine alternatives, reducing HPC counts from 10^{4}–10^{0} cfu/mL. Full-scale testing downstream of an anthracite/sand biofilter was conducted over a period of nearly 250 days. The challenge with UV treatment was that HPC counts slowly climbed in the UV effluent during the nearly 1-year duration of testing. Microbial growth was indicated to come from the main filter gallery pipelines (mortar-lined steel pipe) downstream of the UV reactor as sample taps were regularly backflushed with chlorine solution. For up to 50 days, UV and chlorine disinfection options showed similar performance. From 50 to 150 days, UV effluent counts increased from 10^{1}–10^{2} cfu/mL, while chlorine disinfectants kept HPCs to $<10^{1}$ cfu/mL. Disinfection-effluent counts remained stable at these levels.

WASTEWATER TREATMENT

There are many resources that describe how nitrification is achieved and optimized in wastewater systems. This section is offered to provide a summary of information that may be insightful to drinking water engineers and operators.

Controlled nitrification in wastewater treatment is used to decrease typically much higher ammonia loading (20–40 mg/L NH_3-N) than is experienced in drinking water (typically below 1 mg/L NH_3-N), allowing specific reactor and media designs to achieve full nitrification.

Reactor and media designs are optimized to promote biofilm growth with AOM to produce nitrite and then achieve full nitrification with NOB. Reactors can range from completely mixed (similar to rapid mixing/coagulation in drinking water) and plug-flow (similar to a chlorine or ozone contactor) basins to designs that also include media that allow attachment of microbial colonies. Retention times for these basins can range from 1–5 days, depending on temperature (Grady et al. 1999), and can reach efficiencies that exceed 90 percent (Daigger and Boltz 2011).

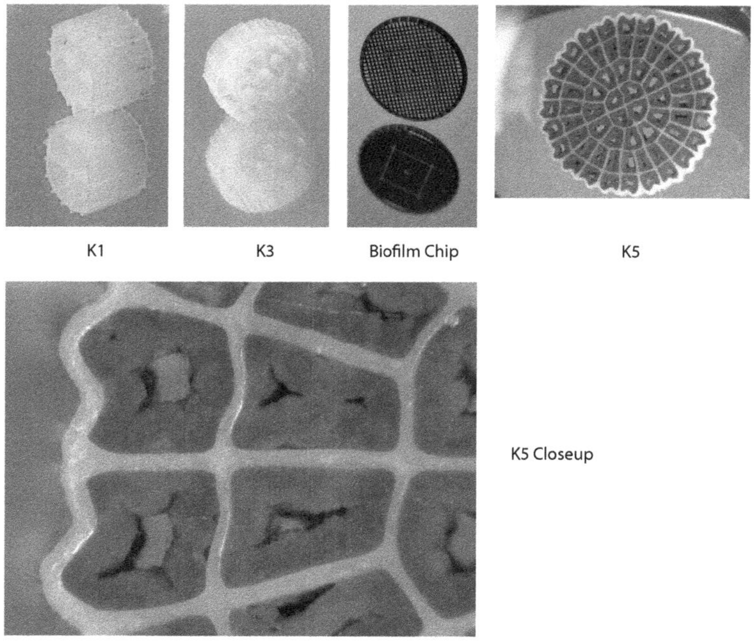

Sources: AnoxKaldnes AB, Alan Werker, and Magnus Christensson.

Figure 2-1 Typical plastic media used in deammonification moving bed biofilm reactor (MBBR) systems with specific surface areas ranging from 500 up to 1,200 m^2/m^3

Some process design advances over the past decades include optimizing recycle streams rich in ammonia and phosphate and addition of nutrients. Some designs optimize growth for specific bacteria and may be fed methanol or other organic carbon sources, operate under anoxic conditions, and use plastic support media (example support media is shown in Figure 2-1). Some of the specific technologies are proprietary and can include the following:

- ANAMMOX: anaerobic ammonia oxidation or deammonification;
- SHARON: stable-reactor, high-activity ammonia removal over nitrate;
- SBR: sequencing batch reactor;
- SHARON + ANAMMOX upflow granulation;
- MBBR: moving bed biofilm reactor (media options illustrated in Figure 2-1);
- MABR: membrane aerated biofilm reactor; and
- SGSSBR: suspended growth sequencing batch reactor.

COMPARING NITRIFICATION IN WATER AND WASTEWATER

Although both the drinking water and wastewater industries were the primary reasons behind modern society's capability to eradicate waterborne disease in many cities worldwide, these two industries have a long history of being separated in management strategies, treatment goals, and operator training and expertise. This is changing in many ways, one of which is the integration of biological processes in drinking water treatment. The institution of biofiltration—with or without purposeful growth of nitrifiers and AOM or successfully achieving partial/full nitrification or denitrification—challenges water supply managers and operations staff to become more knowledgeable in these processes. This chapter has provided applications of these practices in drinking water along with innovations in wastewater biological treatment that may eventually help drive further innovation in future drinking water process development.

There are some key differences and similarities between water and wastewater nitrification. Wastewater practices typically fall within the following categories.

- Wastewater nitrification is often operated to reduce ammonia levels from >10 mg/L NH_3-N, whereas drinking water ammonia levels may typically be below 1 mg/L NH_3-N.
- Nitrifying bacteria are slower growing in drinking water (due to low ammonia and nutrient levels) compared to wastewater applications. This slow growth concept is a key to nitrification control in drinking water distribution systems. It suggests that
  - combining free ammonia with booster chlorination, discussed in Chapter 9, can be effective for nitrification control;
  - presence of chloramine residual during a reservoir fill cycle can inhibit and further lower nitrifiers' growth rates; and
  - the above is supported by research indicating certain AOM are not as impacted by ammonia levels compared to other nitrifiers (Regan et al. 2002), indicating chlorine boosting/combining with free ammonia can be an effective tool to maintain a uniform disinfectant residual throughout the distribution system.
- Drinking water nitrification treatment is like wastewater tertiary submerged attached-growth reactors (low organic carbon levels and bacteria grow on filter media, sediments, and basin walls).
- Many of the differences between wastewater nitrification and drinking water nitrification may be due to the fundamental differences in microbial ecology beyond the scope of this brief review.

KEY TAKEAWAYS

Ammonia control in drinking water sources can be achieved with either physiochemical or biological processes (biological processes intended to remove ammonia should not be confused with biofiltration downstream of ozonation, which also can remove ammonia but operates at a much slower rate). Biological processes can have similarity to wastewater attached-growth reactors and are being increasingly applied in drinking water treatment when conditions can be optimized for AOM growth. Typical biological process operations promote nitrification during low and steady filtration rates. Nitrification can also occur in sedimentation basins but is more likely to need an attached-growth process such as a filter media to successfully decrease ammonia levels.

There have been developments in recent years showing that biological drinking water treatment processes can be operated reliably to remove ammonia. Furthermore, there are many postbiofiltration disinfection options available to water treatment operators to inactivate any sloughing of bacteria such that the distribution system can remain protected against colonization from nitrifiers.

Some additional lessons derived from wastewater treatment indicate that effective nitrification control strategies for drinking water distribution storage reservoirs are to (a) operate a completely mixed tank, (b) minimize biofilm deposits, and (c) maintain a high turnover rate that continuously washes out AOM. A completely mixed tank can provide the highest possible chloramine concentration and lowest possible free ammonia concentration and nitrifier growth rate.

REFERENCES AND BIBLIOGRAPHY

Andersson, A., P. Laurent, A. Kihn, M. Prévost, and P. Servais. 2001. "Impact of Temperature on Nitrification in Biological Activated Carbon (BAC) Filters Used for Drinking Water Treatment." *Water Research*, 35:12:2923–2934.

Bouwer, E.J., and P.B. Crowe. 1988. "Biological Processes in Drinking Water Treatment." *Journal AWWA*, 80:9:82–93.

Chen, A., L. Wang, D.A. Lytle, and T.J. Sorg. 2018. "Arsenic/Iron Removal from Ground Water with Elevated Ammonia and Natural Organic Matter." *Journal AWWA*, 110:3:E2–E17.

Christensson, M., S. Ekström, R. Lemaire, E. Le Vaillant, E. Bundgaard, J. Chauzy, L. Stålhandske, Z. Hong, and M. Ekenberg. 2011. "ANITA™ Mox–A BioFarm Solution for Fast Start-up of Deammonifying MBBRs." Proceedings of the Water Environment Federation 84th Annual Conference and Exposition, Los Angeles, California, October 2011.

Daigger, G.T., and J. P. Boltz. 2011. "Trickling Filter and Trickling Filter–Suspended Growth Process Design and Operation: A State-of-the-Art Review." *Water Environment Research*, 83:5:388–404.

European Union (EU). 2018. European Union Parliament Legislation in Progress. Revision of the Drinking Water Directive (February 1).

EU. Drinking Water Directive. 1998. European Council Directive 98/83/EC relating to the quality of water intended for human consumption (November 3).

Feben, D. 1935. "Nitrifying Bacteria in Water Supplies." *Journal AWWA*, 27:4:439–447.

Grady, C.P.L., Jr., G.T. Daigger, and H.C. Lim. 1999. *Biological Wastewater Treatment* (2nd ed). New York: Marcel Dekker.

Janda, V., and J. Rudovsky. 1994. "Removal of Ammonia from Drinking Water by Biological Nitrification." *Aqua*, 43:3:120–125.

Kihn, A., A. Andersson, P. Laurent, P. Servais, and M. Prévost. 2002. "Impact of Filtration Material on Nitrification in Biological Filters Used in Drinking Water Production." *AQUA*, 51:1:35–46.

Kimura, K., Y. Watanabe, and N. Ohkuma. 2001. "A Novel Biofilm-Membrane Reactor for Advanced Drinking Water Treatment—Pilot Scale Study." *Water Supply*, 1:5–6:157–168.

Kors, L.J., J.H.N. Moorman, A.P.M. Wind, and J.P. van der Hoek. 1998. "Nitrification and Low Temperature in a Raw Water Reservoir and Rapid Sand Filters." *Water and Science Technology*, 37:2:169–176.

Lauderdale, C., C. Alito, J. Hooper, K. Dowdell, and E. Wert. 2018. *Optimizing Biofiltration for Various Source Water Quality Conditions (Project 4555)*. Denver: Water Research Foundation.

Laurent, P., Niquette, P., Kihn, A., Prévost, M., Servais, P., & MacLean, R. G. (1997). Optimizing Nitrification in Biological Filters Used for Drinking Water Treatment. Water Quality Technology Conference, Denver, Colorado, USA

Lytle, D.A., C. Muhlen, and B. Almassalkhi. 2007a. Elevated Natural Source Water Ammonia and Nitrification in the Distribution Systems of Four Water Utilities. AWWA Water Quality Technology Conference. Charlotte, NC. Nov. 4–8.

Lytle, D.A., A.S. Chen, T.J. Sorg, S. Phillips, and K. French. 2007b. "Microbial As(III) Oxidation in Water Treatment Plant Filters." *Journal AWWA*, 99:12:72–86.

Lytle, D.A., D. Williams, C. Muhlen, B. Almassalkhi, and C. White. 2009. Consequences and Reduction of Elevated ammonia in Illinois Groundwaters: U.S. EPA Research Efforts. Presented at ISAWWA/IWEA Annual Meeting, Springfield, Illinois, March 16–19.

Lytle, D.A. 2013. Raw Water Ammonia: Implications on Water Quality, Regulatory Compliance and Management Strategies. Presented at Illinois AWWA Webinar Series, January 23.

Lytle, D.A., C. White, D. Williams, L. Koch, and E. Nauman. 2013. "Innovative Biological Water Treatment for the Removal of Elevated Ammonia." *Journal AWWA*, 105:9:E524–E359

Lytle, D.A., D. Williams, C. Muhlen, M. Pham, K. Kelty, M. Wildman, G. Lang, M. Wilcox, and M. Kohne. 2015. "The Full-Scale Implementation of an Innovative Biological Ammonia Treatment Process." *Journal AWWA*, 107:12: E648–E665.

Mofidi, A.A., and K.G. Linden. 2004. "Disinfection Effectiveness of UV for Bacteria Leaving Biologically Active Filters." *Journal of Water Supply: Research and Technology*, 53:8:553–566.

Mofidi, A.A., & A. Hill. 2023. Personal communication.

Murphy, B., J.T. O'Connor, and T.L. O'Connor. 1997. "Willmar Minnesota Battles Copper Corrosion Part I." *Public Works*, 128:11:65.

Neemann, J.J., R.A. Hulsey, D. Rexing, and E. Wert. 2004. "Controlling Bromate Formation During Ozonation with Chlorine and Ammonia." *Journal AWWA*, 96:2:26–28.

Regan, J.M., G.W. Harrington, and D.R. Noguera, 2002. "Ammonia- and Nitrite-Oxidizing Bacterial Communities in a Pilot-Scale Chloraminated Water Distribution System." *Applied and Environmental Microbiology*, 68:1:73–81.

Rittmann, B.E., P.M. Huck, and E.J. Bouwer. 1989. "Biological Treatment of Public Water Supplies." *Critical Reviews in Environmental Control*, 19:2:119–184.

Rittmann, B.E., and V.L. Snoeyink. 1984. "Achieving Biologically Stable Drinking Water." *Journal AWWA*, 76:10:106–114.

Rogalla, F., P. Ravarini, G. de Larminat, and J. Coutelle. 1990. "Large Scale Biological Nitrate and Ammonia Removal." *Water and Environmental Journal*, 4:4:319–328.

Servais, P., P. Laurent, and G. Randon. 1995. "Comparison of the Bacterial Dynamics in Various French Distribution Systems." *Journal of Water Supply Research and Technology*, 44:1:10–17.

Skadsen, J. 1993. "Nitrification in a Distribution System." *Journal AWWA*, 85:7:95–103.

Stoughtenger, S.S., and E.C. Wert. 2009. "Development of Nitrification in Biological Filters. In *Proc. of the AWWA Water Quality Technology Conference*. Denver: American Water Works Association.

Teefy, S., C. Peña, B. Gentry, I. Najm, A. Mofidi. 2008. Comparison of Bromate Control Strategies: pH Suppression vs. Chloramine Addition. AWWA Water Quality Technology Conference, Denver, Colo.: American Water Works Association.

Teefy, S., I. Najm, S. Sangines, B. Gentry, C. Peña. 2010. Full-Scale Demonstration of Bromate Control with Pre-Chloramination. AWWA Annual Conference and Exhibition, Denver: American Water Works Association.

Tokuno, S. 1997. "Granular Activated Carbon Filtration and Nitrification." *Water Engineering and Management*, 2:22–25.

Tokuno, S. 1999. "Granular Activated Carbon Filtration and Chloramine." *Water Eng. Manage.*, 1:16.

Uhl, W., and R. Gimbel. 2000. "Dynamic Modeling of Ammonia Removal at Low Temperatures in Drinking Water Rapid Filters." *Water Science and Technology*, 41:4–5:199–206.

Upadhyaya, G., Brown, J., Evans, A., Carter, J., Lauderdale, C., O. Schneider. 2017. Biofilter Conversion Guidance Manual (Project 4496). Water Research Foundation, Denver, CO.

US Environmental Protection Agency (USEPA). 1975. *Process Design Manual for Nitrogen Control*. Technology Transfer, Washington, D.C.: USEPA.

White, C. P. 2010. *Molecular Microbial Ecology and Operational Evaluation of a Full-scale and Pilot-scale Biologically Active Rapid Sand Filter for Drinking Water Treatment*. M.S. thesis. Cincinnati, Ohio: University of Cincinnati.

White, C.P., D.A. Lytle, R.W. DeBry, A.F. Galloway. 2009. Design, Evaluation, and Molecular Microbial Ecology of a Biologically Active Filter for Ammonia Oxidation. Water Quality Technology Conference. American Water Works Association, Seattle, WA.

Wilczak, A., J.G. Jacangelo, J.P Marcinko, L.H. Odell, and G.J. Kirmeyer. 1996. "Occurrence of Nitrification in Chloraminated Distribution Systems." *Journal AWWA*, 88:7:74–85.

Wilczak, A., L.L. Hoover, and H.H. Lai. 2003. "Effects of Treatment Changes on Chloramine Demand and Decay." *Journal AWWA*, 95:7:94–106.

Wilczak, A., 2004. Personal communication.

Wert, E.C., J.J. Neemann, D.J. Rexing, and R.E. Zegers. 2008. "Biofiltration for Removal of BOM and Residual Ammonia Following Bromate Control." *Water Research,* 42:1–2:372–378.

Wert, E.C. 2011. Personal communication.

This page intentionally blank.

Chapter 3

Nitrification in Chloraminated Drinking Water Distribution Systems—Occurrence

Ashley Pifer, David G. Wahman, Katherine Alfredo

INTRODUCTION

This chapter discusses available information on nitrification occurrence in chloraminated drinking water distribution systems. Chapter 4 introduces causes and controls of nitrification in chloraminated drinking water distribution systems. Both chapters are intended to serve as an overview, leading to further discussions on nitrification microbiology, monitoring, prevention, response, and engineering improvements in subsequent chapters. The key points from this chapter are summarized in Table 3-1.

Table 3-1 Key points from Chapter 3

Nitrification in chloraminated drinking water distribution systems	• Nitrification is a natural microbial process that occurs in ammonia-containing aquatic systems, including chloraminated drinking water systems. Ammonia serves as a nutrient (substrate) for ammonia-oxidizing microorganisms (AOM) that may include bacteria (AOB), ammonia-oxidizing archaea, and/or complete ammonia oxidizers. Of these three classes of microorganisms, AOB are the most widely studied in drinking water. Chapter 5 includes an in-depth discussion of the microbiology of nitrifying microorganisms. • Except for areas closest to the water treatment plant or distribution system booster disinfection facilities, AOM are ubiquitous in chloraminated distribution systems, according to studies. Low-flow dead-end mains created by design or closed valves, oversized mains, and poorly cycled or mixed water storage facilities are primary areas where nitrifiers can proliferate because chloramine levels can be very low or nondetectable. • If the AOM growth rate (supported by ammonia from chloramine decay and demand reactions, excess ammonia feed, or naturally occurring ammonia) exceeds the rate of AOM inactivation from chloramine or other disinfecting agents, nitrification will proceed in the presence of a disinfectant residual. Low levels of AOM are detected even at chloramine residuals reaching 5 mg Cl_2/L or higher. • Nitrification episodes can become more prevalent when water temperatures exceed 15°C, but some systems experience noticeable nitrification even with water temperatures below 5°C due to increased water age. Nitrification likely occurs to some extent in all chloraminated distribution systems, and if not controlled, it can be a year-round phenomenon. • Nitrifying microorganisms remain present in drinking water distribution systems in the winter. Their metabolic activity may decrease to a point where they no longer create detectable signs of nitrification, but they are still present and will increase activity as soon as water temperature increases. • Nitrification in drinking water distribution systems occurs over a wide pH range (7–10), but studies have demonstrated that higher pH (above pH 9) is less favorable for nitrification. • AOM grow in greater numbers in sediments and pipe/reservoir wall biofilm than in the bulk water. The highest numbers have been found in sediments. • AOB in distribution systems appear more resistant to chloramine disinfection than expected based on inactivation studies performed on suspended bacteria. This is possibly due to biofilm protection, cell aggregation, and the presence of a continuously supplied growth substrate (i.e., free ammonia). • Nitrification can occur in pipes made of any material. Concrete-lined pipes had the lowest levels of AOB and heterotrophic plate count bacteria, possibly related to higher pH/alkaline conditions at concrete pipe surfaces. AOB levels in unlined cast-iron pipes were inconsistent; however, nitrification episodes do occur in unlined cast-iron pipes. • Effective corrosion control programs may aid in nitrification control by minimizing chloramine decomposition and pipe biofilm, sediment, or microniche development. Metals leachable from system components, such as copper and zinc, have been shown to inhibit nitrification. On the other hand, phosphate, a common corrosion inhibitor component, has also been observed to increase nitrification rates.

NITRIFICATION IN CHLORAMINATED DRINKING WATER DISTRIBUTION SYSTEMS

Occurrence of Nitrification in Distribution Systems

As chloraminated water flows through a distribution system, the combined chlorine (chloramine) residual gradually decreases (see Chapter 4 for a discussion on chloramine decay and demand) due to bulk water (decay) reactions and pipe wall/sediment (demand) reactions, releasing free ammonia. The ammonia released through chloramine decay and demand provides a substrate for ammonia-oxidizing microorganisms (AOM), which use ammonia as their energy source, are slow growers, and are also fairly resistant to chloramine disinfection (see Chapter 5 for a discussion on the microbiology of nitrifying microorganisms).

The ammonia-oxidizing bacteria (AOB) *Nitrosomonas europaea,* for example, was found to be more resistant to chloramine than was *Escherichia coli* O157:H7 (Chauret et al. 2008). Because of these characteristics, AOM are well suited to living in chloraminated distribution systems, proliferating in areas with extended water ages (e.g., water storage reservoirs and dead-end mains) where there is likely a continuous availability of ammonia via chloramine decay and demand reactions. Protected environments such as corroded or tuberculated pipes, accumulated sediments, and biofilm serve to shield AOM from disinfectant residual and thereby also favor nitrification. Assuming water age is proportional to distance traveled through the system, it is understood that there is a greater likelihood of nitrification at points farther from the treatment plant or at any location where water stagnates (Harrington et al. 2003).

Based on utility surveys, 30–63 percent of utilities practicing chloramination for secondary disinfection experience periodic nitrification episodes affecting localized or widespread system areas (AWWA Disinfection Systems Committee 2008; Seidel et al. 2005; Kirmeyer et al. 2004; Wilczak et al. 1996). These findings likely reflect the perception of utility personnel about the patterns and frequency of major nitrification episodes, specifically that observable nitrification is sporadic and typically affects only localized areas. However, background nitrification likely occurs in all chloraminated systems, particularly in areas prone to corrosion/tuberculation, biofilm attachment, or solids accumulation. Such areas may include pipe surfaces or tank walls, especially at locations farther away from the treatment plant where a combination of factors (e.g., oversized mains, dead-end configurations, closed valves) allows for growth of AOM.

Some utilities may not realize the extent of nitrification in the distribution system, depending on their monitoring program (e.g., sample numbers, frequency, location, and parameters monitored). Overall, nitrification has been reported to occur in water distribution systems using chloramine throughout the United States, Canada, Europe, and Australia (e.g., Rosenfeldt et al. 2009; Carrico et al. 2008; Pintar et al. 2005; Sathasivan et al. 2005; Lipponen et al. 2002; Cunliffe 1991).

Temperature and pH Impacts on Nitrification in the Distribution System

Water temperature in distribution system pipes and storage facilities (reservoirs) can vary widely (e.g., 0–30°C), both seasonally and spatially. In drinking water distribution systems, nitrification has occurred over a wide (5–34°C) temperature range (Sarker et al. 2013; Pintar et al. 2005; Sung et al. 2005; Odell et al. 1996; Wilczak et al. 1996; Kirmeyer et al. 1995; Cunliffe 1991; Wolfe et al. 1988) but is most often reported at temperatures above 15°C (Wolfe et al. 1990; Skadsen 1993). As water temperature increases, three factors may explain why nitrification increases somewhat exponentially: increasing nitrifying bacteria growth rate (optimum 25–30°C), increasing chloramine (disinfectant) decay/demand, and increasing ammonia (substrate) release from the increasing chloramine decay/demand. Storage facilities (ground storage reservoirs, elevated tanks, etc.) and standpipes may experience significant nitrification during the summer as thermal stratification produces stagnated volumes of relatively warmer water combined with lower chloramine residuals.

Overall, region or climate does not seem to make a difference in repeated occurrence of nitrification episodes (Wolfe and Lieu 2002); however, seasonal variations in water temperature or water use–related system turnover may impact the observed extent and patterns of nitrification. Utilities in the northern United States and in Canada typically experience observable nitrification episodes in the summer or fall only, whereas systems located in more temperate, subtropical, or tropical climates may observe nitrifying activity year-round.

Chloramine decay and free ammonia release rates decrease with increasing pH due to greater chloramine stability at higher pH (i.e., greater than pH 8). See Chapter 4 for a discussion of chloramine chemistry. Maintaining an elevated pH is also a commonly used corrosion control approach. Hence, many chloraminated water systems elect to operate at higher pH ranges for chloramine stability and corrosion control. Nitrifying bacteria's optimal growth rate is pH 7.5–8.0 (Grady and Lim 1980), and it decreases at both lower and higher pH values (see Figures 5-8 and 5-9). Skadsen (2002) reported that an elevated pH of 9.3 versus 8.5 improved the control of nitrification in drinking water distribution systems but did not completely eliminate the problem. Sarker and Sathasivan (2012) investigated the impact of pH on nitrification control in pilot-scale studies and found that increasing pH reduced nitrification. Conversely, in a pilot system with a conventional coagulation process, Harrington et al. (2002) observed that the time for the onset of nitrification was faster at pH 8.5 and pH 8.9 than at pH 7.9. In an associated study, Oldenburg et al. (2002) observed a decrease in *N. europaea* inactivation rates with increasing pH over a range of chloramine concentrations, which suggests that the impact of pH on nitrification is, to some extent, system specific. Overall, nitrification has been reported in drinking water distribution systems with pH levels ranging from 6.5–9.8 (Odell et al. 1996; Wilczak et al. 1996; Kirmeyer et al. 1995).

Effect of Chloramine Concentration

If the AOM growth rate—supported by ammonia from chloramine decay/demand, excess ammonia feed, or naturally occurring ammonia—exceeds the inactivation rate of AOM from residual chloramine, nitrification will proceed in the presence of a chloramine residual. Nitrification can start and progressively increase even in the presence of low-to-moderate chloramine residuals (i.e., 1 mg Cl_2/L or less). As shown by the following examples from several water systems, once nitrifying communities are developed, their growth cannot be easily stopped, even by raising chloramine doses more than 8 mg Cl_2/L. A more detailed discussion of inactivation of AOM is presented in Chapter 5.

Cunliffe (1991) examined 1,184 samples collected from five chloraminated drinking water supplies to determine presence of nitrifying bacteria. Nitrifying bacteria were detected in 64 percent of all samples and in 21 percent of samples with total chlorine residuals greater than 5.0 mg Cl_2/L (Table 3-2); however, the median number of nitrifying bacteria present was several orders of magnitude higher when residuals were <0.3 mg/L. In a water quality survey of 10 drinking water distribution systems, Wilczak et al. (1996) observed greater than 0.050 mg N/L increases in nitrite and nitrate concentrations even

Table 3-2 Frequency of detection of nitrifying bacteria in five chloraminated South Australian water systems using a most probable number procedure

Total Chlorine Residual (mg Cl_2/L)	Number of Samples Tested	Samples Containing Nitrifying Bacteria (%)	Median Number of Nitrifying Bacteria (#/mL)
0.1–0.2	343	88	130
0.3–1.0	156	71	4.1
1.1–2.0	215	64	2.0
2.1–3.0	182	57	0.5
3.1–4.0	134	46	<0.2
4.1–5.0	62	42	<0.2
>5.0	92	21	<0.2
Total	1,184	64	2.0

Source: Adapted from Cunliffe 1991.

though substantial chloramine concentrations (3–6 mg Cl_2/L) were entering the distribution systems.

Kirmeyer et al. (1995), Odell et al. (1996), and Harrington et al. (2002) observed that a chloramine dose above 2.0 mg Cl_2/L may prevent the onset of nitrification. However, once a population of nitrifying bacteria was established, a similar chloramine concentration did not inhibit nitrifying bacteria growth. Fleming et al. (2008) evaluated three full-scale chloraminated systems with nitrification potential curves (a method where the recommended chloramine residual is determined based on the growth and inactivation of nitrifying bacteria) and recommended a minimum chloramine residual of 1.8–2.2 mg Cl_2/L based on system specific operating conditions (i.e., pH 8.1–10.0 and temperature 17–29°C).

Skadsen (1993) observed nitrification—indicated by a decrease in chloramine residual, an increase in heterotrophic plate count (HPC), and/or an increase in nitrite concentration—in a full-scale drinking water distribution system with a chloramine residual of 4.6 mg Cl_2/L. The pH was 8.7–8.8 at the finished water storage facility (reservoir) and may have decreased by 0.5–1.0 pH unit in the distribution system. Signs of nitrification persisted even after increasing the chloramine concentration to 8.0 mg Cl_2/L. The water temperature was 14–25°C, and the chlorine-to-ammonia-to-nitrogen mass ratio (Cl_2:NH_3-N) was 3.6:1. Even after increasing the Cl_2:NH_3-N ratio to greater than 4.5:1, nitrification was still detected in this distribution system. This is not surprising given the significant loading of free ammonia that would have occurred at the chloramine concentrations and ratios reported above. Nitrification was finally stopped by temporarily transitioning the full distribution system from chloramines to free chlorine at 0.7–1.8 mg Cl_2/L. These reports do not contain information about the rates of chloramine decomposition that are critical to chloramine's ability to counteract a nitrification episode.

Bench-scale annular reactor studies by Schrantz et al. (2013) were able to halt an active nitrification event by simultaneously increasing the monochloramine concentration and decreasing the free ammonia concentration in the annular reactor influent. When the annular reactor effluent monochloramine concentrations reached 1.5 mg Cl_2/L, nitrite-oxidizing bacteria (NOB) activity decreased, and AOM and NOB activity could no longer be observed when the effluent monochloramine concentrations were stable at 2.9-3.0 mg Cl_2/L. These results appear to contradict the inability to arrest nitrification in practice by adjusting the chloramine concentration. The authors note that the apparent discrepancy is likely due to the presence of reactive pipe materials in actual systems and the time required to arrest the nitrification event, with monochloramine possibly exceeding the time available to bring the system back into regulatory compliance, which may depend on the severity of the nitrification episode.

Woolschlager et al. (2001) developed a comprehensive disinfection and water quality model that accounts for the ammonia-producing reactions and the dynamics of suspended and attached bacteria, including nitrifiers, within drinking water distribution systems. Results from the model indicated that increasing the chloramine residual to a concentration like that measured at the treatment plant effluent, 3.3 mg Cl_2/L, to inactivate nitrifiers was ineffective because the ammonia released from chloramine decay/demand allowed the AOB growth rate on ammonia to exceed the AOB inactivation rate from chloramines.

Interactions Between Heterotrophic Plate Counts and Nitrifying Bacteria

Heterotrophic bacteria can potentially benefit from interactions with nitrifying bacteria because nitrifying bacteria are known to release soluble microbial products that essentially serve as a food source for heterotrophic bacteria (Rittmann et al. 1994). As an example, Kindaichi et al. (2004) grew a nitrifying biofilm on an inorganic medium (i.e., feeding no organic carbon). The resulting community was 22 percent AOB, 28 percent NOB, and 50 percent heterotrophs. Even though no organic carbon was fed to this system, AOB/NOB could support heterotrophic growth. The interactions between nitrifiers and other bacterial communities are further discussed in Chapter 5.

Nitrification in Bulk Water Biofilms

Although nitrifiers do not require surfaces on which to grow, they will adhere to surfaces if available. For example, in the marine environment, nitrifying bacteria do not remain in the water but are found in deposited sediments. An important feature of nitrifying bacteria is that they can grow in aggregates. Examples of cell aggregates are shown in Chapter 5 (see Figure 5-4). This is significant because it highlights that drinking water nitrification is predominantly a biofilm phenomenon, and this bacterial cell aggregation provides a possible mechanism for some of the resistance to chloramine and free chlorine disinfection. The next section describes biofilms and nitrifier occurrence in more detail.

System Configuration, Operations and Maintenance, Piping Type and Condition

Low-flow lines such as dead ends, pipes sized for fire-flow requirements, and mains with closed valves as well as storage reservoirs are the most susceptible places for nitrification to develop (Cohen et al. 2001). Dead-end pipes should be avoided where possible, or special precautions such as continuously running blow offs or automatic flush hydrants can be taken to preclude excessive water age. The latter options are less favored in systems with limited water flow capacity or source water volumes. A valve-exercising program can be used to help identify unintentionally closed valves. It is not uncommon in water systems for at least 10 percent of the valves to be closed, creating artificial dead ends and potential water quality problems (Kirmeyer et al. 2000). When replacing water mains, utilities should consider downsizing water mains where excessive water age occurs while accounting for necessary fire flow and demand. Additional best practices related to water age are provided in Chapters 7 and 9 of this manual and in AWWA Manual M68, *Water Quality in Distribution Systems*.

Pipe material and corrosion. Nitrification can occur irrespective of pipe material (e.g., plastic, polyvinyl chloride, asbestos–cement, ductile iron, and cast iron), but biofilm development will depend on the material used. Certain pipe materials may provide more favorable conditions for nitrification.

For example, unlined cast-iron pipes or old mortar-lined iron pipes with heavy tuberculation may provide a protective environment for nitrifying bacteria growth, and corrosion byproducts in these pipe types will contribute to chlorine demand (Cohen et al. 2001). However, Stewart and Lieu (1997) determined AOB levels were inconsistent in cast-iron pipes, possibly due to microhabitat differences or inconsistencies in the AOB enumeration methods. Yet, they also found that AOB were found in numbers as high as 100,000/cm^2 (or 10,000/mg dry weight) in distribution system biofilm, suggesting that biofilm may act as a reservoir of AOM in the distribution system.

Stewart and Lieu (1997) further found that concrete-lined pipes had lower AOB and HPC levels compared to cast iron. This may be due to the alkaline nature of concrete and the impact of high pH on AOB. Carrico et al. (2008) conducted a study on the impacts of a one-month free chlorine application to three chloraminated systems in North Carolina. In these studies, cast-iron pipe was implicated in nitrification recurrence in two of the three systems, and water quality improved after replacement of cast-iron pipe with cement-lined ductile iron pipe.

Zhang and Edwards (2010) supported the idea that corrosion-related metals (such as zinc and copper) and nutrients (such as phosphate and inorganic and organic carbon) can affect nitrification. For example, addition of copper at 0.1 mg/L or zinc at 1 mg/L decreased ammonia conversion at low alkalinity (10 mg/L-$CaCO_3$), but the same impacts were not observed at high alkalinity (500 mg/L-$CaCO_3$), possibly due to complexation or scale-formation decreasing the amount of free copper present. Implementation of an effective corrosion control program may help minimize pipe biofilm, improve chloramine residuals, decrease HPCs, and aid in overall nitrification control. Utilities considering changes to disinfection or corrosion control practices, which change the water quality within the system, are cautioned to fully explore these options to avoid unintended consequences. Chapter 1 summarizes previous research related to nitrification and corrosion interactions in drinking water distribution systems.

Water age and reduced inactivation. The frequent detection of nitrifying microorganisms at the ends of distribution systems could be due to a combination of decreased water flows (increasing water age) and lower chloramine residuals (reduced inactivation). When lower chloramine residuals exist, AOM and NOB detection in the bulk water may be a result of organisms shifting from the biofilm to the bulk water. Srinivasan et al. (2008) found that the ratio of total bulk water bacteria to total biofilm bacteria increased with decreasing free chlorine residuals. This chapter further discusses water age in the section on storage facilities.

Sediment and biofilm. Accumulated sediment and biofilm can protect microorganisms from bulk water chloramine residual. Higher AOB concentrations were detected in storage reservoir and pipe sediment materials than in pipe biofilm samples (Wolfe et al. 1990). AOB levels in pipe sediment were around 10 most probable number (MPN)/mg dry weight, but their levels in pipe wall scrapings were <0.2–7.4 MPN/mg dry weight. Liu et al. (2019) investigated free chlorine and monochloramine penetration into a 2 cm-thick sediment from a drinking water distribution system storage tank. They found that limited penetration occurred over months of disinfectant application, indicating that a protective environment developed in the sediment.

The important factor contributing to AOB survival in the distribution system and reservoirs is their presence in the sediment and biofilm material, presumably because of the larger amounts of essential nutrients and protection from disinfectant (Wolfe et al. 1990). In another study, Lipponen et al. (2002) surveyed 15 drinking water distribution systems in Finland for the presence of AOB and NOB, finding high concentrations of these bacteria, especially in pipe sediments. The surveyed sites had only traces of total chlorine residual, and the estimated MPN/g dry weight AOB concentrations in the pipe sediment were in the thousands to hundreds of thousands per milligram. The NOB MPN/mg dry weight of pipe sediment were in the hundreds concentration range. The water samples contained a maximum of 390,000 MPN/L AOB and 300,000 MPN/L NOB, which would indicate a severe nitrification episode despite the low temperatures (5–19°C, average 12°C). No further operational details of the surveyed systems were presented, except for a low chloramine residual entering the systems below 0.7 mg Cl_2/L (a typical disinfectant residual for European practice).

Biofilm growth may contribute to the persistence of nitrifying microorganisms in drinking water systems. Camper et al. (1996) reported that a significant fraction of the bacteria growing in distribution systems was associated with biofilm. If the chloramine demand exerted by the biofilm or substratum is sufficiently large, the disinfectant will only penetrate the outer part of the biofilm, and ammonia will be released to the biofilm interior. In addition to protection from chloramine, the biofilm can provide protection during periods of free chlorination that may lead to the recurrence of nitrification when a switch is made back to chloramines. Carrico et al. (2008) determined that biofilm was not reduced by free chlorination (HPC/acridine orange direct count basis). Rosenfeldt et al. (2009) found that free chlorination lowered cell counts, but they increased back to previous levels when the system switched back to chloramine. Lee et al. (2011) and Lee et al. (2018) directly measured free chlorine and monochloramine penetration in a laboratory-grown nitrifying biofilm and found that minimal free chlorine biofilm penetration occurred even after applying it to a biofilm that had been fully penetrated by monochloramine. This implies that free chlorination may only inactivate the outer biofilm, leaving the inner biofilm unaffected by free chlorine, depending on the exposure time, flushing procedure, bulk water concentration, and system specifics (i.e., pipe materials). In addition, Pressman et al. (2012) directly measured monochloramine penetration into biofilm exposed to different free ammonia concentrations and found that a greater free ammonia concentration during biofilm exposure to monochloramine resulted in a faster recovery of the biofilm when monochloramine was removed.

Nitrification in Chloraminated Water Storage Facilities

As introduced in Chapter 1, nitrification can be a challenge in finished water storage facilities, which are used to equalize water demands and pumping, reduce distribution system pressure fluctuations, and provide reserves for firefighting, power outages, and other emergencies. Excessive water age in many storage facilities, particularly if accompanied by chloramine decay/demand and sediment and biofilm accumulations, is probably the most important factor related to overall water quality deterioration.

Excessive water age is caused by underutilization, where water is not regularly or substantially cycled through pipes or storage facilities and short-circuiting within poorly mixed storage facilities, leaving water to stagnate in uncirculated areas of storage (e.g., upper areas of thermally stratified storage facilities). If storage short-circuiting is severe, water age may range from a few minutes to weeks or months within the storage structure.

Uncirculated storage (i.e., floating storage located at too low an elevation in a pressure zone to regularly exchange water or systems with pumping schedules that keep tanks full without allowing cycling) or poorly mixed/cycled storage located remotely or sequentially in systems may show nitrifying activity first due to increased water age. Representative sampling of water in storage facilities is essential to characterize water quality throughout the storage contents.

Few guidelines are available on water turnover rates. The US Environmental Protection Agency's Area-Wide Optimization Program (AWOP) operational guideline for storage tank turnover is to maintain average turnover times less than 5 days but encourages systems to establish storage facility–specific turnover time targets based on water quality. Kirmeyer et al. (1999) collected available information on water turnover rates by conducting interviews with state regulators in the United States and collecting available guidelines from the literature, finding that

- the state of Georgia and the country of Scotland recommended a 50-percent daily turnover,
- the state of Virginia recommended 33-percent daily turnover,
- the state of Ohio recommended 25-percent daily turnover,
- a maximum 1- to 3-day turnover was recommended in Switzerland, and
- 5–7 days was recommended in Germany.

None of these guidelines have been specifically established for nitrification control. Even with good turnover, poor mixing can result in pockets of older water with lower disinfectant residuals, and even with good mixing, low residuals can result from excessive detention time (Grayman et al. 2000). Overall, turnover should be based on maintaining a desired water quality in the finished water storage facility.

EXAMPLES OF NITRIFICATION AND AMMONIA-OXIDIZING MICROORGANISMS/NITRITE-OXIDIZING BACTERIA DETECTION IN CHLORAMINATED DRINKING WATER DISTRIBUTION SYSTEMS

The following case studies collected across the globe show how different distribution system conditions can affect AOM/NOB occurrence and nitrification.

South Australian Water System

Cunliffe (1991) depicted a nitrification episode for a site located near the end of a chloraminated system in South Australia (Figure 3-1). The total chlorine residual decreased 1.5–2.5 mg Cl_2/L to below 0.1 mg Cl_2/L. The numbers of nitrifying bacteria and concentrations of oxidized nitrogen, nitrite, and nitrate all increased.

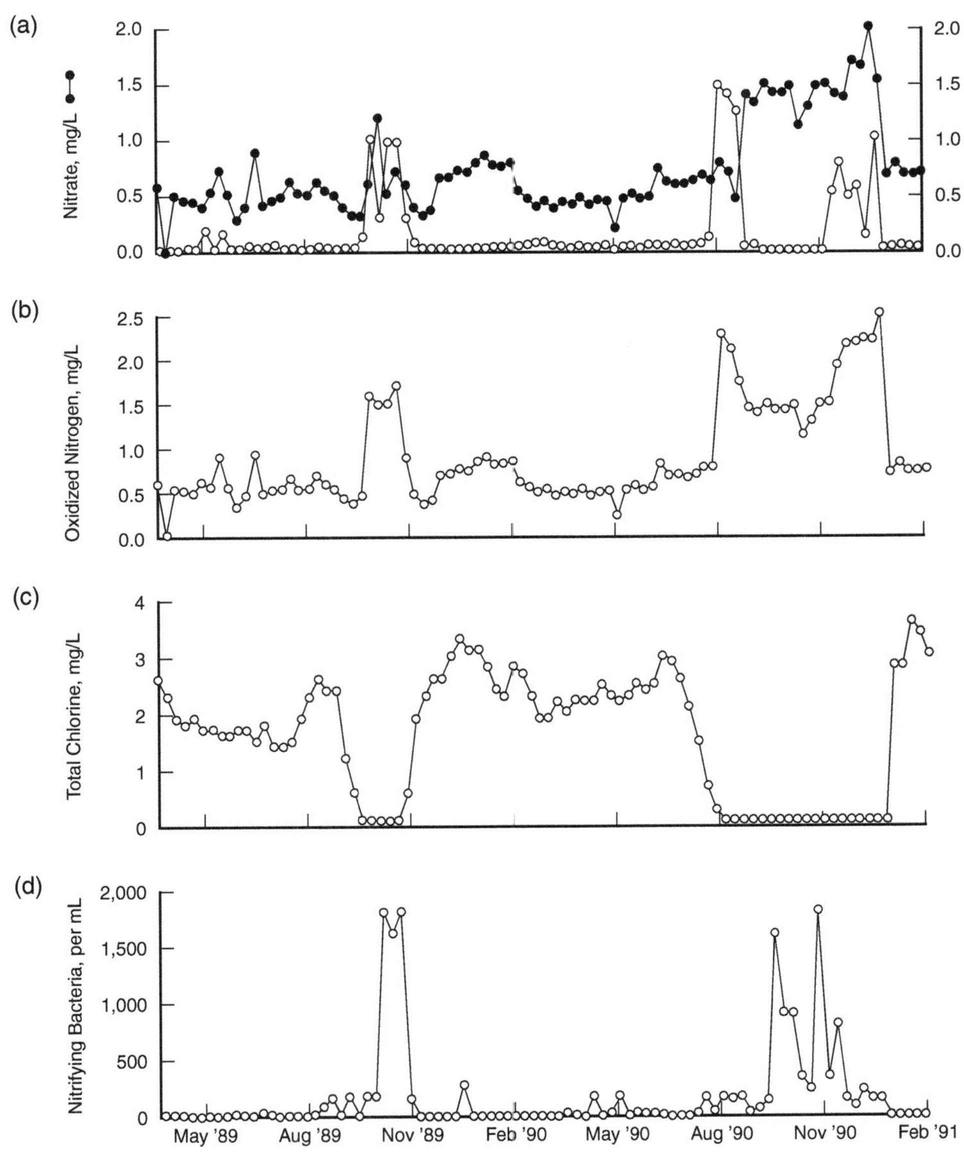

Source: Cunliffe 1991

Figure 3-1 Nitrification episode in a South Australian distribution system. Relationship between (a) nitrite and nitrate concentrations, (b) oxidized nitrogen concentrations, (c) total chlorine residual, and (d) number of nitrifying bacteria.

Typically, nitrite concentrations increase relatively quickly but are not always sustained. When nitrite is detected, AOM are typically well established in biofilm attached on material surfaces. Therefore, once AOM are established, it is conceivable that nitrite levels can increase noticeably within hours (Stewart and Lieu 1997). The AOB concentrations, determined using a culture-based method and shown in Figure 3-1d, started increasing about one month before the jump in nitrite levels and chloramine residual loss. Enumeration of AOM can be a long process (see Chapter 5 for discussion of isolation and enumeration techniques), and while molecular methods for AOM can provide more accurate results, field indicators such as nitrite provide real time results. HPC R2A bacteria counts may be used in some systems for diagnostic purposes. Experiences with HPC R2A as a diagnostic tool are mixed, and AOB enumeration has not been widely implemented. Some utilities have tested adenosine triphosphate (ATP) methods, which provide a gross parameter of microbial activity. Both HPC R2A and ATP measurements may require frequent sampling to develop baselines and detect trends. See Chapter 6 for additional discussion on these techniques for nitrification monitoring.

The AOB detection frequency and median number in the South Australian water distribution system referenced above decreased as the total chlorine residual increased, but nitrifying bacteria were still detected in 21 percent of samples that contained more than 5.0 mg Cl_2/L of total chlorine, although at very low levels. Table 3-3 presents the detection frequency in the chloraminated water system between 1988 and 1990 (although no temperature or pH conditions were given). Nitrifying bacteria were detected in 64 percent of the samples collected (Cunliffe 1991).

Surveys of individual systems showed that there is an association between increasing distance from the chloramine dosing station (i.e., increasing water age) and the frequency of bacteria detection. The results from a survey performed in 1988 and 1989 are shown in Table 3-3 (Cunliffe 1991). The percentage of positive samples for AOB and bacterial concentrations increased with the distance in the distribution system despite maintaining a chloramine residual at or above 2 mg Cl_2/L.

South Asian Drinking Water Distribution System

Cruz et al. (2020) characterized biofilm and bulk water samples from locations in a South Asian chloraminated drinking water distribution system. Biofilms were collected from harvested sections of cement-lined cast-iron, ductile-iron, and stainless-steel pipes, as well

Table 3-3 Occurrence of nitrifying bacteria in a chloraminated distribution system in Australia

Distance from Dosing Station (km)	Positive Samples (%)	Median Number of Bacteria (#/mL)	Median Chloramine Residual (mg Cl_2/L)
Before	91	2.5	<0.1
1.5	8	<0.2	6.4
27	0	<0.2	3.9
42	5	<0.2	5.6*
83	53	0.2	2.2
101	33	<0.2	2.5
110	35	<0.2	2.3
125	63	0.9	2.5
129	100	3.5	2.2
129	67	1.3	2.3
143	88	3.5	2.0

*Supplementary chloramination occasionally applied at this location
Source: Adapted from Cunliffe 1991

as asbestos–cement pipes. The pipes had been in service for 6–60 years. The sampling locations were <1–18.3 km from the water supply source. Monochloramine, free ammonia, nitrite, and nitrate were measured in bulk water samples and ranged from 1–2 mg/L-Cl_2, 0.06-0.19 mg/L-N, <0.01–0.06 mg/L-N, and 0.13–1.0 mg/L-N, respectively. The microbial communities in the biofilms were assessed using polymerase chain reaction (PCR), and nitrification activity tests were used to determine the MPNs for AOM and NOB.

Cruz et al. (2020) used statistical methods to group samples into two clusters, low diversity (LD) and high diversity (HD), based on the number and distribution of microbial species present. PCR and MPN data indicated that of the microbes related to nitrification, samples in the LD cluster were dominated by AOB (24.9–68.8% of sequences). Nitrifiers in samples from the HD cluster had AOB, AOA, and NOB, and NOB predominated in seven of the eight samples. The LD cluster included predominantly pipes that had been in service less than 35 years (10 of 11 pipes), while six of the eight pipes in the HD cluster had been in service 35 years or more. The LD cluster was correlated to higher monochloramine residuals (1.7 ± 0.2 mg/L-Cl_2), and the HD cluster was correlated to higher concentrations of total phosphorus, nitrate, and total trihalomethanes and lower pH. These differences are in line with differences in bacterial detections reported by Cunliffe (1991).

Examples of Water Storage Facility Nitrification and AOM Occurrence

Metropolitan Water District of Southern California, 1985–1990. Several reports (Wolfe et al. 1990; Ike et al. 1988; Wolfe et al. 1988) describe nitrification episodes that occurred from 1985 to 1987 in Garvey and Orange County reservoirs of the Metropolitan Water District of Southern California (MWDSC) system. MWDSC switched from free chlorine to chloramine on June 3, 1985, and at this time, MWDSC maintained a 1.5 mg Cl_2/L total chlorine residual, 3:1 Cl_2:N mass ratio (which is low compared to the 4.5:1–5:1 Cl_2:N mass ratios recommended today), 8.2 pH, and water temperatures that ranged between 12°C (winter) to 27°C (summer). Respectively, the Garvey and Orange County reservoir capacities were 1,600 and 200 acre-ft, hydraulic detention times ranged from 4–16 days (average 9 days) and 2–9 days (average 4 days), and both had synthetic elastomeric covers (see Figure 3-2).

Wolfe et al. (1988) reported that the Garvey Reservoir experienced a nitrification episode in September 1985 when the total chlorine residual and ammonia levels decreased, HPC levels increased, and nitrite reached 0.4 mg N/L. A detention time increase of 100 percent in the preceding weeks may have led to the nitrification episode. The Garvey Reservoir was ultimately taken out of service on September 28, 1985, and breakpoint chlorinated in October 1985. Subsequently, the chlorine-to-nitrogen ratio of this reservoir's influent was increased from 3:1 to 4:1, possibly preventing the reservoir from experiencing nitrification in 1986. It should be noted that higher (4.5:1 to 5:1) chlorine-to-nitrogen ratios are recommended today to further minimize free ammonia.

The Orange County reservoir did not experience nitrification in 1985 (Wolfe et al. 1988) but did in 1986. Over August 13–14, 1986, the total chlorine residual in the reservoir decreased from 1.4–0.8 mg Cl_2/L, and reservoir bulk water nitrite was 0.11 mg N/L (below 0.005 mg N/L in reservoir influent). Reservoir bulk water AOB levels were 280–510 MPN/mL, but AOB were not detected (i.e., below 0.2 MPN/mL) in the reservoir influent, indicating that AOB were actively growing in the reservoir. Likewise, HPC levels were approximately 10–30 times higher in the reservoir bulk water (1,830–12,300 cfu/mL) than influent (145–260 cfu/mL). The reservoir was taken out of service on August 14, 1986, and subsequently breakpoint chlorinated.

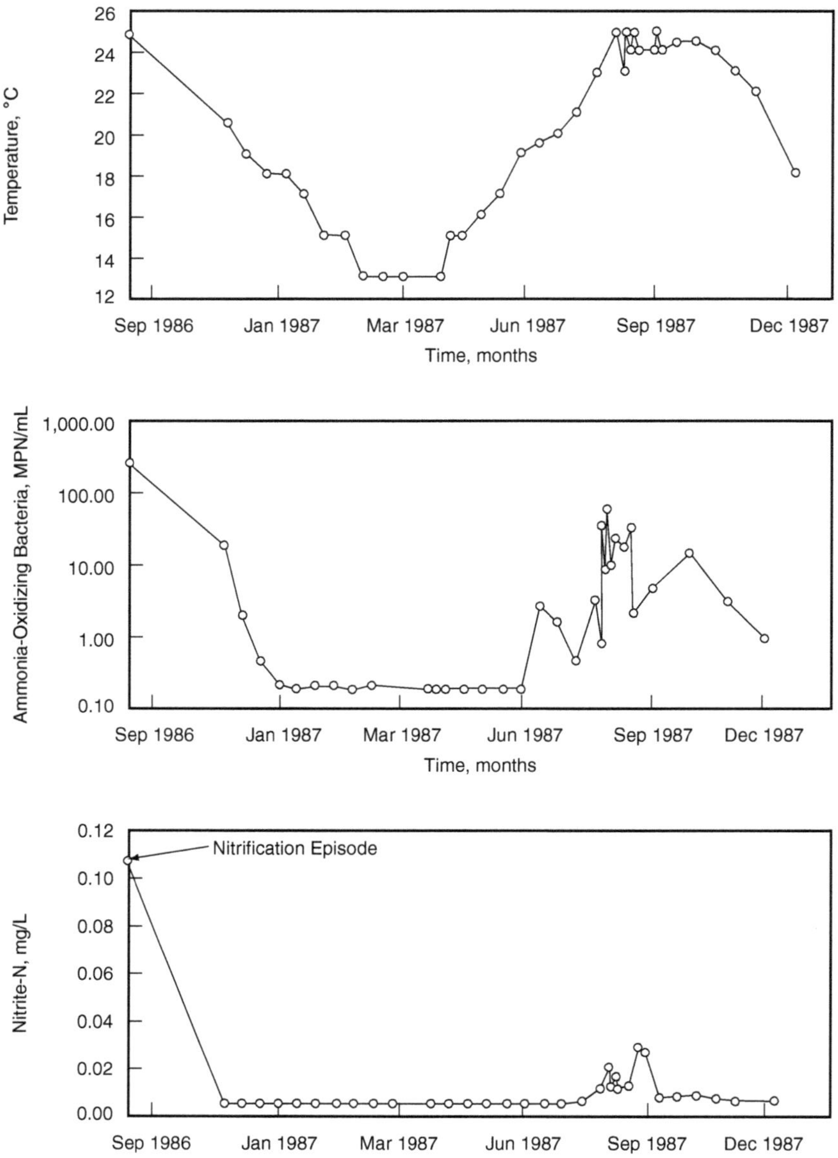

Source: Ike et al. 1988

Figure 3-2 Seasonal relationship between temperature, ammonia-oxidizing bacteria, and nitrite in the Orange County reservoir in California.

As a result of the nitrification episodes in the Garvey and Orange County reservoirs in 1985 and 1986, MWDSC implemented a sampling program in its covered reservoirs (Wolfe et al. 1988). Ike et al. (1988) present Orange County reservoir monitoring results from August 1986–December 1987. The AOB seasonal occurrence at the top of the water column (Figure 3-2) ranged from a high of 500 MPN/mL during the August 1986 nitrification episode (Wolfe et al. 1988) to below 0.2 MPN/mL during the winter months.

On June 18, 1987, an AOB increase occurred (18°C water temperature and 400 cfu/mL HPC), but no nitrite was detected. By August 5, 1987, within six days, the nitrite level had increased to 0.03 mg N/L, the chloramine residual decreased from 1.5 mg Cl_2/L to below 1.0 mg Cl_2/L, and the AOB level was 22 MPN/mL. In general, Ike et al. (1988) found that an AOB increase was detected when the water temperature was over 18°C and when the HPC was 350–500 cfu/mL. Nitrite was usually not detected until an AOB level of about

1 MPN/mL was reached. Detention time was observed to affect nitrification in the Orange County reservoir as nitrite and AOB numbers decreased when the detention time was decreased to 2 or 3 days, implying that AOB had limited impact on the bulk water within these shorter detention times (Ike et al. 1988).

During the nitrification event in the summer of 1987 (August 5 and September 3), Ike et al. (1988) conducted detailed sampling throughout the Orange County reservoir's depth for biological, chemical, and gravimetric parameters (Table 3-4). In the water column, AOB, nitrite, and temperature levels were the highest at the surface where chloramine residual was the lowest, but HPC levels appeared consistent throughout the water column. Greater levels of AOB, HPC, and nitrite within the reservoir compared to the reservoir influent indicated that AOB were actively growing in the reservoir.

Ike et al. (1988) also enumerated AOB in biofilm and sediment samples collected from the underside of the cover and from the bottom and sides of the reservoir. The bottom sediment appeared to have more AOB per area than the cover biofilm. For this system, Wolfe et al. (1990) also reported that the Orange County reservoir sediment contained higher AOB levels (67 MPN/mg dry weight) than those found in pipe sediment (10.4–12.1 MPN/mg dry weight) or pipe wall scrapings (below 0.2–7.4 MPN/mg dry weight). The bottom sediment was approximately 1 cm thick, brown in color, flocculant-like in consistency, and much of it was organic as indicated by its 51-percent total volatile solids content. In contrast, the biofilm material beneath the reservoir cover was very thin and difficult to remove (Table 3-4).

Supplementing information provided by Ike et al. (1988), Wolfe et al. (1990) also presented results from the MWDSC system over 1985–1987. Sampling locations included a treatment plant, the Garvey and Orange County reservoirs, and two finished water

Table 3-4 Biological and chemical analyses of samples collected from within a covered California finished water storage reservoir on August 5 and September 3, 1987

Site	Date	Temperature (°C)	Total Chlorine Residual (mg Cl_2/L)	Nitrite (μg N/L)	Ammonia-oxidizing bacteria (MPN/mL)	Heterotrophic plate count (cfu/mL)
Reservoir influent	Aug 5	23.5	1.5	<5	<0.18	9
	Sep 3	24.5	1.4	<5	<0.18	26
Reservoir effluent	Sep 3	24.5	1.5	12	10.2	390
Bulk water at 0.3 m (1 ft) depth (surface)	Aug 5	27.0	0.8	31	22	1,900
	Sep 3	24.5	0.9	19	1.9	1,100
Bulk water at 5 m (16 ft) depth (middle)	Aug 5	24.0	1.2	18	0.69	1,200
	Sep 3	24.5	1.3	18	0.18	1,400
Bulk water at 10 m (32 ft) depth (bottom)	Aug 5	24.0	1.1	16	6.7	1,200
	Sep 3	24.5	0.8	18	0.19	680
Cover biofilm	Aug 5				1,500*	94,000*
	Sep 3				52*	ND
Floor sediment	Aug 5				420,000*	3,800,000*
	Sep 3				39,000	700,000*
	Sep 3				67†	1,400†
Side wall sediment	Aug 5				200,000*	550,000*

Source: Adapted from Ike et al. 1988
* MPN/cm^2
† MPN/mg dry weight
MPN, most probable number; ND, not determined

pipelines. AOB (MPN basis) were found throughout the system, detected only when the water temperature was above 16–18°C, highest in the summer and below detection limits in the winter for both reservoir inlets and water columns, and generally much higher in the reservoir bulk water than reservoir influents. Few AOB were observed when HPCs were less than approximately 350 cfu/mL, and in the Orange County reservoir, nitrite levels generally remained below 0.01 mg N/L until AOB exceeded 10 MPN/mL. During the summer months, AOB numbers in both reservoir water columns were approximately 10 times greater than the reservoir influents, indicating that these organisms were able to survive and grow in the presence of 1.2–1.5 mg Cl_2/L of chloramine with an initial 3:1 Cl_2:N mass ratio. Wolfe et al. (1990) suggested that chloraminated systems may have to increase the residuals substantially above 1.5 mg Cl_2/L to control nitrifier growth.

Southern California utility #1. In the summer of 1994, Stewart and Lieu (1997) sampled a 2.75-million gallon (MG) coated steel tank in southern California. The utility blended groundwater with MWDSC's finished surface water. The chloramine (2.0 mg Cl_2/L) and nitrite (0.003 mg N/L) levels were similar throughout the tank. Although AOB were not detected in the bulk water, high levels of AOB and HPC bacteria colonized the biofilm layers attached to the tank sides, which were not smooth, as the paint coating was chipped in some areas. AOB levels on the tank wall surface ranged from a low of 11 MPN/cm^2 (middle level) to a high of 860 MPN/cm^2 (bottom level). The sludge sample also had high AOB levels (4,000 MPN/mg dry weight). Surface attachment (i.e., biofilm) offers bacteria protection from inactivation and serves as a source for seeding the bulk water.

Southern California #2. Baribeau et al. (2001) evaluated three covered reservoirs (one steel and two concrete) in southern California to characterize the suspended and attached biomass populations. The total chlorine residual in the reservoirs typically fluctuated 1–2 mg Cl_2/L, and the temperature varied between 12 and 26°C. Coupons used to examine biofilm were made of the same material as the reservoir in which they were deployed (i.e., steel or concrete) and were placed at two depths, corresponding to whether the coupon was constantly submerged (bottom) or intermittently submerged (top). AOB were detected in all reservoirs, in both bulk water and biofilm samples, and at all sampling depths.

Overall, bulk water sampling indicated higher AOB and HPC levels in the upstream steel reservoir than the two downstream concrete reservoirs, and AOB concentrations in the bulk water samples typically varied by 0–10 MPN/mL. AOB were detected within two weeks following coupon placement in the reservoirs, and higher HPC and AOB levels were measured in the biofilm coupons collected from the reservoir bottom (constantly submerged) than top (intermittently submerged). Reservoir cycling was not sufficient to prevent nitrification.

In one of the reservoirs, a severe nitrification episode occurred in winter when the water temperature was only 13°C–14°C. Free chlorination at 2.0 mg Cl_2/L effectively decreased biomass concentrations in both bulk water and biofilm samples. After free chlorination, AOB were not detectable in the bulk water samples and were reduced by 2–3.5 orders of magnitude in the biofilm samples. Even with the AOB decreases seen from free chlorination, experience has shown that nitrification may reoccur within a period of a few weeks or months (Baribeau et al. 2001).

Pennsylvania. In addition to chloramine residual loss and an HPC increase, nitrification in water storage reservoirs may lead to customer complaints. Organic compounds released by nitrifying bacteria and ammonia have been associated with taste and odor problems (Bouwer and Crowe 1988). Burlingame and Brock (1985) reported customer taste and odor complaints that occurred during nitrification episodes in storage tanks (10 MG) in Pennsylvania. Strong musty odors and a chemical taste were detected in the tank effluents in September 1983 after the chloramine residual decreased to nondetectable levels and HPC counts exceeded 500 cfu/mL. The tanks were drained and refilled, which restored chloramine residuals and eliminated the taste and odor problem. Both tanks began to display a similar loss of residual in 1984; therefore, cycling was improved, and booster chloramination was conducted at the tanks. Both measures resulted in maintenance of a residual above 1.0 mg Cl_2/L and no taste and odor complaints throughout 1984. An evaluation of historical data indicated that the recurring taste and odor events only occurred after chloramine residuals declined below 0.5 mg Cl_2/L and bacterial counts correspondingly increased. Subsequent monitoring found the nitrite level to increase significantly when the chloramine residual disappeared during the summer.

Note on the enumeration of nitrifying microorganisms in the examples. Past studies presented within this chapter include MPN levels for nitrifying microorganisms. It is important to note that, given the low recovery efficiency of the culture-dependent MPN techniques used (Wolfe and Lieu 2002), actual levels of AOM are likely to be much greater than reported. For example, Brion and Billen (2000) calculated that nitrifying bacteria MPN counts converted to biomass were 2–3 orders of magnitude lower than the estimates deduced from the measurements of potential activity in wastewater samples. Also, MPN counts from soil and sediment samples were only 0.1–5 percent of the estimated populations that would be required to produce the observed nitrifying activity (Belser and Mays 1982). Hoefel et al. (2005) found that AOB detected by culture-independent techniques were several orders of magnitude greater than those from the culture-dependent MPN, and Rosenfeldt et al. (2009) determined that total culturable cells were 0.2–4.6 percent of the total cells present in two chloraminated distribution systems. Estimates derived from nitrifying activity measurements based on the oxygen consumption rate in wastewaters (similar to BOD_5 technique) were deemed more realistic and in agreement with treatment models (Brion and Billen 2000). Given these discrepancies, MPN levels for nitrifying bacteria reported in this chapter for drinking water distribution systems should be considered underestimated and used only in a relative way. Chapter 5 discusses the details of methods for nitrifying microorganism isolation and enumeration that are currently available, as well as their recovery efficiencies.

KEY TAKEAWAYS

Nitrifying microorganisms and low (background) levels of nitrification are likely present in the majority of chloraminated systems. As water flows and ages within a distribution system, the chloramine residual decreases due to bulk water decay reactions and pipe wall/sediment demand reactions, releasing free ammonia. The ammonia that is released provides a substrate for AOB, which are ubiquitous in chloraminated distribution systems, as well as complete ammonia-oxidation bacteria and AOA, which have been detected in some chloraminated systems. AOB and AOA activity produces nitrite, which is used as substrate for NOB. Nitrification has been reported in distribution systems throughout the range of pH, temperature, and monochloramine residuals common to drinking water systems. Certain conditions, including pH of 7.5–8.0, temperature >15°C, low monochloramine residuals (<1.5 mg/L-Cl_2), and excess free ammonia are favorable conditions for nitrifying microorganisms to proliferate and more likely lead to significant levels of nitrification. Nitrification is typically associated with attached growth nitrifying bacteria in biofilm or sediment on the floor and walls of a reservoir or pipeline. Bulk water nitrification (i.e., suspended growth) is expected to occur only if this biofilm (attached growth) nitrification remains uncontrolled. Therefore, the level of nitrification would be the greatest in areas with extended water ages (i.e., water storage reservoirs and dead-end mains) and protective environments (i.e., corroding pipes/tubercles, sediments, and biofilm).

DISCLAIMER

The US Environmental Protection Agency through its Office of Research and Development collaborated in the production of this chapter. It has been subjected to the agency's peer and administrative review and has been approved for external publication. Any opinions expressed in this chapter are those of the authors and do not necessarily reflect the views of the agency; therefore, no official endorsement should be inferred. Mention of trade names or commercial products does not constitute endorsement or recommendation for use.

REFERENCES AND BIBLIOGRAPHY

AWWA Disinfection Systems Committee. 2008. "Committee Report: Disinfection Survey, Part 2—Alternatives, Experiences, and Future Plans." *Journal AWWA*, 100:11:110–124.

Baribeau, H., C.A. Kinner, R. De Leon, and P.A. Rochelle. 2001. "Nitrification in Steel and Concrete Drinking Water Reservoirs: A Case Study." In *Proc. of AWWA Annual Conference*. Denver: American Water Works Association.

Bautista-de los Santos, Q.M., J.L. Schroeder, M.C. Sevillano-Rivera, R. Sungthong, U.Z. Ijaz, W.T. Sloan, and A.J. Pinto. 2016. "Emerging Investigators Series: Microbial Communities in Full-scale Drinking Water Distribution Systems—A Meta-analysis." *Environmental Science: Water Research and Technology*, 2:4:631–644.

Belser, L.W., and E.L. Mays. 1982. "Use of Nitrifier Activity Measurements to Estimate the Efficiency of Viable Nitrifier Counts in Soils and Sediments." *Applied and Environmental Microbiology*, 43:4:945–948.

Bouwer, E.J., and P.B. Crowe. 1988. "Biological Processes in Drinking Water Treatment." *Journal AWWA*, 80:9:82–93.

Brion, N., and G. Billen. 2000. "Wastewater as a Source of Nitrifying Bacteria in River Systems: The Case of the River Seine Downstream from Paris." *Water Research*, 34:12:3213–3221.

Burlingame, G.A., and G.L. Brock. 1985. Water Quality Deterioration in Treated Water Storage Tanks. In *Proc. of the AWWA Annual Conference*. Denver, Colo.

Camper, A., W. Jones, and J. Hayes. 1996. "Effect of Growth Conditions and Substratum Composition on the Persistence of Coliforms in Mixed-Population Biofilms." *Applied and Environmental Microbiology*, 62:11:4014–4018.

Carrico, B.A., F.A. Digiano, N.G. Love, P. Vikesland, K. Chandran, M. Fiss, and A. Zaklikowski. 2008. "Effectiveness of Switching Disinfectants for Nitrification Control." *Journal AWWA*, 100:10:104–115.

Chauret, C., C. Smith, and H. Baribeau. 2008. "Inactivation of *Nitrosomonas Europaea* and Pathogenic *Escherichia Coli* by Chlorine and Monochloramine." *Journal of Water and Health*, 6:3:315–322.

Cohen, Y.K., C. Smith, and H. Baribeau. 2001. Nitrification: Causes, Prevention, and Control. *Opflow*, 27:12:6–12.

Cruz, M.C., Y. Woo, H.C. Flemming, and S. Wuertz. 2020. "Nitrifying Niche Differentiation in Biofilms from Full-Scale Chloraminated Drinking Water Distribution System." *Water Research*, 176:115738.

Cunliffe, D.A. 1991. "Bacterial Nitrification in Chloraminated Water Supplies." *Applied and Environmental Microbiology*, 57:11:3399–3402.

Fleming, K.K., G.W. Harrington, and D.R. Noguera. 2008. "Using Nitrification Potential Curves to Evaluate Full-Scale Drinking Water Distribution Systems." *Journal AWWA*, 100:10:92–103.

Friedman, M., G.J. Kirmeyer, M. LeChevallier, S. Seidl, and J. Routt. 2010. *Criteria for Optimized Distribution Systems*. Denver: Water Research Foundation.

Grady, C.P.L., Jr., and H.C. Lim. 1980. *Biological Wastewater Treatment*. New York: Marcel Dekker.

Grayman, W.M., L.A. Rossman, C. Arnold, R.A. Deininger, C. Smith, J.F. Smith, and R. Schnipke. 2000. *Water Quality Modeling of Distribution System Storage Facilities*. Denver: Awwa Research Foundation.

Harrington, G.W., D.R. Noguera, C.C. Bone, A.I. Kandou, P.S. Oldenburg, J.M. Regan, and D. Van Hoven. 2003. *Ammonia from Chloramine Decay: Effects on Distribution System Nitrification*. Denver: Awwa Research Foundation.

Harrington, G.W., D.R. Noguera, A.I. Kandou, and D.J. Vanhoven. 2002. "Pilot-Scale Evaluation of Nitrification Control Strategies." *Journal AWWA*, 94:11:78–89.

Hoefel, D., P.T. Monis, W.L. Grooby, S. Andrews, and C.P. Saint. 2005. "Culture-Independent Techniques for Rapid Detection of Bacteria Associated with Loss of Chloramine Residual in a Drinking Water System." *Applied and Environmental Microbiology*, 71:11:6479–6488.

Ike, N.R., R.L. Wolfe, and E.G. Means. 1988. "Nitrifying Bacteria in a Chloraminated Drinking Water System." *Water Science and Technology*, 20:11–12:441–444.

Kasuga, I., H. Nakagaki, F. Kurisu, and H. Furumai. 2010. "Predominance of Ammonia-Oxidizing Archaea on Granular Activated Carbon Used in a Full-Scale Advanced Drinking Water Treatment Plant." *Water Research*, 44:17:5039–5049.

Kindaichi, T., T. Ito, and S. Okabe. 2004. "Ecophysiological Interaction between Nitrifying Bacteria and Heterotrophic Bacteria in Autotrophic Nitrifying Biofilms as Determined by Microautoradiography-Fluorescence in Situ Hybridization." *Applied and Environmental Microbiology*, 70:3:1641–1650.

Kirmeyer, G., K. Martel, G. Thompson, L. Radder, W. Klement, M. LeChevallier, H. Baribeau, and A. Flores. 2004. *Optimizing Chloramine Treatment*. Second ed. Denver, Colo.: Awwa Research Foundation.

Kirmeyer, G.J., M. Friedman, J. Clement, A. Sandvig, P.F. Noran, K.D. Martel, D. Smith, M. LeChevallier, C. Volk, E. Antoun, D. Hiltebrand, J. Dyksen, and R. Cushing. 2000. *Guidance Manual for Maintaining Distribution System Water Quality*. Denver, Colo.: Awwa Research Foundation.

Kirmeyer, G.J., L. Kirby, B.M. Murphy, P.F. Noran, K.D. Martel, T.W. Lund, J.L. Anderson, and R. Medhurst. 1999. *Maintaining Water Quality in Finished Water Storage Facilities*. Denver: Awwa Research Foundation.

Kirmeyer, G.J., L.H. Odell, J. Jacangelo, A. Wilczak, and R. Wolfe. 1995. *Nitrification Occurrence and Control in Chloraminated Water Systems*. Denver: Awwa Research Foundation.

Lee, W.H., D.G. Wahman, P.L. Bishop, and J.G. Pressman. 2011. "Free Chlorine and Monochloramine Application to Nitrifying Biofilm: Comparison of Biofilm Penetration, Activity, and Viability." *Environmental Science & Technology*, 45:4:1412–1419.

Lee, W.H., J.G. Pressman, and D.G. Wahman. 2018. "Three-Dimensional Free Chlorine and Monochloramine Biofilm Penetration: Correlating Penetration with Biofilm Activity and Viability." *Environmental Science & Technology*, 52:4:1889–1898.

Lipponen, M.T.T., P.J. Martikainen, R.E. Vasara, K. Servomaa, O. Zacheus, and M.H. Kontro. 2004. "Occurrence of Nitrifiers and Diversity of Ammonia-Oxidizing Bacteria in Developing Drinking Water Biofilms." *Water Research*, 38:20:4424–4434.

Lipponen, M.T.T., M.H. Suutari, and P.J. Martikainen. 2002. "Occurrence of Nitrifying Bacteria and Nitrification in Finnish Drinking Water Distribution Systems." *Water Research*, 36:17:4319–4329.

Liu, H., D.G. Wahman, and J.G. Pressman. 2019. "Evaluation of Monochloramine and Free Chlorine Penetration in a Drinking Water Storage Tank Sediment Using Microelectrodes." *Environmental Science & Technology*, 53:16:9352–9360.

Nagymáté, Z., Z.G. Homonnay, and K. Márialigeti. 2016. "Investigation of Archaeal and Bacterial Community Structure of Five Different Small Drinking Water Networks with Special Regard to the Nitrifying Microorganisms." *Microbiological Research*, 188:80–89.

Odell, L.H., G.J. Kirmeyer, A. Wilczak, J.G. Jacangelo, J.P. Marcinko, and R.L. Wolfe. 1996. "Controlling Nitrification in Chloraminated Systems." *Journal AWWA*, 88:7:86–98.

Oldenburg, P.S., J.M. Regan, G.W. Harrington, and D.R. Noguera. 2002. "Kinetics of *Nitrosomonas Europaea* Inactivation by Chloramine." *Journal AWWA*, 94:10:100–110.

Pievac, P., C. Schauberger, L.Poghosyan, C.W. Herbold, M.A.H.J. van Kessel, A. Daebeler, M. Steinberger, M.S.M. Jetten, S. Lücker, M. Wagner, and H. Daims. 2017. "*AmoA*-Targeted Polymerase Chain Reaction Primers for the Specific Detection and Quantification of Comammox *Nitrospira* in the Environment." *Frontiers in Microbiology*, 8:1508.

Pintar, K.D.M., W.B. Andersson, R.M. Slawson, E.F. Smith, and P.M. Huck. 2005. "Assessment of a Distribution System Nitrification Critical Threshold Concept." *Journal AWWA*, 97:7:116–129.

Pinto, A.J., D.N. Marcus, U.Z. Ijaz, Q.M. Bautista-de Lose Santos, G.J. Dick, and L. Raskin. 2015. "Metagenomic Evidence for the Presence of Comammox *Nitrospira*-like Bacteria in a Drinking Water System." *mSphere*, 1:1:e00054-15.

Pressman, J.G., W.H. Lee, P.L. Bishop, and D.G. Wahman. 2012. "Effect of Free Ammonia Concentration on Monochloramine Penetration Within a Nitrifying Biofilm and Its Effect on Activity, Viability, and Recovery." *Water Research*, 46:3:882–894.

Regan, J.M., A.-Y. Cho, S. Kim, and C.D. Smith. 2007. *Monitoring Ammonia-Oxidizing Bacteria in Chloraminated Distribution Systems*. Denver: Awwa Research Foundation.

Regan, J.M., G.W. Harrington, H. Baribeau, R.D. Leon, and D.R. Noguera. 2003. "Diversity of Nitrifying Bacteria in Full-Scale Chloraminated Distribution Systems." *Water Research*, 37:1:197–205.

Rittmann, B., J.M. Regan, and D.A. Stahl. 1994. "Nitrification as a Source of Soluble Organic Substrate in Biological Treatment." *Water Science & Technology*, 30:6:1–8.

Rosenfeldt, E.J., C. Baeza, and D.R.U. Knappe. 2009. "Effect of Free Chlorine Application on Microbial Quality of Drinking Water in Chloraminated Distribution Systems." *Journal AWWA*, 101:10:60–70.

Santillana, G.E., H.J. Smith, M. Burr, and A.K. Camper. 2016. "Archaeal Ammonium Oxidation Coupled with Bacterial Nitrite Oxidation in a Simulated Drinking Water Premise Plumbing System." *Environmental Science: Water Research & Technology*, 2:4:658–669.

Sarker, D.C., and A. Sathasivan. 2012. Nitrification control by adjusting pH in severely nitrified bulkwaters. *Water Science and Technology-Water Supply*, 12:5:683–690.

Sarker, D.C., A. Sathasivan, C.A. Joll, and A. Heitz. 2013. "Modelling Temperature Effects on Ammonia-oxidising Bacterial Biostability in Chloraminated Systems." *Science of the Total Environment*, 454–455:88–98.

Sathasivan, A., I. Fisher, and G. Kastl. 2005. "Simple Method for Quantifying Microbiologically Assisted Chloramine Decay in Drinking Water." *Environmental Science & Technology*, 39:14:5407–5413.

Scott, D.B., M.I. Van Dyke, W.B. Anderson, and P.M. Huck. 2015. "Influence of Water Quality on Nitrifier Regrowth in Two Full-scale Drinking Water Distribution Systems." *Canadian Journal of Microbiology*, 61:12:965–976.

Schrantz, K.A., J.G. Pressman, and D.G. Wahman. 2013. "Simulated Distribution Nitrification: Nitrification Index Evaluation and Viable Ammonia-oxidizing Bacteria." *Journal AWWA*, 105(5):E242–E254.

Seidel, C.J., M.J. McGuire, R.S. Summers, and S. Via. 2005. "Have Utilities Switched to Chloramines?" *Journal AWWA*, 97:10:87–97.

Skadsen, J. 1993. "Nitrification in a Distribution System." *Journal AWWA*, 85:7:95–103.

Skadsen, J. 2002. "Effectiveness of High pH in Controlling Nitrification." *Journal AWWA*, 94:7:73–83.

Srinivasan, S., G.W. Harrington, I. Xagoraraki, and R. Goel. 2008. "Factors Affecting Bulk to Total Bacteria Ratio in Drinking Water Distribution Systems." *Water Research*, 42:13:3393–3404.

Stehr, G., B. Böttcher, P. Dittberner, G. Rath, and H.P. Koops. 1995. "The Ammonia-Oxidizing Nitrifying Population of the River Elbe Estuary." *FEMS Microbiology Ecology*, 17:3:177–186.

Stewart, M.H., and N.I. Lieu. 1997. "Nitrification in Chloraminated Drinking Water and Its Association with Biofilms." In *Proc. of AWWA Water Quality Technology Conference*. Denver, Colo.: American Water Works Association.

Sung, W., X. Huang, and I.W. Wei. 2005. "Treatment and Distribution System Effects on Chloramine Decay, pH, Nitrification, and Disinfection By-Products: Case Study." *Journal of Water Resources Planning and Management*, 131:3:201–207.

van der Wielen, P.W.J.J., S. Voost, and D. van der Kooij. 2009. "Ammonia-Oxidizing Bacteria and Archaea in Groundwater Treatment and Drinking Water Distribution Systems." *Applied and Environmental Microbiology*, 75:14:4687–4695.

Wang, H., C.R. Proctor, M.A. Edwards, M. Pryor, J.W. Santo Domingo, H. Ryu, A.K. Camper, A. Olson, and A. Pruden. *Environmental Science & Technology*, 48(18): 10624–33.

Wilczak, A., J.G. Jacangelo, J.P. Marcinko, L.H. Odell, and G.J. Kirmeyer. 1996. "Occurrence of Nitrification in Chloraminated Distribution Systems." *Journal AWWA*, 88:7:74–85.

Wolfe, R.L., and N.I. Lieu. 2002. "Nitrifying Bacteria in Drinking Water." In *Encyclopedia of Environmental Microbiology*, Vol. 4. G. Bitton, ed. New York: John Wiley & Sons, Inc.

Wolfe, R.L., N.I. Lieu, G. Izaguirre, and E.G. Means. 1990. "Ammonia-Oxidizing Bacteria in a Chloraminated Distribution System: Seasonal Occurrence, Distribution, and Disinfection Resistance." *Applied and Environmental Microbiology*, 56:2:451–462.

Wolfe, R.L., E.G. Means III, M.K. Davis, and S.E. Barrett. 1988. "Biological Nitrification in Covered Reservoirs Containing Chloraminated Water." *Journal AWWA*, 80:9:109–114.

Woolschlager, J., B. Rittmann, P. Piriou, L. Kiene, and B. Schwartz. 2001. "Using a Comprehensive Model to Identify the Major Mechanisms of Chloramine Decay in Distribution Systems." *Water Supply*, 1:4:103–110.e

Zhang, Y. and M. Edwards. 2010. "Nutrients and Metals Effects on Nitrification in Drinking Water Systems." *Journal AWWA*, 102:7:56–66

Chapter 4

Overview of Causes and Control of Nitrification in Chloraminated Drinking Water Distribution Systems

Amlan Ghosh, David G. Wahman, Katherine Alfredo

INTRODUCTION

This chapter provides an integrated overview of nitrification causes and controls in chloraminated drinking water distribution systems. In-depth discussions of nitrification microbiology, monitoring, prevention, response, and engineering improvements are provided in previous and subsequent chapters. A summary of nitrification causes, best practices for prevention, and response practices is also included in Chapter 6, "Nitrification," of AWWA Manual M68, *Water Quality in Distribution Systems*.

CONDITIONS PROMOTING AND LIMITING GROWTH OF NITRIFYING MICROORGANISMS IN DRINKING WATER DISTRIBUTION SYSTEMS

Nitrifying microorganisms (including ammonia-oxidizing bacteria [AOB] and nitrite-oxidizing bacteria [NOB]) are ubiquitous in nature and can proliferate under favorable conditions in chloraminated distribution systems. This is due to free ammonia being inherently present through chloramine application. Typically, ammonia is added in slight excess during chloramine formation and subsequently released during chloramine demand and decay reactions, all of which result in the presence of free ammonia in the distribution system. In cases where naturally occurring free ammonia is present in source water, water systems intending to perform free chlorine disinfection may unintentionally form chloramines and/or have free ammonia in the treated water if breakpoint chlorination has not been achieved. In addition to the necessary ammonia substrate, factors favoring nitrification include source water quality, subsequent treatment, and distribution system operating conditions. Key points from this chapter are summarized in Table 4-1. Note that detailed discussion of Table 4-1 points is provided in subsequent chapter sections with examples and references.

After chloraminated water leaves the treatment facility, water quality and distribution system configuration, condition, and operation affect the chloramine residual stability and potential for nitrification within the system. Distribution systems that do not maintain

Table 4-1 Key points from Chapter 4

Free ammonia—essential nutrient for nitrification	• Minimization of free ammonia is an important strategy to control nitrification. Free ammonia can occur in chloraminated systems (or in free chlorine systems that have free ammonia in source water and have not achieved complete breakpoint through treatment) from excess ammonia (from source water and/or fed in treatment) or released via chloramine demand and/or decay reactions in distribution systems.
Chloramine demand and decay reactions—key sources of free ammonia and root cause of nitrification	• Chloramine demand and decay in the distribution system release free ammonia that becomes the substrate (i.e., food or nutrients) for nitrifying microbial growth. • Raw and finished water quality, treatment, and distribution system factors determine/affect chloramine residual stability (and ammonia excess or release) and nitrifier growth rates. • Chloramine concentration decreasing over time through its inherent chemistry (e.g., autodecomposition) will always occur and is termed *decay*. • Chloramine concentration decreasing over time from reacting (often relatively quickly) with other constituents (organic and inorganic) will occur if the constituents are present and is termed *demand*. • Difficulty maintaining sufficient chloramine residuals at system extremities and/or at areas of low flow or high hydraulic residence times may indicate the following: (1) excessive chloramine demand or decay potential in the water entering the distribution system, which could be caused by natural organic matter (NOM), nitrite, bromide, low pH, etc.; (2) water has not been appropriately treated or stabilized posttreatment to satisfy chloramine demand (e.g., nonoptimized biological treatment process following ozonation); (3) inappropriate chlorine-to-ammonia ratio during chloramine formation at the water treatment plant prior to entry to the distribution system; (4) excessive detention time in the distribution system; and/or (5) poor distribution system cleanliness (i.e., buildup of sediments or biofilm that can create chloramine demand and/or shield nitrifiers from disinfectant).

(continued)

Table 4-1 Key points from Chapter 4 (*Continued*)

	• Chloramine demand and decay should be measured on a routine basis in the raw and treated water after each unit process to understand treatment impacts, plant effluent, and planning stages (bench or pilot scale) to verify proposed treatment change impacts. In addition to total chlorine and monochloramine, water quality parameters, including free ammonia, temperature, pH, and alkalinity, should also be monitored in the raw, process, and treated water for a complete understanding of treatment process impacts on chloramine demand.
Source water quality, and/or treatment as chloramine demand and decay factors	• Systems with multiple sources should select water sources that have either lower free ammonia in the source water or are more favorable to chloramine stability and with less demand to help control nitrifier growth. However, other water quality parameters (e.g., regulated contaminants, corrosivity) and treatment targets should also be considered while selecting the most appropriate source. • Systems should consider adjusting their chloramine target seasonally to account for seasonal changes in water quality and distribution system characteristics. • Source waters of naturally higher pH and lower alkalinity may be more favorable to chloramine stability and nitrification prevention. • Source water bromide can contribute to chloramine decay. • Raw water source oxygenation in the lake or reservoir can help eliminate anoxia and lower the overall oxidant demand (chlorine, ozone, peroxide, chlorine dioxide, etc.) during water treatment. The resulting chloramine residual may be more stable and limit growth of nitrifying microorganisms in the distribution system. • Source water chloramine demand (e.g., NOM, reduced iron, and reduced manganese) should be removed at the treatment plant to the greatest degree possible prior to chloramine formation as they exert chloramine demand in the bulk water, on downstream pipe walls, or in sediment (e.g., reduced manganese and reduced iron). • Total organic carbon (TOC) removal by clarification (coagulation/flocculation/sedimentation), adsorption, biological filtration, membrane filtration, ion exchange, or others will minimize the demand exerted by natural organic substances. • Ozonation, if applied appropriately and followed by biofiltration, may decrease chloramine demand due to TOC and oxidant demand changes. However, poorly operated ozonation may result in an increase in chloramine demand. • Hydrogen peroxide, used at higher doses for advanced oxidation, may react with chlorine or chloramine during disinfection, unless it is quenched before disinfectant application. • Filter media should be prevented from passing into the distribution system, as media may exert additional chloramine demand. • Chlorite entering or forming in the distribution system because of chlorine dioxide primary disinfection treatment has been reported by some water systems as effective for nitrification prevention. Note that chlorine dioxide followed by ozonation converts all chlorite to chlorate ion and therefore negligible chlorite-related nitrification control benefit would be expected. • Sodium chlorite fed directly into the distribution system has been reported effective by some water systems for preventing nitrification. • Maintaining higher pH in the distribution system reduces the possibility of initiation of nitrification. However, operating pH should be evaluated on a case-by-case basis to balance chloramine stability with corrosion control and other regulatory requirements.

(continued)

Table 4-1 Key points from Chapter 4 (*Continued*)

Increased water age in the distribution system	• Water distribution systems can provide ideal growth environments for certain types of nitrifying microorganisms (e.g., *N. oligotropha*). For example, darkness, long detention times (water age), warmer water, and a low-level continuous supply of free ammonia provided by degrading chloramine via demand and decay reactions. • The quiescent conditions of finished water storage facilities and piping dead-ends can be associated with long residence times and lower chloramine residuals, which could allow nitrifying microorganisms to proliferate. Similarly, pipe corrosion can cause particle accumulation and tuberculation, thereby providing physical areas for nitrifiers to grow and while being protected from disinfectant exposure, shear, and washout (via normal flows or flushing).
Distribution system operations and maintenance practices for control of nitrification	• Minimizing water age through operational means and maintaining system cleanliness and proper corrosion control are all prerequisites of chloraminated system optimization and nitrification control. Distribution system water age can be minimized by managing storage tank operating levels, storage tank mixing and cycling, managing distribution system hydraulics by directing water to low flow areas through valving, optimizing pumping schedules, and regular flushing. • System cleanliness can be maintained through a storage-facility cleaning program and pipe flushing or physical cleaning. • Some systems preferentially blend or select different finished waters in the distribution system (maximizing use of those more favorable to chloramine stability) to control nitrifier growth. • Ensuring distribution system integrity is maintained; repairing pipes and disinfecting them properly before they are put back in service, along with minimizing cross-connections and contamination, are important strategies to preserve chloramine stability and limit nitrification.
Distribution system configuration or design improvements for control of nitrification	• Water age is fixed to a certain extent due to system configuration relative to water use. Water storage reservoir or tank inlet and outlet configuration, fill/draw cycling practices and/or inadequate mixing may result in stagnant water zones within storage facilities and elevated water age in downstream distribution. These can be eliminated through improved storage cycling, inlet and outlet modifications, and/or active mixing. • Distribution system design that avoids dead ends, and/ or includes blowoffs or other flushing methods, is helpful for higher bulk water turnover, reduced water age, and improved water quality. • Pipeline materials can impact chloramine demand and biofilm growth. • Chloramine boosting via precisely controlled application of chlorine and ammonia into the distribution system at points in areas experiencing relatively lower chloramine residuals can counteract the effects of water age and pipe walls by reconverting free ammonia into chloramine and eliminating the ammonia-oxidizing bacteria food source.

water quality with stable chloramine residuals and low ammonia and are not designed or operated for good water turnover, and system cleanliness may not be well optimized for nitrification control.

Nitrification causes and controls are centered around availability or minimization of free ammonia that may be present in source water or can occur from excess ammonia fed at treatment facilities or released via chloramine demand and/or decay reactions through distribution systems. Free ammonia is the essential nutrient for nitrifying microorganisms. Where free ammonia is routinely present in water systems, nitrifiers are expected to be present. The extent to which nitrification becomes problematic (i.e., chloramine residual loss, increased nitrite) depends on a utility's ability to optimize chloramine application at the treatment plant and distribution system operations, while accounting for demand and decay reactions.

Typically, nitrification results from a combination of several factors or conditions; therefore, all possible factors should be evaluated in chloraminated distribution systems and in systems considering chloramination. Tables 4-2–4-5 summarize a range of possible causes and controls of nitrification and could be used as general guidelines for nitrification control with the understanding that the conditions or factors listed may or may not apply to a given system, and factors causing nitrification and associated control measures are interrelated. Site-specific evaluations are necessary to implement effective nitrification prevention and control practices. The concepts summarized in Tables 4-2–4-5 are also discussed in detail in the subsequent chapters.

Systems preparing for chloramine conversion should prepare a nitrification detection and control program that should be implemented when beginning chloramination. In general, distribution system problem areas with low flow, turnover, or demand and difficulties maintaining free chlorine often continue to generate water quality challenges after chloramine conversion. Hence, work done to identify and address low-flow areas prior to beginning chloramination may prevent future problems with residual loss and nitrification when using chloramines. Optimization of chloraminated distribution system operations typically involves progressive improvements in chloramine formation, water quality monitoring, and operational tools for nitrification prevention and control. Where necessary, it may take engineering improvements of the source, treatment, and/or distribution systems to resolve persistent trouble areas. Because of differences between free chlorine and chloramine reactivity and residual stability, systems previously requiring booster free chlorination may find that those applications may no longer be needed upon switching to chloramine; however, they may be useful in boosting chloramine residuals, although the locations of these stations may be different.

CHLORAMINE CHEMISTRY—AS MAJOR CAUSE OF NITRIFICATION

Chloramine Demand and Decay Definitions and Factors

The concentration of chloramine applied (i.e., ammonia-nitrogen reacted with free chlorine to form predominantly monochloramine—the preferred species—along with very minimal excess residual-free ammonia, accounting for source water ammonia, if present) influences the amount of ammonia available to support nitrification in the downstream distribution system. Only free ammonia (NH_3), which predominates at higher pH, can be used as a nutrient by nitrifiers. As long as ammonia is combined in monochloramine (NH_2Cl) or as ammonium (NH_4^+), which predominates at lower pH, it does not support nitrification.

Understanding chloramine chemistry, especially chloramine demand and decay reactions, is critical to minimizing free ammonia, minimizing nitrification, and maintaining the chloramine residual in the distribution system. Chloramine formation and subsequent decay occur through a complex series of chemical reactions (Jafvert and Valentine 1992). Both demand and decay reactions reduce the chloramine concentration and release free ammonia; therefore, the agent responsible for controlling AOB growth (the monochloramine residual) is decreasing through the distribution system, while the agent responsible for supporting AOB growth (ammonia) is increasing (Harrington et al. 2003).

Once formed, the chloramine (monochloramine) can react (e.g., chloramine residual decreases, releasing ammonia) by two general mechanisms: demand reactions and decay reactions. In this chapter, these terms are defined as the following:

***Chloramine demand* refers to constituents in the water or on surfaces in contact with the water that react with chloramine** (e.g., nitrite, bromide, natural organic matter, and internal pipe surfaces themselves). Chloramine demand reactions can be fast or slow, depending on the constituents with which it is reacting.

***Chloramine decay* refers to the inherent chloramine autodecomposition that decreases chloramine concentrations over time.** Chloramine decay (autodecomposition) is generally slow.

Table 4-2 summarizes the key chloramine demand and decay factors, their potential to cause or control nitrification, and the feasibility of implementation of related nitrification controls. Detailed examples and references are included in following chapter sections.

Chloramine residual level. The presence of a chloramine residual can mitigate nitrification by inactivating nitrifiers. However, the release of ammonia (i.e., the food source initiating nitrification) through chloramine decay and demand can exacerbate nitrification. The stage 1 Disinfectants and Disinfection Byproduct Rule (D/DBPR) established a maximum residual disinfectant level for total chlorine of 4.0 mg Cl_2/L as a running annual average of distribution system residuals. The Surface Water Treatment Rule requires that a detectable disinfectant residual be maintained in at least 95 percent of samples collected throughout the distribution system monthly (or heterotrophic plate count measurements not exceeding 500 colony-forming units per mL).

Higher free and total chlorine residual levels are required by several state regulatory agencies, as discussed in Chapter 6. Target chloramine residual is system specific. Some utilities have increased the target residual over the years, either seasonally or year-round, to better control nitrification. Higher than detectable minimum levels for chloramine residuals (e.g., 0.5 mg Cl_2/L) have been required or recommended by some states and recommended in utility research-based guidance (Friedman et al. 2010, see discussion in

Table 4-2 Causes and control of nitrification in the distribution system: chloramine demand and decay

Parameter or Condition	Cause of Nitrification	Control of Nitrification	Feasibility of Implementation for Nitrification Control
Water age	As water age increases, more time is provided for chloramine demand and decay reactions to reduce chloramine residual and release free ammonia, promoting conditions favorable to nitrification.	To the extent possible, water age should be minimized in a system to limit the time provided for chloramine demand and decay reactions to occur.	Several strategies can be employed for minimization of water age, ranging from simple operational changes to costly infrastructure improvements. Operational changes may include adjustment of valve positions (open/closed), implementing hydrant flushing during low demand periods for bulk water turnover, tank/reservoir cycling and level optimization, pumping modifications, or taking certain tanks offline seasonally during low demand periods. Infrastructure improvements may include looping or (re)sizing lines to manage water quality, implementing automatic flush stations, etc.

(continued)

Table 4-2 Causes and control of nitrification in the distribution system: chloramine demand and decay (*Continued*)

Parameter or Condition	Cause of Nitrification	Control of Nitrification	Feasibility of Implementation for Nitrification Control
Chloramine demand	Chloramine demand reactions with both organic and inorganic constituents release free ammonia and decrease chloramine concentrations.	Chloramine demand should be reduced as much as possible during treatment by coagulation, pre- or intermediate oxidation, and filtration and by maintaining distribution system cleanliness.	A portion of the demand can be reduced through conventional treatment and chlorination. Chloramine demand can also be reduced by removal of accumulated sediments/deposits through routine cleaning of water mains and tanks.
Chloramine decay	Monochloramine decays over time as a function of pH, alkalinity and temperature, releasing free ammonia.	Decay decreases as pH increases. Avoid pH drops in the system.	A stable pH should be maintained in the water leaving the treatment plant and throughout the distribution system. Unlike temperature, both pH and alkalinity can be controlled. Lower water age also assists with maintaining pH. However, pH targets and the tolerance for variability in the distribution system should be developed such that optimal corrosion-control conditions are maintained.
Chloramine residual entering system	Low chloramine residual may not be sufficient to reach the ends of the distribution system and inactivate nitrifiers.	Increase chloramine residual entering the distribution system to maintain an adequate residual at the extremities of the distribution system. Boost chloramine residuals at key locations in the distribution system as appropriate.	Treated water with a higher chloramine concentration entering the distribution system helps limit microbial growth. Consecutive systems may need to boost chloramine residuals at key locations in the distribution system.
Chlorine-to-ammonia ratio entering system	Overfeeding ammonia at the treatment plant and booster chloramination stations because of poor process control results in suboptimal monochloramine formation and may result in free ammonia in the treated water, which can be a major cause of nitrification.	Minimizing free ammonia entering the system is extremely important. If possible, maintain Cl_2:N weight ratios of 4.5:1–5:1 at the entry point to form a stable monochloramine residual.	The chlorine-to-ammonia ratio must be controlled based on process monitoring data (e.g., total chlorine, monochloramine, and free and total ammonia residual).
Organic material (commonly measured as total organic carbon [TOC])	High concentrations of organic materials in the water exert chloramine demand and promote free ammonia release.	Removing organic materials (measured by TOC removal) in treated water minimizes the initial chloramine demand.	TOC may be removed through conventional and enhanced coagulation. Further reduction may be achieved with granular activated carbon, high-pressure membrane filtration, or ion exchange but may be cost prohibitive in the absence of other treatment drivers.

(continued)

Table 4-2 Causes and control of nitrification in the distribution system: chloramine demand and decay (*Continued*)

Parameter or Condition	Cause of Nitrification	Control of Nitrification	Feasibility of Implementation for Nitrification Control
Temperature	Nitrifiers grow faster as water temperature increases (i.e., typically above 15°C). Warm water in piping and storage exacerbates nitrification. Chloramine degradation is faster as water temperature increases, leading to lower residual and higher free ammonia levels.	Cold water in winter may limit nitrification for many systems, but nitrifiers exhibit slow metabolic activity even at low temperatures and can rapidly begin the nitrification process as soon as temperature becomes favorable.	While water temperature may not be precisely controllable, strategies including blending of source waters, shading reservoirs, distribution system flushing, storage tank mixing, and cycling may assist with managing water temperatures. Higher frequencies for monitoring water quality parameters during periods of high temperature may also assist in the early detection and control of nitrification.
pH	Chloramine inactivates ammonia-oxidizing bacteria slowly as pH increases, but chloramine is more stable as pH increases.	Some utilities operate at high pH (above 9.0) to better control nitrification. High pH is also beneficial for chloramine stability and corrosion control in some cases.	Higher pH should be implemented based on corrosion-control evaluations if precipitation/scaling or resulting increase in disinfection byproduct formation is not problematic. Any other regulatory requirements related to pH of treated water should also be considered.
Bromide	Elevated bromide concentrations catalyze chloramine decay.	Bromide is typically not removed by conventional treatment processes.	Optimizing pH may decrease catalytic effect of bromide on chloramine demand.
Iron, manganese, sulfide	Inorganics such as iron, manganese, and sulfide exert chlorine demand.	Chlorine demand from inorganics should be accounted for when adding chlorine and ammonia for monochloramine formation.	Before adding the appropriate ratio of chlorine and ammonia for monochloramine formation, inorganics should be addressed at the water treatment facility through preoxidation and filtration or sequestration (if necessary).
Alkalinity	Alkalinity is consumed during nitrification, and low alkalinity can impair nitrification. Low alkalinity may contribute to wide pH shifts observed in the distribution system.	Higher alkalinity water would be more desirable to maintain pH, chloramine residual, and corrosion control.	Addition of lime or soda ash can be implemented for pH and alkalinity control, but pH/ alkalinity adjustment should be performed while considering corrosion control and other regulatory implications.

Chapter 6). Targeting minimum detectable or substantially higher chloramine residuals will be inadequate to control nitrification and related chloramine residual loss if (1) free ammonia continues to be present to sustain nitrifier growth (see recommended free ammonia levels and supporting discussion in Chapter 6), (2) the nitrification byproduct (nitrite) is present, or (3) other constituents are present in water that exert a chloramine

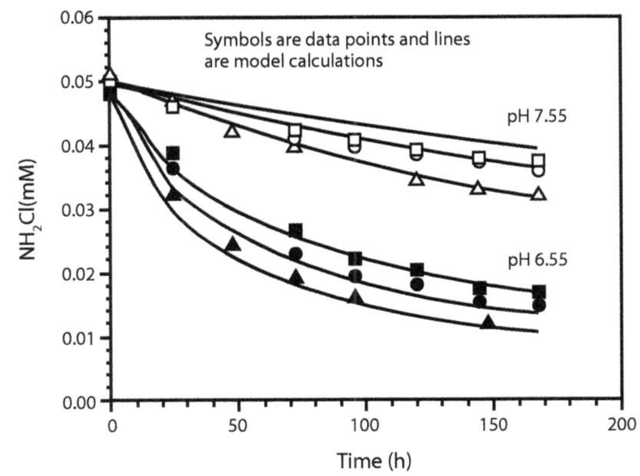

Source: Vikesland et al. 2001

Figure 4-1 Monochloramine decay as a function of Cl/N molar ratio. Cl/N = 0.5 (□,■), Cl/N = 0.6 (▪,▪), and Cl/N = 0.7 (△,▲). Open symbols are for pH ≈ 7.5 and filled symbols are for pH ≈ 6.5 $[NH_2Cl]_0$ = 0.05 mM, $C_{T,CO3}$ = 4 mM, μ = 0.1 M, *T* = 25°C.

demand. In each system, chloramine demand and decay rates, combined with factors affecting water age, microenvironments (solids and biofilm accumulations, corrosion tubercles), and distribution system operation determine the chloramine residual level sufficient to control nitrification and maintain residuals through the system.

Chlorine (as Cl_2) to ammonia (as N) weight ratio. Guidance for the chlorine as Cl_2 to ammonia as N (Cl_2:N) weight ratio historically has ranged from 3:1 to 5:1. The stoichiometrically ideal free chlorine (Cl_2) to ammonia nitrogen (as N) weight ratio for monochloramine formation is 5.07:1 (1 mol of chlorine to 1 mol of ammonia). This ideal ratio assumes no demand and only monochloramine formation. Because the fastest chloramine formation occurs at pH 8.4 (Kirmeyer et al. 2004), system chloramine residuals would best be maintained at pH >8.3. The Cl_2:N mass ratio should be balanced at the treatment plant to limit excess free ammonia and maximize monochloramine formation. It is important for operations staff to understand that the Cl_2:N ratio is based on the actual free chlorine available (i.e., residual) to react with free ammonia at a given point in the treatment train or in the system if boosting.

Any free ammonia occurring naturally in the source water will need to be factored into dosing ratios as well. As Cl_2:N increases, chloramine decay rate increases (Figure 4-1), and conversely, as chloramine decays, the Cl_2:N ratio decreases. Therefore, measures may be needed to monitor and recombine liberated free ammonia in the distribution system by booster chlorination to maintain the Cl_2:N weight ratio near 5:1 throughout the system. In general, higher Cl_2:N weight ratios (up to 5:1) are normally needed at the entry to the distribution system because there will undoubtedly be some demand and loss of chlorine residual in the distribution system.

Maintenance of the ratio very close to 5:1 may be difficult. If chlorine dosing exceeds the available ammonia and demand, odorous dichloramine formation and/or residual loss may occur. This can lead to breakpoint reaction-associated tastes and odors and/or increased disinfection byproduct (DBP) formation.

Chloramine demand. Several chloramine demand reactions can occur within drinking water distribution systems, including reactions with nitrite (Wahman and Speitel 2012; Vikesland et al. 2001; Margerum et al. 1994; Johnson and Margerum 1991); bromide (Vikesland et al. 2001); natural organic matter (Duirk et al. 2005; Duirk 2003, 2006; Duirk et al. 2002; Vikesland et al. 1998), microorganisms, biofilm, and soluble microbial products (Maestre et al. 2016; Wahman et al. 2016; Speitel et al. 2014; Maestre et al. 2013; Adhikari et al. 2012; Krishna et al. 2012); pipe surfaces (Zhang and Andrews 2013; Vikesland and Valentine 2000, 2002a, 2002b); and hydroxylamine, a chemical intermediate produced by AOB (Wahman and Speitel 2015; Wahman et al. 2014). All these reactions can be generalized by the following reaction:

$$\text{Demand (e.g., nitrite, bromide, NOM)} + NH_2Cl \rightarrow NH_3 + Cl^- + \text{products} \quad \text{(Eq. 4-1)}$$

For Equation 4-1, 1 mol of free ammonia is released for each mole of monochloramine reacted. Therefore, the reduction of monochloram+ine demand reactions (i.e., minimizing nitrite, bromide, NOM, and temperature and maximizing pH) can minimize ammonia release and nitrification potential to an even greater extent than measures targeted at minimizing decay (e.g., addressing water age, temperature, and pH).

Chloramine decay. Although more stable than free chlorine, chloramine is inherently unstable, and its fate is controlled by a complex set of reactions that result in ammonia release and active chlorine reduction represented by the model developed by Jafvert and Valentine (1992) and slightly revised by Vikesland et al. (2001), which can be summarized by two reactions (Eq. 4-2 and 4-3) that both yield free ammonia (i.e., the food for microorganisms initiating nitrification):

$$3NH_2Cl \rightarrow N_2 + NH_3 + 3H^+ + 3Cl^- \quad \text{(Eq. 4-2)}$$

$$4NH_2Cl + 3H_2O \rightarrow 3NH_3 + NO_3 + 5H^+ + 4Cl^- \quad \text{(Eq. 4-3)}$$

In addition to ammonia, the nitrogen bound in monochloramine is preferentially oxidized to nitrogen gas (Eq. 4-2) with smaller quantities oxidized to nitrate (Eq. 4-3) (Vikesland et al. 1998). If Eq. 4-2 dominates, 1 mol of ammonia is produced for every 3 mol of monochloramine lost. Temperature, pH, inorganic carbon, and chlorine-to-ammonia ratio affect chloramine decay rates as discussed further in this chapter. Maximizing chloramine stability will minimize decay-related ammonia release and nitrification potential. To assist in understanding the important factors related to chloramine decay, web-based applications have been developed that allow chloramine formation and decay simulations to be conducted (Wahman 2018).

Organic materials. Elevated levels of organic materials (e.g., TOC and NOM) present in source water, and in the distribution system water after treatment, can exert a chloramine demand (i.e., react with chloramine resulting in chlorine residual loss and ammonia release). The concept of biostability was discussed in Srinivasan and Harrington (2007). Biostability is the ability to limit regrowth of microorganisms in drinking water, and it depends on the concentration of disinfectant residual as well as the concentration of substrates that are required for the regrowth of microorganisms. Presence of higher concentrations of organic materials (especially biodegradable organic materials) in the treated water affects the biostability curve by increasing the concentration of substrates.

pH and inorganic carbon (alkalinity). pH is one of the most important factors affecting monochloramine formation and decay. Monochloramine formation is fastest at a pH of 8.4, and monochloramine stability increases with increasing pH. Therefore, establishing the proper pH is essential not only for corrosion control and water stability overall but also for maintaining chloramine residual and limiting nitrification. Monochloramine decay is generally acid-catalyzed; therefore, water pH (Figures 4-1 to 4-3) and the presence of inorganic carbon (Figure 4-3) impact monochloramine stability (Vikesland et al. 2001). Dissolved inorganic carbon is related to alkalinity and is the sum of carbon dioxide, carbonic acid, bicarbonate ion, and carbonate ion. Higher pH favors chloramine stability, minimizing decay-related ammonia release—the key nutrient for nitrification. Maintaining high pH (e.g., 9.0 or higher) has been recognized as a key factor

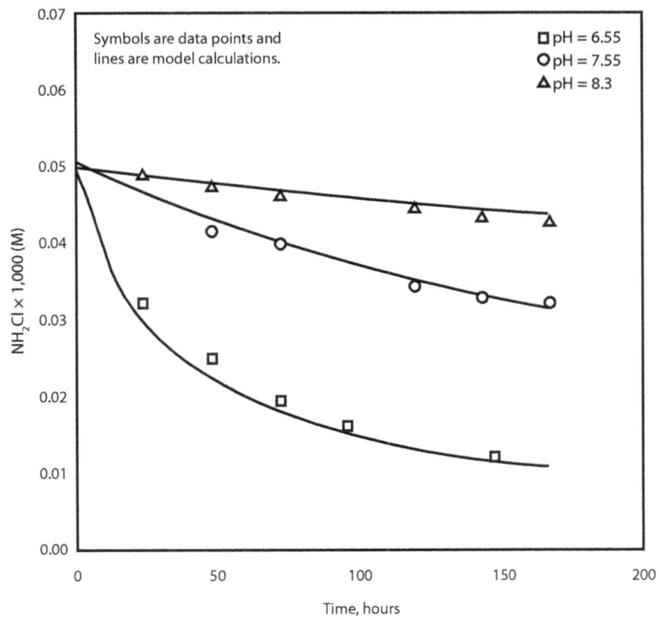

Source: Valentine et al. 1998

Figure 4-2 Effect of pH on monochloramine decay (autodecomposition) at 25°C; 4 mg/L Cl_2 = 0.056 mM NH_2Cl.

in successful nitrification control programs (Friedman et al. 2010, Wilczak et al. 2010). Because nitrification consumes alkalinity and reduces pH while simultaneously producing chloramine-demanding nitrite, it is essential to take appropriate measures to monitor, prevent, and/or control localized nitrification.

Temperature. The chloramine decay rate increases as temperature increases (Figure 4-2). The water temperature in distribution system mains and storage facilities can vary seasonally based on source water, ambient air temperatures, and sun exposure, which are out of utility control. However, utilities using chloramination need to understand how temperature can impact chloramine decay and nitrification. For example, utilities often report seasonal nitrification episodes (AWWA Disinfection Systems Committee 2008) with nitrification more likely (but not exclusively) during the summer or warmest seasons (Pintar et al. 2000). Additionally, thermal stratification of stored water can result in higher water temperatures and higher water age, which can be avoided through addition of mixing and/or operational changes to storage tanks to minimize chloramine decay and subsequent nitrification. As previously stated, chloramine demand and decay (ammonia release) reactions are faster in warmer temperatures, theoretically providing more ammonia to support nitrification, which can also proceed faster at warmer temperatures due to increased biological growth rates (e.g., Sarker et al. 2013). The optimal water temperature range for nitrifying microorganism growth is 25–30°C; however, nitrification can occur at temperatures as low or lower than 5°C in systems with long detention times (Pintar et al. 2000).

Bromide. The presence of bromide does not directly cause nitrification. However, at high enough concentrations, bromide may have a significant effect on chloramine demand and therefore, on ammonia release (Figures 4-3–4-6). At bromide concentrations below 0.1 mg/L, the impact of bromide is minimal (Vikesland et al. 2001). However, at higher concentrations, the presence of bromide may increase the rates of chloramine decay and ammonia release due to the formation of bromine-substituted haloamines (i.e., bromamines or bromochloramine), which are less stable than monochloramine. Pope (2006) studied the

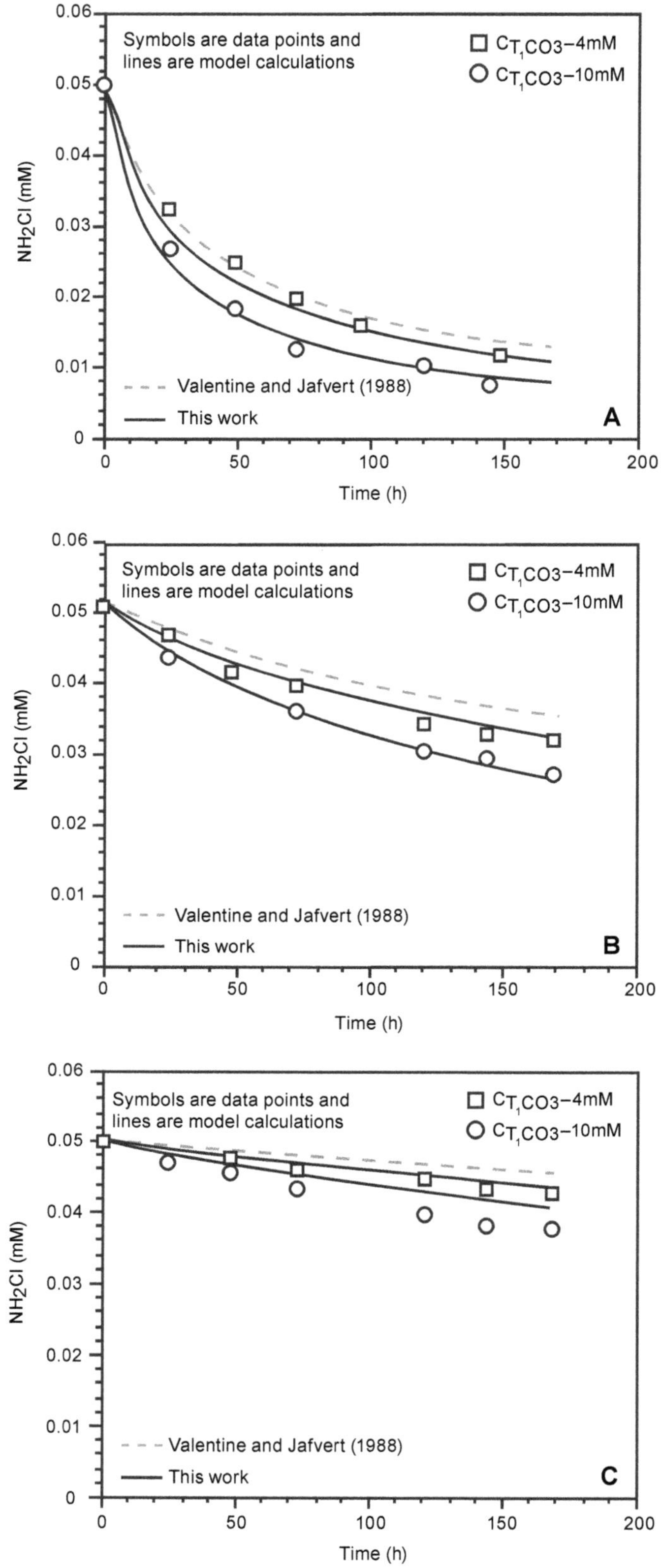

Source: Vikesland et al. 2001

Figure 4-3 Effect of total carbonate concentration on monochloramine decay at (A) pH ≈ 6.6, (B) pH ≈ 7.6, and (C) pH ≈ 8.3; Cl/N = 0.7 mol mol^{-1}, μ = 0.1 M, *T* = 25°C

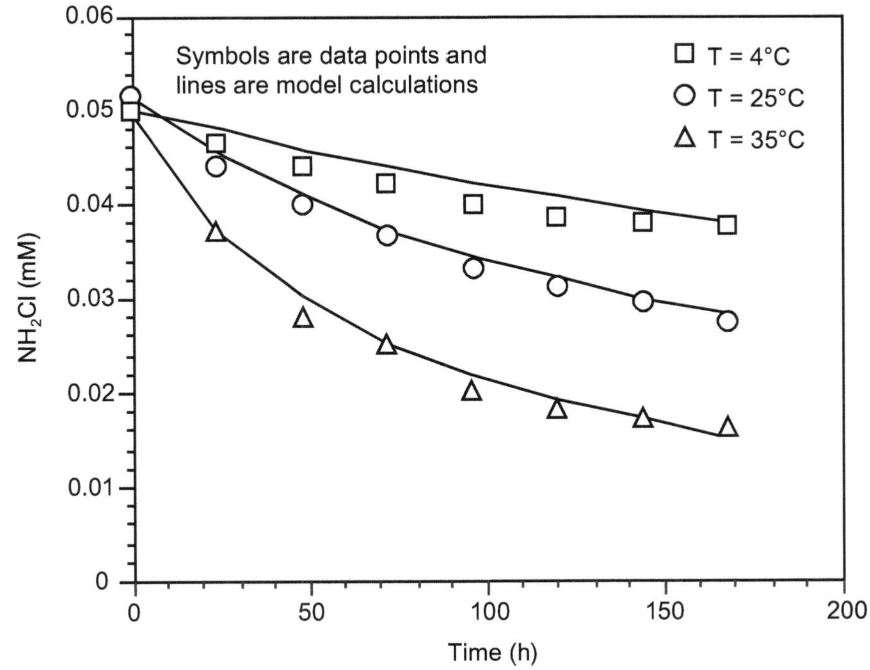

Source: Vikesland et al. 2001

Figure 4-4 Effect of temperature on monochloramine decomposition; Cl/N = 0.7 mol mol^{-1}, pH = 7.5, $C_{T,CO3}$ = 10 mM, μ = 0.1 M

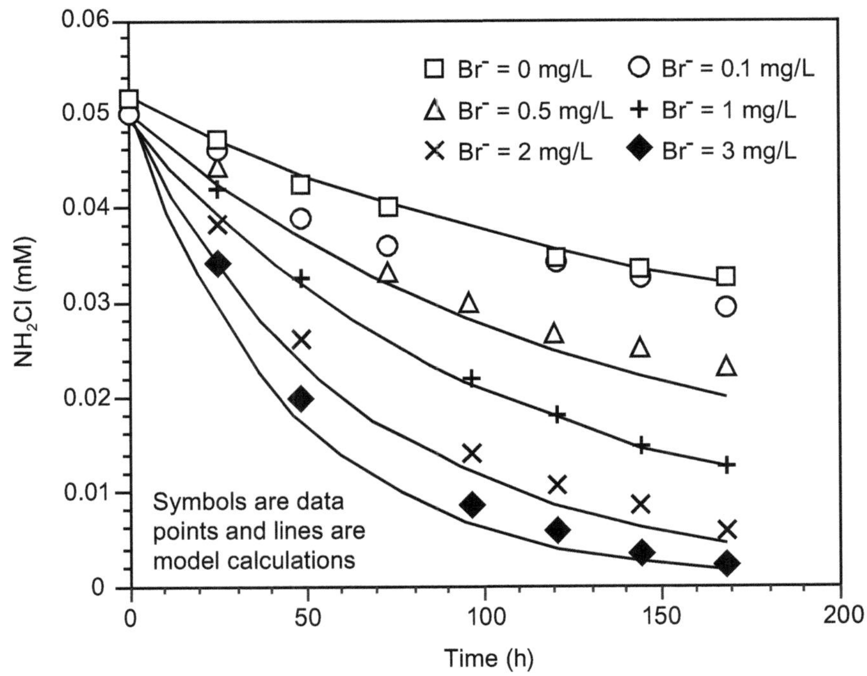

Source: Vikesland et al. 2001

Figure 4-5 Effect of 0–3 mg/L bromide on monochloramine decomposition at pH ≈ 7.5; Cl/N = 0.7 mol/mol, $C_{T,CO3}$ 4 mM, μ = 0.1 M, *T* = 25°C

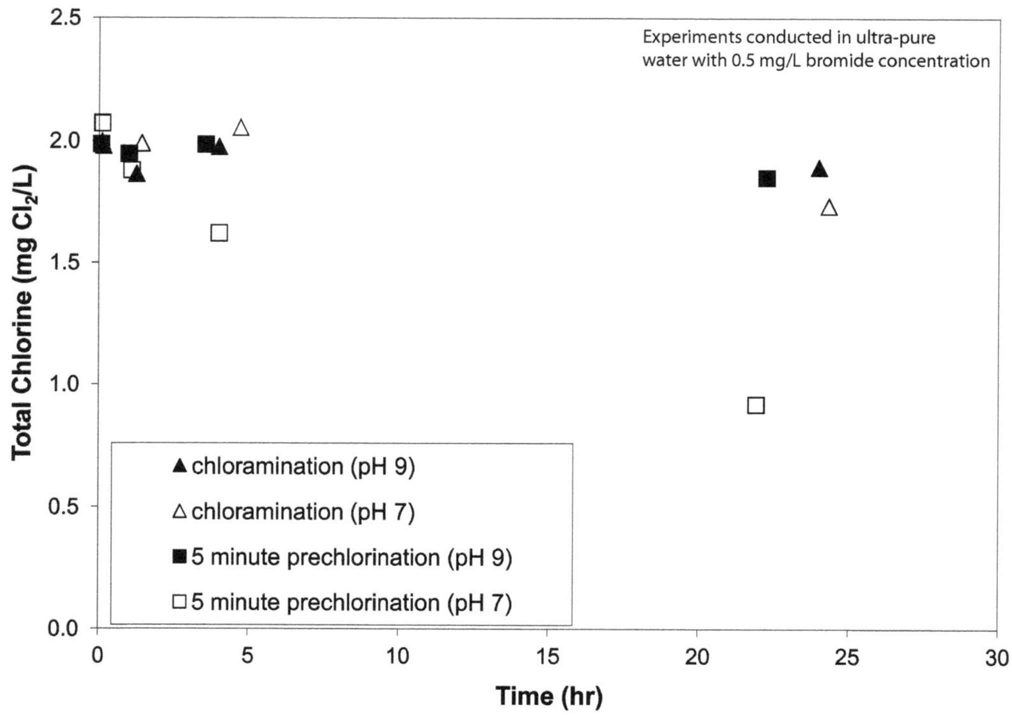

Source: Adapted from Pope 2006

Figure 4-6 Impact of bromide on chloramine decay

impact of pH and free chlorine contact time prior to ammonia addition on chloramine stability in ultrapure water (Figure 4-5). These data indicate that pH and chloramine application conditions (i.e., preformed chloramine versus short periods of exposure to free chlorine prior to chloramine formation) play an important role in chloramine stability in the presence of high bromide concentrations.

Iron, manganese, sulfide. The presence of inorganics such as iron, manganese, or sulfide affects the chlorine demand of water. Prior to application of chlorine and ammonia for monochloramine formation, these inorganics should be removed from water through treatment processes such as oxidation and filtration or sequestered.

SOURCES AND TREATMENT AS NITRIFICATION CAUSES AND CONTROLS

Source Water Quality and Treatment Operations

It is important that source water quality and treatment processes are optimized for chloramine stability and nitrification control. Source water quality characteristics and the subsequent type and amount of treatment applied can inherently affect chloramine stability and distribution system nitrification potential. While treatment systems are designed to achieve multiple water quality objectives (beyond nitrification control), their impacts on chloramine stability and nitrification control should be considered. In addition to optimization of chloramine application, the removal of particulate matter and organic materials like TOC and assimilable organic carbon (AOC) during treatment while minimizing corrosivity is beneficial to nitrification control. Attention must be paid to organic material removal (including organic carbon and organic nitrogen), oxidation, and disinfection sequence to ensure that oxidant demand is satisfied within the treatment train and minimized in the distribution system. In general, chloramine stability increases for water that is lower in TOC, AOC, and biodegradable dissolved organic carbon (BDOC) and higher in pH. Bench-scale testing to evaluate chloramine decay and demand is recommended prior to chloramine conversion and in conjunction with nitrification control studies. A series of tables is presented in this chapter summarizing conditions that could cause or contribute to nitrification (Tables 4-3–4-5). Table 4-3 below summarizes the causes and controls of nitrification through various water treatment processes, while Tables 4-4 and 4-5 address causes and controls of nitrification due to distribution system design and distribution system operations and maintenance (O&M), respectively.

Conventional and enhanced coagulation (TOC and NOM removal). Conventional coagulation or enhanced coagulation (with depressed pH) followed by sedimentation can remove a significant portion of reactive TOC and may have beneficial effects on chloramine stability overall. The impact of NOM removal on chloramine demand will depend on the

Table 4-3 Causes and control of nitrification in the distribution system: water treatment

Treatment Type/ Unit Process	Cause of Nitrification	Control of Nitrification	Other Considerations
Type of filtration (conventional filtration, unfiltered systems, or alternative filtration processes)	Sediment deposition in the distribution system may be greater in systems served by unfiltered supplies requiring more frequent cleaning.	Unfiltered waters are typically from high-quality sources and maintain chloramine residual well. However, because they are unfiltered systems, small amounts of sediment may be present, which can accumulate and deposit in distribution systems over time. Conventional clarification (coagulation, flocculation, sedimentation) and/or softening with filtration (inline, direct, media, micro- or ultrafiltration) remove a large portion of oxidant demand, including chloramine demand. Nanofiltration removes more natural organic matter and oxidant demand.	It is important to verify that conventional treatment and/or alternative filtration processes produce stable chloramine residual. Consider the pH needs throughout the treatment plant (e.g., the optimal pH for coagulation with aluminum salts is much less than the optimal pH for chloramine formation or corrosion control).
Oxidation and primary disinfection	Ozonation should be followed by appropriate filtration or biodegradation steps to reduce subsequent chloramine demand. Advanced oxidation processes (e.g., ozone or ultraviolet [UV] with high hydrogen peroxide doses) may exert chlorine demand and negatively impact chloramine stability unless it is managed appropriately, and/ or compensated for with appropriate postchlorine dose.	Primary disinfection with chlorine and chlorine dioxide will satisfy a portion of oxidant demand and produce a more stable chloramine. UV disinfection will have no impact on chloramine demand because it will not oxidize the demand-causing substances, but if chloraminated water is passed through a UV reactor, a portion of total chlorine will be destroyed.	It is important to verify the impacts of alternative oxidants and their subsequent process units on chloramine stability in the project planning stage.

(continued)

Table 4-3 Causes and control of nitrification in the distribution system: water treatment (*Continued*)

Treatment Type/ Unit Process	Cause of Nitrification	Control of Nitrification	Other Considerations
Chlorine dioxide and chlorite ion	Like all chemical oxidants, chlorine dioxide used in place of free chlorine for disinfectant concentration is expected to increase assimilable organic carbon, even though the increase is likely to be lower than with an equivalent dose of chlorine.	Chlorite ion, a regulated byproduct of chlorine dioxide disinfection, may selectively inactivate nitrifiers even at low levels (0.1–0.2 mg/L). Chlorine dioxide improves chloramine stability like chlorine does.	Chlorite ion is not naturally present in water and is either an impurity or a byproduct of chlorine dioxide application. Chlorite has its own maximum contaminant level. Verify chlorine dioxide's effect on chloramine stability.
Biological treatment	Some biological filters, especially those with granular activated carbon filtration and biomass growth medium, may cause an increase in chloramine demand due to not fully optimized biological processes.	Optimized biological treatment may decrease regrowth of microorganisms and chloramine demand by consuming assimilable/ biodegradable organic material.	Proper contact times for the biological processes and careful process control should be implemented.
	Uncontrolled nitrification at the treatment plant may result in seeding the system with nitrifiers and excessive biomass released from biofilters may be responsible for microbial regrowth in the distribution system.	Complete and controlled nitrification at the treatment plant may control fluctuations of raw water ammonia.	Provide either a physical or disinfectant barrier at the end of treatment train to remove microorganisms and byproducts of their growth.
Corrosion control	Lack of optimum corrosion-control treatment at the treatment plant may lead to pipeline tuberculation and precipitates/sediments, biofilm, and persistent nitrification in distribution systems. Lower pH, as recommended with some corrosion inhibitors, may favor chloramine degradation and nitrification.	Optimum corrosion control is a prerequisite for maintaining water quality in the distribution system and long-term successful nitrification control. Optimizing corrosion-control treatment is also a requirement of US Environmental Protection Agency's Lead and Copper Rule Revisions. Chloramine is more stable at higher pH.	Establishing optimum corrosion control may require long-term studies and adjustments. Alkalinity, pH, disinfectant residuals, and corrosion inhibitors are factors in corrosion and nitrification control.
Orthophosphate	PO_4-P added for corrosion control may promote nitrification, especially in low flow areas, and may contribute to chloramine demand. However, addition of phosphate is still beneficial for the control of metals release from the distribution system and premise plumbing.	Optimizing PO_4-P residual in the distribution system could make phosphorus the limiting nutrient for regrowth, thereby assisting in nitrification control. However, optimal phosphorus concentration needs to be determined with careful consideration of corrosion-control treatment.	Phosphorus present in the source water is effectively removed by coagulation. Change of corrosion-control treatment (i.e., from orthophosphate to high pH) should be carefully evaluated so as not to compromise lead control.

nature and reactivity of NOM. Processes downstream of coagulation and sedimentation (i.e., intermediate oxidation with ozone or other oxidants, including free chlorine, and any excess free ammonia added to form chloramine) may have an overriding effect on chloramine stability and associated nitrification potential.

Studies by Wilczak et al. (1999) were conducted at a plant treating reservoir raw water with moderate TOC (5 mg C/L) and alkalinity (120 mg $CaCO_3$/L). No impact on chloramine demand was observed because of improved TOC removal in pilot tests of enhanced coagulation with alum and ferric chloride salts. Very little difference in chloramine demand was observed when TOC was removed from the settled water (without ozonation) by a small, activated carbon column; however, tests on ozonated and ozonated/biofiltered waters indicated a greater impact of TOC removal on chloramine stability (Wilczak et al. 1999). Bone et al. (1999) found little difference in chloramine demand and ammonia release rates between waters after conventional and enhanced coagulation (3.4 versus 2.9 mg C/L residual DOC). Harrington et al. (2002) found that, for the same water supply, enhanced coagulation may delay the onset of nitrification and nitrification episodes may be less frequent.

Wilczak et al. (2003) investigated chloramine stability for a plant treating reservoir raw water with moderate TOC (e.g., 4 mg C/L) and alkalinity (e.g., 100 mg $CaCO_3$/L). This plant used conventional coagulation/flocculation/sedimentation followed by dual media filtration (anthracite/sand). Disinfection consisted of intermediate ozone applied after sedimentation, followed by free chlorine postfiltration, and then ammonia for distribution chloramination. Figure 4-7 shows results of chloramine demand testing conducted by this utility, where waters from various treatment stages were dosed to provide an initial 2.5 mg Cl_2/L chloramine residual (pH 8.9, 20°C) and then sampled over several days to measure chloramine residual. Coagulation and sedimentation eliminated a substantial amount of the chloramine demand, as shown by the settled water compared with the raw water (Figure 4-7). However, samples collected later in the treatment train after ozonation

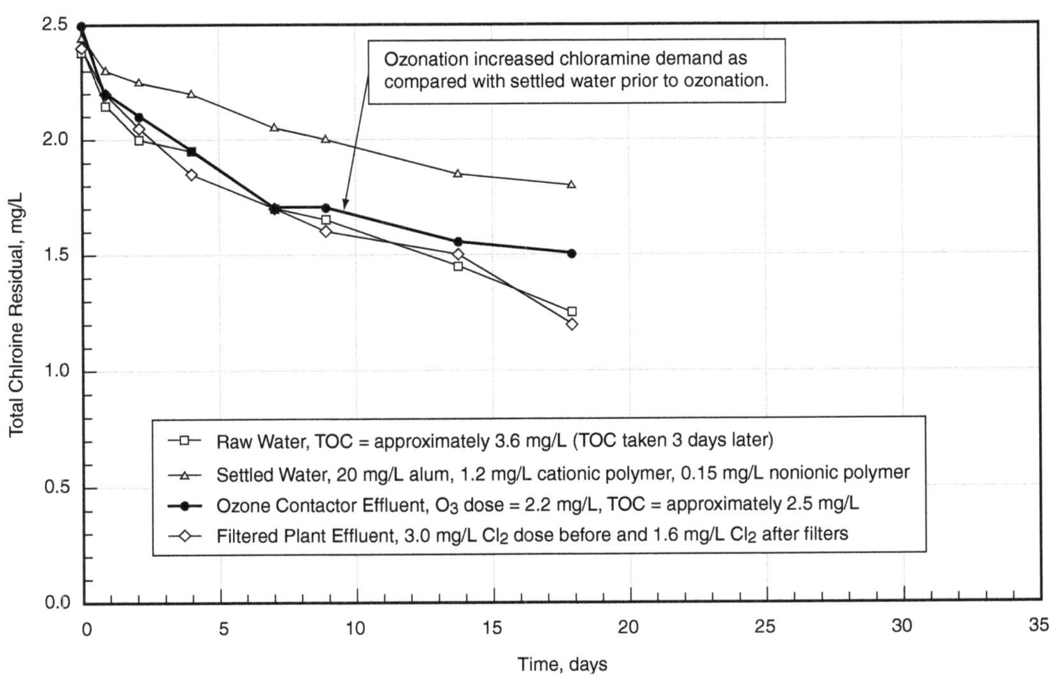

Source: Wilczak et al. 2003

Figure 4-7 Impact of coagulation and ozonation on chloramine demand. Chloramine demand/decay profile: 20°C, pH = 8.9.

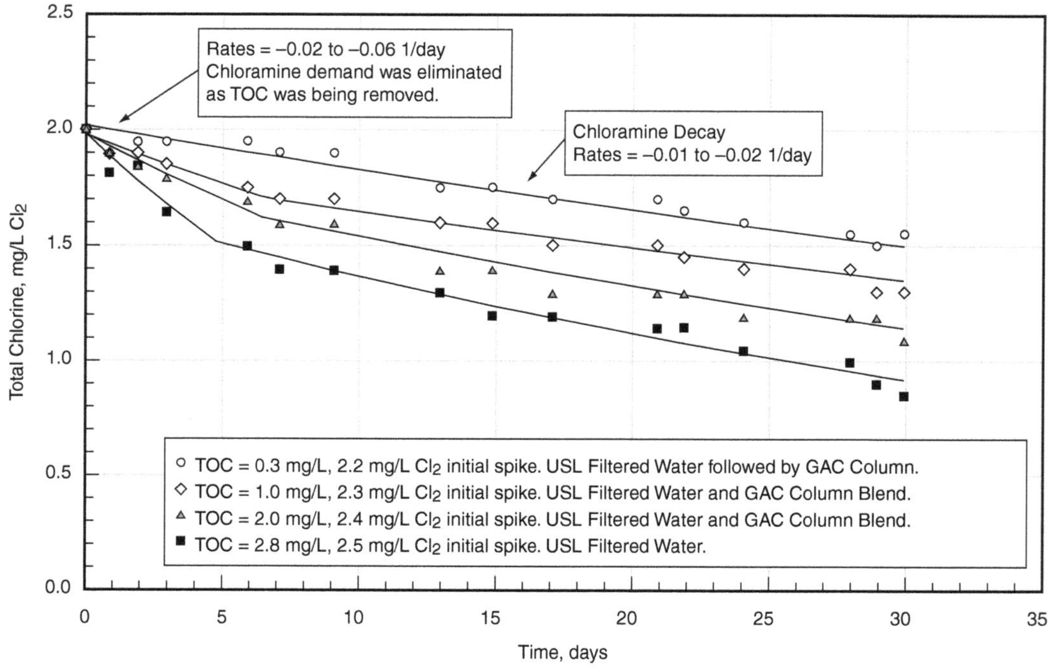

Source: Wilczak et al. 2003

Figure 4-8 Effect of TOC removal by granular activated carbon adsorption on chloramine demand. Ozonated, filtered effluent and GAC effluent blends; 20°C, pH = 8.5.

and chlorination showed chloramine demand curves similar to those of the raw water, supporting that oxidant addition created a TOC more reactive with chloramine. Monitoring total chloramine loss for various treatment samples, held in uniform conditions, may be useful for operators to better understand the impact of each treatment process on overall chloramine stability.

Results from tests to predict impacts of TOC removal on chloramine demand of ozonated/granular activated carbon (GAC)-filtered water are shown in Figure 4-8. The initial chloramine demand due to NOM was completely eliminated in the water sample experiencing the greatest TOC removal by (GAC) adsorption (Wilczak et al. 2003).

Some utilities may wish to investigate the use of more advanced and expensive NOM removal strategies such as GAC adsorption, ion exchange, and nanofiltration (NF). These would require significant capital investment, and there may be more cost-effective ways of achieving the same goal unless there are other reasons to use these processes (e.g., removal of microcontaminants of concern).

Management of the source water quality through source water protection methods, where possible, may decrease raw water TOC and the need for expensive NOM removal processes. Chloramine is quite stable in some coagulated/settled waters, and the addition of oxidation processes for other purposes should be carefully investigated for their impact on chloramine demand and decay (Wilczak et al. 2003). More cost-effective alternatives to controlling nitrification through removal of organic materials are the creation of optimal conditions for chloramine formation that promotes residual stability and minimizes excess free ammonia leaving the treatment plant, optimized pH, decreased chloramine demand at the treatment plant, booster chloramination in the distribution system, reduction of water age through elimination of excess storage, and elimination of dead zones. Alternatively, installation of GAC or NF to remove TOC or other DBP precursors may eliminate the need for chloramination, depending on the volume of water treated, system size, and ability to maintain free chlorine residual in the distribution system without forming excessive amounts of DBPs.

Effect of oxidants/disinfectants. Oxidation satisfies a portion of the chloramine demand, lowering the overall disinfectant residual loss. Bone et al. (1999) showed that a longer contact time with free chlorine (45 min versus 1 min) stabilized chloramine residuals and less free ammonia was released. Note that longer free chlorine contact time may also increase DBP formation, such as total trihalomethanes and five haloacetic acids.

Effect of pre-/intermediate-ozonation on chloramine demand. The use of pre- or intermediate ozonation can have both positive and negative influences on chloramine demand and, correspondingly, nitrification. When appropriately applied and residual quenched, there should not be any ozone residual leaving the ozone contactor. The ozone should reduce the oxidant demand, thereby reducing downstream chlorine/chloramine demand. Also, ozone can improve removal of organics from water by downstream biological or adsorptive filtration processes, thereby decreasing treated water chlorine/chloramine demand and increasing biological stability of water.

However, unless the ozonation process is optimized such that oxidant demand is reduced and ozone residual quenched, application of ozone can also have negative influences on chloramine demand. Intermediate ozonation of coagulated and settled water increased the total chlorine demand at two water treatment plants (Figure 4-7). Ozonation applied to settled water at approximately 1:1 ozone/TOC mass ratio increased chloramine demand (Wilczak et al. 2003).

A combination of ozone and hydrogen peroxide (typical ozone-to-peroxide mass ratio of 5:1 or greater) was used to control taste and odor at a water treatment plant (Wilczak et al. 2003). Residual hydrogen peroxide was removed in the filter media and did not react with chlorine applied after the filters. Chloramine stability was significantly degraded when higher doses of hydrogen peroxide (ozone-to-peroxide mass ratio of 4:1) were used to manage elevated tastes and odors in the summer of 2001. Not only did the overall chlorine use increase at the plant from around 4.6–6.6 mg Cl_2/L to react with the residual peroxide in the filter influent (no peroxide residual was present in the effluent), but chloramine stability was also impacted. A rapid loss in total chlorine residual occurred in a section of the distribution system, indicating a different reaction was taking place. The section of the distribution system supplied by the plant experienced rapid chloramine residual loss followed by nitrification in storage reservoirs.

Presence of chlorite ion from chlorine dioxide byproduct, or chlorite application in distribution systems. Chlorite ion (a byproduct of chlorine dioxide disinfection) has been observed to selectively inactivate AOB (e.g., Rungvetvuthivitaya et al. 2014; McGuire et al. 1999). In some instances, utilities using chlorine dioxide for primary disinfection and chloramine for residual maintenance have reported that chlorite can be a nitrification inhibitor (Zhu et al. 2010). This is an area worthy of further review and study where utilities have converted to chlorine dioxide and chloramine to comply with DBP regulations. Some utilities have applied sodium chlorite directly in distribution systems to study its effect in nitrification-prone areas. Some studies report nitrification prevention benefits (Blute et al. 2013), while others report mixed results (McGuire et al. 2009). Zhu et al. (2010) reported that chlorite was more effective when applied prior to onset of nitrification as a preventive approach in the distribution system, and less effective at controlling nitrification episodes already in progress. The stage 1 D/DBPR regulated maximum contaminant level for chlorite of 1.0 mg/L is also a possible deterrent to deliberately applying chlorite for nitrification control.

Effect of activated carbon on chloramine. Monochloramine reacts with activated carbon (granular or powdered) as follows:

$$NH_2Cl + H_2O + C^* \rightarrow NH_3 + HCl + CO^* \qquad \text{(Eq. 4-4)}$$

$$2NH_2Cl + CO^* \rightarrow N_2 + C^* + 2HCl + H_2O \qquad \text{(Eq. 4-5)}$$

where C^* and CO^* represent the carbon surface and oxidized carbon surface, respectively. Monochloramine reacts with activated carbon in a catalytic (as opposed to adsorptive) process (Komorita and Snoeyink 1985). A catalytic process is one where chemical reaction rates are altered by a substance (the catalyst is activated carbon in this case) which is not itself changed during the chemical reaction. Dichloramine, if present, reacts faster with activated carbon than monochloramine (Komorita and Snoeyink 1985). Dechloramination can occur with activated carbon potentially for long periods of operation (as it is a catalytic process not limited by adsorptive capacity of activated carbon [White 1999]) and in packed-beds the process reaches steady-state following a period of higher removal.

Fairey et al. (2007) studied monochloramine reduction from five commercially available GACs in laboratory-scale column studies. Steady-state monochloramine destruction was achieved in each of the carbons studied, although steady-state levels varied widely. Some treatment plants pass chloramine through GAC filters, but this practice is undesirable due to disinfectant destruction and application of ammonia to the filter bed (unless nitrification at the treatment plant is desired and controlled, as is sometimes the case. See Chapter 2).

Point-of-use (POU) filters using GAC or activated carbon-block media are commonly found in many households, schools, and buildings. In recent years, the provision of certified activated-carbon filters to remove lead at the POU (either for an individual household or faucet) is a common stopgap measure taken by local governments and utilities as lead service lines and lead fixtures are replaced or concentrations of lead release decline to below the action level for lead. Both Newark, N.J., and Flint, Mich., promoted POU filter distribution to residents when elevated lead was observed in their systems, and Washington, D.C., installed POU filters on fountains and taps in all public schools to reduce lead exposure. The popularity of POU filters with GAC or activated carbon-block media in households is not just limited to lead reduction, as activated carbon can remove unpleasant tastes and odors as well as other contaminants from water. In the United States alone, the use of POU filters increased in homes from 27 percent in 1995 to 41 percent in 2000 (Raucher et al. 2004). In 2018, approximately 60 percent of a surveyed population used a household filter (NSF International, n.d.). However,despite ubiquitous use, installation within chloraminated systems can pose a problem. Recent research found increased nitrite concentrations in some POU filters that far exceeded 2 mg/L N (Alfredo et al. 2019). In work tracking nitrite concentrations of filters actively installed in schools, microbiological activity and nitrite concentrations were greatest at the schools supplied with the best water quality, meaning the filters, while successful in removing lead, were detrimental to other aspects of water quality (Alfredo et al. 2020). The increase in nitrite concentrations coincided with measured microbial community shifts, suggesting that the process was dynamic and ever changing—thus complicating monitoring since a single water sample could easily misrepresent the true risk from nitrite exposure.

Filtration via inert and/or biologically active granular media. Filtration would be expected to be neutral or beneficial to chloramine stability due to its likelihood of removing particulates, which can include microbial agents and chemically reactive organics and inorganics. Removal of these contaminants will likely result in reduction of chloramine demand.

The city of Arlington, Tex., began experiencing atypical water quality problems beginning in 1999 after ozonation and biofiltration were added to the existing conventional treatment process. These problems included high chloramine residual decay rates, system-wide nitrification, and elevated manganese levels in the finished water. This study reviewed literature and existing plant and distribution system data and collected additional data to

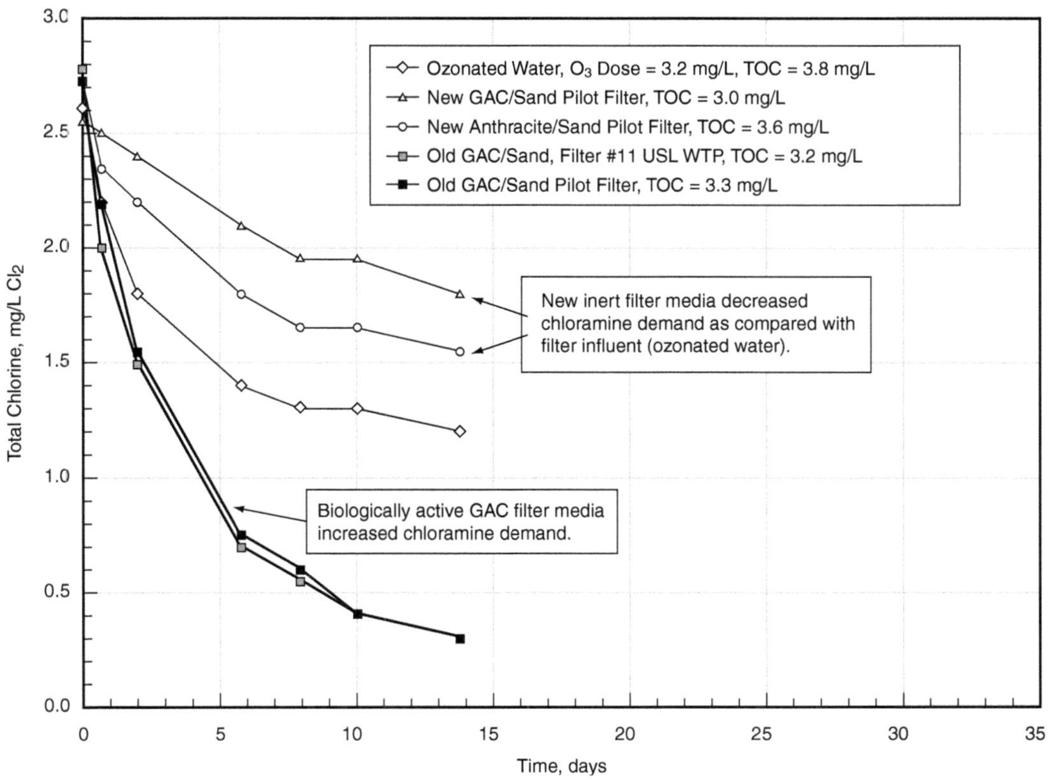

Source: Wilczak et al. 2003

Figure 4-9 Effect of inert and biologically active filtration on chloramine demand. Ozonated and filtered water; 20°C, pH = 8.5.

determine whether modifications to the plant and distribution system operations would resolve these water quality problems. Results demonstrated that chloramine stability was improved with increased free chlorine contact time at a pH near 8.3. Improved biofiltration by enhancing and optimizing biological growth through nutrient addition improved chloramine stability, as well as eliminated manganese breakthrough at the filters. Careful monitoring and minimizing water age in the distribution system also helped to reduce water quality issues (Vokes 2007).

Filtration of ozonated water through biologically active GAC/sand media at two treatment plants in California (Wilczak et al. 2003) substantially increased the chloramine demand despite lower TOC and lower BDOC (Figure 4-9). Free chlorine was first applied in the treatment train after the filters that contained sand and biologically active 11-year-old GAC originally installed to control tastes and odors. Pilot filters were set up to investigate the impact of the filter media on chloramine demand: (a) biologically active GAC/sand media taken from one of the full-scale filters, (b) new GAC/sand media, and (c) new anthracite/sand media. New GAC or anthracite/sand did not exhibit any negative impact on chloramine demand, as shown by the excellent stability of chloramine in both pilot filter effluents (Figure 4-9). The application of a more intensive air scour during backwash in the pilot filters did not eliminate the increased chloramine demand observed in the biologically active pilot GAC/sand filter effluent. No GAC fines were visible on the 0.45-μm filter paper when filtering 5 L of filter effluent, but a slight brown film, possibly bacterial cell fragments, was visually observed. The hypothesis of suspended material shed from the filters and exerting chloramine demand was confirmed by subsequent membrane filtration studies as discussed in the next section (Wilczak et al. 2003).

Membrane filtration. In some cases, membrane filtration or other advanced NOM removal processes may be an excellent option to improve disinfectant stability, minimize DBPs, eliminate regrowth of microorganisms, and remove other contaminants of concern. Wilczak et al. (2003) conducted tests with settled, ozonated, and biologically filtered water samples that were passed through membrane modules and chloraminated. Ultrafiltration (UF) was expected not to impact chloramine stability, whereas NF was expected to improve disinfectant stability.

Chloramine in NF permeate was very stable regardless of whether settled, ozonated, or filtered water was processed (Figure 4-10). The decay rate in NF permeate was comparable to that in Milli-Q® ultrapure water (Millipore Corporation, Billerica, Mass.) obtained as a control in that no initial demand was observed in these test waters. These results indicated that utilities that employ NF may find that free chlorination in the distribution system becomes a viable alternative to chloramine. Chloramine in the UF permeate was not as stable, and no overall improvement was achieved with either settled or ozonated water. While reverse osmosis (RO) is typically not implemented if the treatment objective is solely NOM removal, RO, like NF, will remove DBP precursors and improve disinfectant stability.

Optimized corrosion control. Optimized corrosion control can reduce chloramine demand by reducing the reaction between chloramine and corrosion byproducts and minimizing pipe wall microenvironments and sediments that promote favorable conditions for nitrifying microorganisms. It should be noted, however, that systems electing to use phosphate-based corrosion inhibitors or sequestering agents and relying on pH near neutral (e.g., 7.4–7.6) may experience higher chloramine decay rates, releasing free ammonia that may increase nitrification potential compared to using a higher pH (e.g., >8.3) where chloramines have greater stability (Figure 4-10).

MONITORING FOR PREVENTION AND EARLY DETECTION OF NITRIFICATION

Monitoring is key for prevention, early detection, and mitigation of nitrification. In recent years, several state regulatory agencies have implemented distribution system-monitoring requirements for chloraminated water systems with the objective of nitrification prevention. A discussion of nitrification-related monitoring requirements adopted by state regulatory agencies is included in Chapter 6. Nitrification control plans or action plans should be developed using nitrification-monitoring parameters, frequencies, and locations. These parameters, frequencies, and locations are selected on a site-specific basis and tailored to achieve predetermined water quality objectives. Chapter 6 ("Monitoring for Nitrification Prevention and Control") of this manual describes monitoring parameters, frequencies, locations, and analytical methods that need to be considered for nitrification monitoring (including online analyzers) and data management.

DISTRIBUTION CONFIGURATION AND OPERATIONS AS NITRIFICATION CAUSES AND CONTROLS

Distribution System Design

Distribution system design aspects that may impact nitrification include distribution system configuration and its impact on water age, pipe materials, methods to boost chloramine residual, and storage reservoir design (see Table 4-4).

Distribution system configuration, size, and zones. Systems designed with excessive finished water storage capacity (higher capacity than necessary for meeting peak demands and fire flows) are more difficult to operate for nitrification control if the increased finished water capacity results in increased water age (i.e., stagnant areas in piping and/or

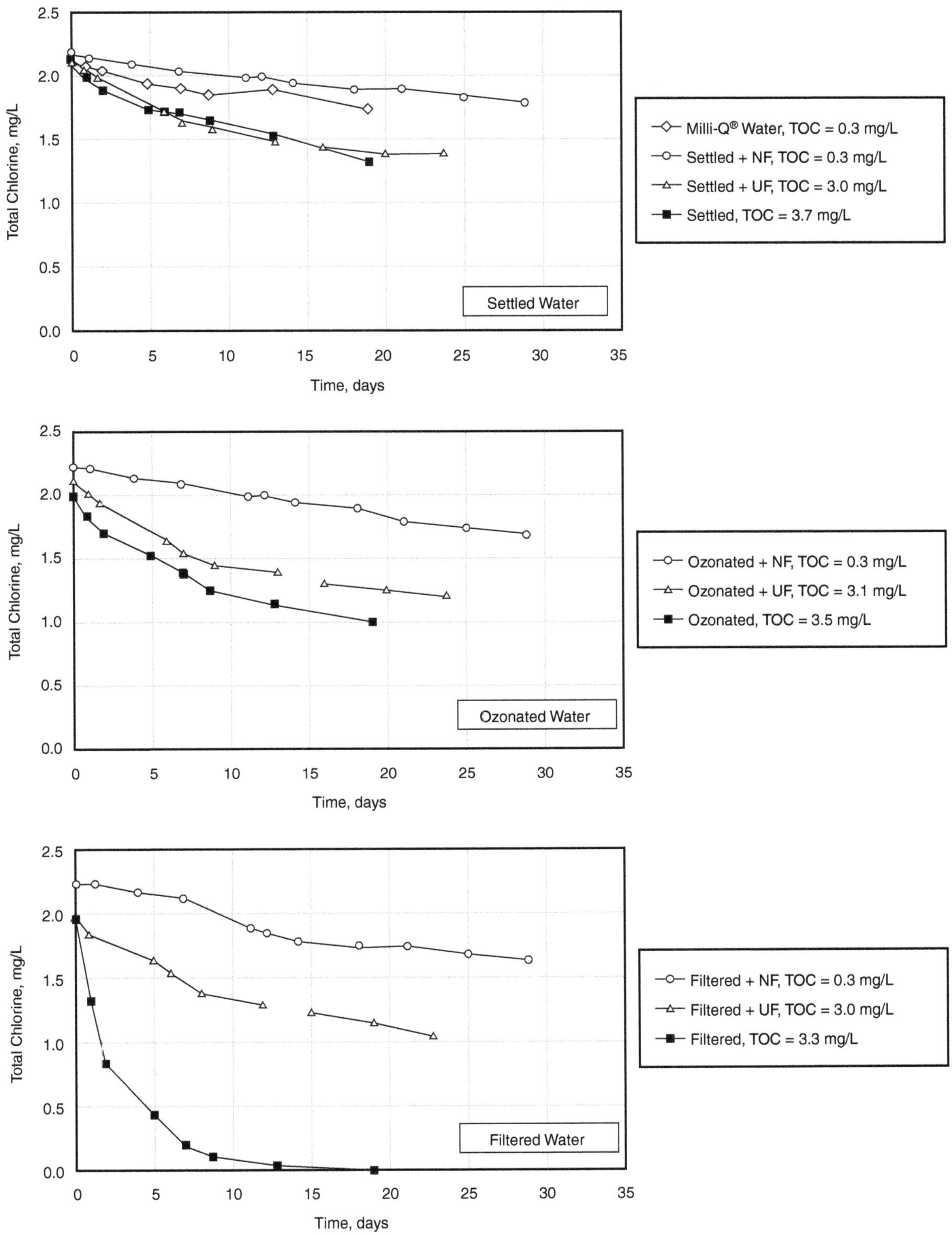

Source: Wilczak et al. 2003

Figure 4-10 Effect of membrane filtration on chloramine demand; settled, ozonated, and biofiltered (old granular activated carbon/sand) water, pH = 8.7; 20°C (no free Cl_2 contact time)

Table 4-4 Causes and control of nitrification: distribution system design and configuration

Parameter or Condition	Causes of Nitrification	Control of Nitrification	Feasibility of Implementation for Nitrification Control
System configuration, size	Oversized finished water storage reservoirs and water mains are primary locations for nitrifier growth. Storage reservoirs where water is not mixed or cycled sufficiently may stratify, resulting in higher water age, which can lead to nitrification.	Proper cycling and/or seasonal storage removal from service can limit water age. Reservoirs with low water mixing/cycling and dead-end pipes could be taken out of service.	Can be limited by system design and potentially expensive to modify. Temporary or permanent decommissioning of redundant storage can be relatively easy but must consider minimum storage capacity for fire protection.
Pressure zone configurations	Sequential pressure zones can increase water age. Most systems reporting nitrification have long and complex distribution systems.	Adjust the timing of pumping operations to bring fresh water up through sequential pressure zones. If this is not possible, boosting chloramine may be necessary to maintain residual in sequential pressure zones.	If it is not possible to modify pressure zones due to hydraulic constraints, operational practices can be implemented instead to manage water age.
Pipe configurations	Dead ends cause water stagnation and long detention times and may allow increased water temperature in shallow pipes in the summer.	Looped pipe networks help water circulate and eliminate dead ends. Valve exercise programs help locate inadvertently closed valves.	It is expensive to modify, so valve exercise and flushing (for bulk water turnover) are typically implemented instead.
Pipeline materials	Unlined cast-iron pipes and other iron pipes may corrode more easily and exert chloramine demand, form tubercles, and shield nitrifiers.	High-quality water, eliminating stagnation, lining pipelines, and proper corrosion control limit the impact of pipeline material.	It is expensive to replace pipes, so operational practices (such as unidirectional flushing) are typically implemented instead.
Finished water storage mixing	Poor mixing or baffling in finished water storage reservoirs causes water stagnation and localized older water through thermal stratification or hydraulic dead zones. Lack of mixing also makes water quality monitoring by analyzers and grab samples difficult to interpret.	Complete mixing is the desired condition for distribution system finished water storage reservoirs. This can be achieved with a combination of pumping/deep cycling, variations to fill/draw duration or intervals, active mixing devices, and/or modifications to inlet/outlet. Active mixing can also be incorporated as part of boosting chloramination system.	Engineering expertise may be needed to fully evaluate and implement active mixing in storage tanks. If structural modifications are needed, they will result in higher cost and facility downtime, although newer model mixers can typically be installed without structural modifications or taking the tank offline. Initial capital cost for mixer installation and potential operating cost depending on the design should be considered.
Finished water storage inlet/outlet	Common inlet/outlet may contribute to stagnation and stratification. Nitrification can occur also with separate inlet/outlet if water in storage is not mixed or sufficiently cycled.	Separating inlet and outlet or adding active mixing may help with water circulation. Directional inlet and outlet valves or an inlet orifice restriction plate can be less costly than piping changes.	High momentum or longer duration pumping can increase turbulence in inlet water and could be implemented instead of piping changes.

(continued)

Table 4-4 Causes and control of nitrification: distribution system design and configuration (*Continued*)

Parameter or Condition	Causes of Nitrification	Control of Nitrification	Feasibility of Implementation for Nitrification Control
Chloramine residual boosting in distribution system	Free ammonia from chloramine degradation can initiate nitrification.	Automated chloramination control system minimizing free ammonia and maintaining higher chloramine residuals (e.g., action targets) in the distribution system appears to help control nitrification. This may be especially valuable in storage reservoirs within the distribution system where residual is low.	Boosting could be a feasible long-term strategy. Boosting with chlorine and ammonia is more desirable because it allows control of chloramine concentration while minimizing free ammonia, but it involves two chemicals and the need to balance them. Chemical storage in many and/or remote locations may be challenging.

storage). To help address this issue, permanent or seasonal removal of storage from service or lowering of storage volume can be a relatively easy improvement if emergency flow and fire storage requirements allow it. Deep cycling of tanks periodically can also assist in reducing water age and preventing nitrification. Low-flow lines (e.g., dead ends and oversized pipes) are the most susceptible pipe configurations for nitrification development. Where necessary, redesign and modification of storage facilities, pressure zones, pipelines, and operations can provide long-term improvements.

Distribution system pipe materials. It is likely that nitrification can occur regardless of pipe material (i.e., plastic, polyvinyl chloride [PVC], asbestos–cement, ductile iron, or cast iron), but certain pipe materials like unlined cast-iron pipes prone to tuberculation may provide more favorable nitrification conditions. Pipe tuberculation provides an environment that is more conducive to the growth of microorganisms, including nitrifiers. Also, pipe corrosion will contribute to chloramine demand, accelerating disinfectant loss.

Internal mixing in finished water storage facilities. Storage tanks should be well mixed to avoid hydraulic dead zones that may provide areas for chloramine degradation and sediment or biofilm accumulation, which can harbor microorganisms (including nitrifiers) and contribute to nitrification. Mixing also reduces thermal stratification and ensures uniform water quality in the storage tank. This mixing may be achieved through various means: higher-momentum pumping, mixing devices, inlet/outlet separation, and inlet modifications to increase fill turbulence. Mahmood et al. (2009) developed a multifactorial computational fluid dynamics and Excel-based mixing model of storage operational (i.e., cycling volume, flow, fill duration) and configuration (i.e., inlet size, location, and flow direction) factors to evaluate potential for storage facilities' impact on water age and water quality. Carlomagno et al. (2005) described the evaluation of internal mechanical mixing options to improve chloramine residual dispersion and minimize nitrification in storage. Comparison of water quality from the top of the tank to the tank to the tank outlet water quality can also indicate if the tank is stratified.

Chloramine residual boosting in the distribution system. Utilities often successfully use booster chlorination applied at key points in the distribution system to effectively manage distribution system disinfection and DBPs with chloramine and free chlorine (AWWA Disinfection Systems Committee 2008). Properly designed and controlled chlorine boosting can be used to convert chloramine decomposition-related free ammonia back to chloramine in the distribution system, thereby helping to minimize nitrification potential (Coley et al. 2010; Wilczak et al. 2003). Automated chloramination control systems for water storage tanks have been helpful in managing nitrification in the distribution system.

Distribution System Operations and Maintenance

Distribution system O&M are related to minimizing water age through operational means, maintaining system cleanliness, performing adequate monitoring, and optimizing corrosion control. In addition, some utilities go through a periodic free chlorination of system areas or blend different water sources, if feasible, as a means of controlling nitrification. If blending is performed, chloraminated water should be blended with chloraminated water only. If chloraminated water is blended with water containing free chlorine, a stable monochloramine residual will be difficult to maintain, and breakpoint chlorination may also occur. The distribution system O&M that can be used to control nitrification are shown in Table 4-5.

Distribution system and storage operation. Detention time or water age plays an important role in nitrification development. Areas with longer detention times are more

Table 4-5 Causes and control of nitrification: distribution system operation and maintenance

Parameter or Condition	Causes of Nitrification	Control of Nitrification	Feasibility of Implementation for Nitrification Control
Water age due to operations	Excessive water age is one of the primary causes of nitrification. Desired water age is site specific and a function of many factors.	Appropriate pumping schedules, reservoir cycling, seasonal reservoir outages, and targeted flushing can decrease detention time.	A control program requires understanding of distribution system hydraulics and adoption of proactive, innovative approaches to minimize water age.
Water storage operating levels	Water systems that are operated at full storage capacity for maximum fire protection and water demands result in increased water age and deteriorated water quality.	Chloraminated water storage tank levels need to either be lowered or operation should allow for significant water cycling.	Decrease of water storage and associated water age is a balancing act between water quality and supply.
Water demand	Decreasing water demand due to conservation increases water age.	Continuous bleeding or automated flushing of long dead-end or oversized pipelines is helpful.	Nitrification control could be considered for long dead-end pipes if allowed by conservation requirements.
System flushing	Periodic or episodic flushing will not be sufficient in areas receiving nitrifying water or where nitrification biofilm has already established.	Systematic preventive flushing reduces deposits and helps identify and eliminate closed valves and dead ends. Automatic flushing stations can more consistently improve water turnover in low flow mains.	Routine flushing (for bulk water turnover) or unidirectional flushing (to reduce chloramine demand) programs provide the most benefit in terms of improving water quality. Automatic flushing hydrants are relatively inexpensive and effective for small pipes. Large water volumes may be needed.
System cleanliness	Sediment deposits at the bottom of storage tanks and biofilms in tank and pipe walls cause disinfectant demand and promote ongoing nitrification.	A storage facility cleaning and maintenance program is crucial for maintaining chloramine residual and nitrification prevention. Roofing must also maintain integrity.	A control program is relatively inexpensive, but large finished water storage facilities can be expensive to clean.

(continued)

Table 4-5 Causes and control of nitrification: distribution system operation and maintenance (*Continued*)

Parameter or Condition	Causes of Nitrification	Control of Nitrification	Feasibility of Implementation for Nitrification Control
Periodic free chlorination of parts or entire system	Free chlorination, especially frequent chlorination of nitrifying water storage reservoirs, is not a long-term solution and is an indication of other parameter(s) or condition(s) not being optimized.	Periodic free chlorination of the entire system or specific portions may successfully control nitrification in the short term. This strategy is currently used; however, there may be a future trend away from this approach.	A control program is easy to implement as a small-scale and successful short-term strategy, but it's not as good in the long term. It may cause high disinfection byproduct levels and taste and odor complaints. The conversion process from chloramine to free chlorine and the reconversion from free chlorine to chloramine need to be carefully controlled to avoid water quality upsets, such as alteration of corrosion-control conditions.
Blending or selecting source water	Blending water with different quality may provide both benefits and challenges with respect to nitrification, depending on how the blending is performed.	Blending or selecting source water with lower natural organic matter, lower nutrients, lower temperature, or higher pH may help control nitrification. Chloraminated water should not be blended with water containing free chlorine.	Control may be implemented, but the utility needs to characterize source waters and determine significant factors. It may be easy to implement if an alternative water source with appropriate characteristics is available and hydraulics allow for it.

likely to have lower disinfectant residuals, higher ammonia concentrations, and sediment accumulation; therefore, they are more likely to experience nitrification. Because storage facilities can become hydraulically or thermally stratified if not sufficiently cycled and mixed, adjustment of water storage operating levels, water demand, and pumping schedules is also necessary.

In temperate, tropical, and subtropical climates, water temperatures in uncirculated storage areas can become very warm. Storage facility studies to assess effectiveness of mixing and cycling often utilize spatially collected temperature measurements from within storage (Fisher et al. 2009). Adjustments to water storage operating levels and spacing of fill/draw cycles appropriately to coincide with water demand and pumping schedules are necessary to minimize water age. Mahmood et al. (2009) developed a spreadsheet model to demonstrate how various interrelated storage configurations and operating schemes affect mixing of stored water. Wong et al. (2005) and Mahmood et al. (2005) conducted extensive studies of storage mixing status and needs. Wong et al. (2005) provided results of extensive studies of storage facilities in planning for chloramination. Since those studies and after about 6 years of chloramination, mixers have been found beneficial where installed, and additional installations were planned. Some storage facilities were removed from service seasonally and others utilized in zonal free chlorination strategies (Wilczak et al. 2012).

Flushing. Flushing is one of the first but often temporary control measures for nitrification. If done correctly, flushing can replace the nitrification-degraded water (e.g., no disinfectant, elevated nitrite) that has a higher total chlorine residual (e.g., >1.5 mg Cl2/L) and lower nitrite/nitrate levels (<0.010 mg N/L). Systematic flushing (spot flushing or unidirectional flushing) can also be a preventive measure if applied effectively and regularly. Flushing moves the affected water and sediment out of the system rather than downstream to other areas, spreading the problem.

After the total chlorine residual increases and nitrite level decreases below the alert levels, flushing can be stopped. However, follow-up monitoring is needed to determine if nitrification activity has progressed to a point where chloramine residuals are again quickly dissipating (i.e., within a day). If this occurs, additional nitrification control measures will be needed. Mains with insufficient flow may benefit from strategically located, automated flushing stations to flush targeted areas more efficiently and regularly (Gupta 2015; Gupta and Akagi 2014). Pryor and Lovins (2009) reported that extensive, increased flushing was needed annually for nitrification control, including numerous automatic flush hydrants. Flushing may not be a viable option in areas prone to drought and arid conditions (i.e., places with source volume limitations). It is further noted that conventional flushing is generally not adequate to completely remove biofilm that is attached to the internal surfaces of pipes and storage facilities. Unidirectional flushing can be an effective tool for nitrification control; however, it is important that it is executed properly; otherwise, it can exacerbate the issue. In addition to flushing, mechanical pipe-cleaning strategies such as ice pigging are also options.

Storage inspection and cleaning programs. Sediment and biological deposit minimization within water storage facilities is a prerequisite to nitrification control. The deposits of residual turbidity, coagulants, corrosion byproducts, and occasionally filter media serve as a support environment for the growth of nitrifying microorganisms. Accumulated sediment (Liu et al. 2019) and biofilm (Lee et al. 2018, 2011; Pressman et al. 2012) can protect nitrifying microorganisms from bulk water chloramine residual. Therefore, a storage inspection and cleaning program is an essential (but not the only) part of maintaining chloramine residual inside storage facilities.

Periodic system wide or zonal secondary disinfectant change from chloramine to free chlorine. Some utilities periodically conduct a system-wide conversion to free chlorine by temporarily ceasing to apply ammonia, or they batch chlorinate storage facilities or areas by applying excess free chlorine to prevent or stop nitrification episodes (Krishna et al. 2014; AWWA Disinfection Systems Committee 2008; Skadsen 1993). Changing the disinfectant residual from combined to free chlorine might be necessary when nitrification control methods have shown to be unsuccessful and nitrification returns at high frequency. If done properly, a residual-type change (from chloramine to free chlorine) can be an effective measure to temporarily resolve nitrification and restore disinfectant residuals, especially in storage tanks and small isolated system areas. Utilities using chloramination report varied durations for routine free chlorination switching. Some utilities using chloramination routinely switch to free chlorine residuals during annual flushing programs to minimize concerns with hazards to aquatic life from chloramine in flushing waters and as a nitrification control measure. Switching to free chlorine will inherently increase DBPs during the free chlorination, may cause upsets to corrosion-control conditions, and may create the potential for chlorinous taste and odor during the switchover and during free chlorine use. It is generally understood that nitrification will likely return upon returning to chloramine unless underlying causes are addressed.

Blending or preferential finished water source selection. Blending finished water sources with different water quality characteristics or preferential selection of best alternative sources on a seasonal or continuous basis may provide more stable water and fewer favorable growth conditions for nitrifying microorganisms and thereby better nitrification control (Wilczak et al. 2012). However, finished water with chloramine residual should not be blended with finished water containing a free chlorine residual. If blending of water with free chlorine and chloraminated water cannot be avoided, a detailed system-specific evaluation is needed, and the chemical feed control systems need to be designed appropriately to ensure a stable chloramine residual in the blended water.

KEY TAKEAWAYS

Typically, nitrification results from a combination of several cumulative and/or competing factors, all of which may need to be addressed to some extent in a chloraminated distribution system to substantially and consistently prevent or reduce nitrification. The relative impacts of the nitrification causative factors are undoubtedly interrelated, and if some factors or issues are addressed, others may become less significant.

The root causes of nitrification are the availability of free ammonia as a substrate for nitrifying microorganisms allowing them to proliferate and an absence of disinfectant residual (chloramine) at levels sufficient to inactivate nitrifiers. Free ammonia may be present in the source water and present in treated water if complete breakpoint chlorination has not been achieved or may be available in chloraminated systems through ammonia feed excess or through ammonia release from chloramine via demand and decay reactions. Understanding chloramine demand and decay for a given water supply is key to understanding approaches used to prevent or control nitrification.

The fundamental basis of nitrification control or inhibition is for the rate of inactivation of nitrifying microorganisms (by chloramine disinfectant residual) to exceed their rate of growth (utilizing ammonia as a nutrient). Chloramine demand and decay reactions, which decrease disinfectant levels and release ammonia, are important because the chloramine concentration dictates inactivation rate and free ammonia concentration dictates nitrifier growth rate. Temperature, pH, and chloramine demand (e.g., NOM, nitrite, manganese, and bromide) are important because they influence chloramine stability and the free ammonia release rate that determines inactivation and growth rates of nitrifying microorganisms. Key points from this chapter are summarized in the following.

- Minimizing excess free ammonia during chloramine formation and from chloramine demand and decay reactions is key to nitrification control.
- Where feasible, selection of raw or finished water supply sources with water quality less favorable to nitrification (e.g., lower TOC and AOC, higher pH, lower disinfectant demand overall, lower temperature, etc.) can be beneficial for overall nitrification control.
- Removal of chloramine-demanding substances (e.g., NOM, manganese, and nitrite) via treatment processes and biological and particulate accumulation minimization via effective distribution system O&M activities are all important approaches to minimizing nitrification because all allow more effective chloramine concentrations and lower free ammonia concentrations as chloramine degrades. Therefore, removal of chloramine-demanding substances from source water and prevention or remediation of accumulations within the system will increase the potential for chloramine inactivation of nitrifiers and limit growth.
- Higher pH (>8.3) favors chloramine stability, minimizes potential for ammonia release via chloramine demand and decay reactions, and maximizes residual availability for inactivation of nitrifying microorganisms. Lower pH, sometimes recommended when using corrosion inhibitors, may increase chloramine degradation potential and limit inactivation of nitrifying microorganisms while increasing nutrients for their growth. Very high pH (>9.4), feasible in soft waters, may further reduce growth rate of nitrifying microorganisms and prevent nitrification.
- Corrosion-control treatment decisions for ongoing compliance with the Lead and Copper Rule need to consider nitrification control implications in systems using or considering chloramine for regulated DBP control.

- Oxidants applied to source water or in pretreatment, along with various types of clarification processes (e.g., coagulation/sedimentation and filtration), can affect NOM microbial assimilability and/or chemical reactivity and therefore may affect chloramine formation, stability, and nitrification potential.
- Bromide from source waters can contribute to chloramine degradation and hence may promote nitrification.
- Higher water temperature leads to faster microbial inactivation and growth rates and faster chloramine decomposition rates (i.e., lower chloramine and higher free ammonia concentrations). The net effect can be that the temperature increase will boost growth rate to a more significant degree than the inactivation rate and, at the same time, accelerate chloramine decomposition. This has implications for nitrification episodes occurring more in summer versus winter and in warmer areas of thermally stratified storage facilities or warmer climates overall.
- Sediments and biofilm accumulations in pipe networks and in storage facilities will contain chloramine-demanding materials, which lead to lower chloramine concentrations and higher free ammonia concentrations. Such environments can also shield nitrifiers from disinfection. Therefore, sediments will foster faster growth rates and slower inactivation rates, suggesting that sediment and biofilm removal by flushing and cleaning of storage facilities or pipes will help prevent or delay the onset of nitrification and may help remediate nitrification episodes when they occur.
- Longer residence times can increase the time and extent of chloramine degradation and ammonia release, allowing more growth and less inactivation of nitrifying microorganisms. The net effect is that longer residence times favor nitrification growth over inactivation.
- Chloramine booster stations and automated chloramine control in water storage tanks can increase the chloramine and decrease the free ammonia concentrations by addition of chlorine to recombine free ammonia (if present) with chlorine or by adding chlorine and ammonia, thereby increasing the rate of inactivation and decreasing the rate of growth.
- System-wide or localized conversion to free chlorine inactivates nitrifiers and removes the nutrients for their growth. However, nitrification will eventually return upon return to chloramination unless underlying causes are addressed. Timing of free chlorination is critical to ensure effectiveness through nitrification season.
- Chlorite, occurring as a chlorine dioxide DBP or applied directly, has yielded mixed results with regard to nitrification control, so additional study is warranted.

Ongoing research and the sharing of utility operational experiences are needed to further understand the interrelated and often competing factors affecting nitrification causes and controls. This will become even more important as more utilities consider and convert to chloramine to comply with DBP regulations and/or where utilities are experiencing weather-related changes like warmer water temperature, longer warm seasons, increased or altered source organics, and/or water scarcity that can translate to increasing conservation requirements and water age issues. Increases in water age can also be due to decreased consumer water use associated with the economy or weather changes year to year. For successful nitrification prevention and control, utilities need to understand and address chloramine formation and stability as well as nitrifier growth and inactivation factors for their particular systems.

DISCLAIMER

The US Environmental Protection Agency through its Office of Research and Development collaborated in the production of this chapter. It has been subjected to the agency's peer and administrative review and has been approved for external publication. Any opinions expressed in this chapter are those of the authors and do not necessarily reflect the views of the agency; therefore, no official endorsement should be inferred. Mention of trade names or commercial products does not constitute endorsement or recommendation for use.

REFERENCES AND BIBLIOGRAPHY

Adhikari, R.A., A. Sathasivan, and K.C.B. Krishna. 2012. "Effect of Biofilms Grown at Various Chloramine Residuals on Chloramine Decay." *Water Science and Technology—Water Supply*, 12:4:463–469.

Alfredo, K., K.L. Chen, and M. Schmelling. 2019. "Stagnation, Use, and Age: Understanding Nitrification in Point-of-use (POU) Carbon-block Filters." Proceedings of the Water Quality and Technology Conference, Dallas, TX.

Alfredo, K., J. Lin, A. Islam, and Z-W. Wang. 2020. Impact of Activated Carbon-block Point-of-use (POU) Filters on Chloraminated Water Quality. *AWWA Water Science*.

American Water Works Association (AWWA) Manual M68: *Water Quality in Distribution Systems*. 2017. Denver: AWWA.

AWWA Disinfection Systems Committee. 2008. "Committee Report: Disinfection Survey, Part 1—Recent Changes, Current Practices, and Water Quality." *Journal AWWA*, 100:10:76–90.

Blute, N., D. Pedersen, X.Y. Wu, and E. Means. 2013. "Demonstration testing of nitrification control using chlorite in Irvine, Calif." *Journal AWWA*, 105:4:E207–E216.

Bone, C.C., G.W. Harrington, P.S. Oldenburg, and D.R. Noguera. 1999. "Ammonia Release from Chloramine Decay: Implications for the Prevention of Nitrification Episodes." In *Proceeding of the AWWA Annual Conference*: American Water Works Association.

Boozarpour, M., A. Wilczak, A. DeGraca, J. Weintraub, and Garrett, S. 2007. "Impact of Chloramine on San Francisco's Water Quality Three Years after Conversion from Free Chlorine." In *2007 AWWA Annual Conference Proceedings*.

Carlomagno, B., A.R. Wong, M. Quady, C. Smith, and W.M. Grayman. 2005. "Evaluation of Mixing Improvements in Distribution System Reservoirs for Nitrification Control and Prevention." In *2005 AWWA Annual Conference Proceedings*.

Coley, R., J. Reynolds, and C. Johnson. 2010. "Nitrification Control in the Florida Keys." In *Proceedings of the AWWA Water Quality Technology Conference*: American Water Works Association.

Duirk, S.E. 2003. *Mechanisms and Modeling of Monochloramine Loss, Natural Organic Matter Oxidation, and Disinfectant by-Product Formation in Water*. University of Iowa.

Duirk, S.E., B. Gombert, J. Choi, and R.L. Valentine. 2002. "Monochloramine Loss in the Presence of Humic Acid." *Journal of Environmental Monitoring*, 4:1:85–89.

Duirk, S.E., B. Gombert, J.P. Croué, and R.L. Valentine. 2005. "Modeling Monochloramine Loss in the Presence of Natural Organic Matter." *Water Research*, 39:14:3418–3431.

Fairey, J.L., G.E. Speitel, Jr., and L.E. Katz. 2007. "Monochloramine Destruction by GAC: Effect of Activated Carbon Type and Source Water Characteristics." *Journal AWWA*, 99:7:110–120.

Fisher, I., A. Sathasivan, P. Chuo, and G. Kastl. 2009. "Effects of Stratification on Chloramine Decay in Distribution System Service Reservoirs." *Water Research*, 43:5:1403–1413.

Friedman, M., G.J. Kirmeyer, M. LeChevallier, S. Seidl, and J. Routt. 2010. *Criteria for Optimized Distribution Systems*. Denver: Water Research Foundation.

Gupta, K. 2015. "Nitrification Monitoring and Control at the Portland Water Bureau." Webcast. *Water Research Foundation.*

Gupta, K., and Y. Akagi. 2014. "Nitrification Monitoring and Control: A Utility Case Study." In *Water Quality Technology Conference*: American Water Works Association. New Orleans

Harrington, G.W., D.R. Noguera, C.C. Bone, A.I. Kandou, P.S. Oldenburg, J.M. Regan, and D. Van Hoven. 2003. *Ammonia from Chloramine Decay: Effects on Distribution System Nitrification*. Denver: Awwa Research Foundation.

Harrington, G.W., D.R. Noguera, A.I. Kandou, and D.J. Vanhoven. 2002. "Pilot-Scale Evaluation of Nitrification Control Strategies." *Journal AWWA*, 94:11:78–89.

Jafvert, C.T., and R.L. Valentine. 1992. "Reaction Scheme for the Chlorination of Ammoniacal Water." *Environmental Science & Technology*, 26:3:577–586 .

Johnson, D.W., and D.W. Margerum. 1991. "Non-Metal Redox Kinetics: A Reexamination of the Mechanisms of the Reaction between Hypochlorite and Nitrite Ions." *Inorganic Chemistry*, 30:25:4845–4851.

Kirmeyer, G., K. Martel, G. Thompson, L. Radder, W. Klement, M. LeChevallier, H. Baribeau, and A. Flores. 2004. *Optimizing Chloramine Treatment*. 2nd edition. Denver: AWWA Research Foundation.

Komorita, J.D., and V.L. Snoeyink. 1985. "Monochloramine Removal from Water by Activated Carbon." *Journal AWWA*, 77:1:62–64.

Krishna, K.C.B., A. Sathasivan, and D.C. Sarker. 2012. "Evidence of Soluble Microbial Products Accelerating Chloramine Decay in Nitrifying Bulk Water Samples." *Water Research*, 46:13:3977–3988.

Krishna, K.C.B., A. Sathasivan, and G. Kastl. 2014. "Effectiveness of Breakpoint Chlorination to Reduce Accelerated Chemical Chloramine Decay in Severely Nitrified Bulk Waters." *Chemosphere*, 117:692–700.

Lee, W.H., J.G. Pressman, and D.G. Wahman. 2018. "Three-Dimensional Free Chlorine and Monochloramine Biofilm Penetration: Correlating Penetration with Biofilm Activity and Viability." *Environmental Science & Technology*, 52:4:1889–1898.

Lee, W.H., D.G. Wahman, P.L. Bishop, and J.G. Pressman. 2011. "Free Chlorine and Monochloramine Application to Nitrifying Biofilm: Comparison of Biofilm Penetration, Activity, and Viability." *Environmental Science & Technology*, 45:4:1412–1419.

Liu, H., D.G. Wahman, and J.G. Pressman. 2019. "Evaluation of Monochloramine and Free Chlorine Penetration in a Drinking Water Storage Tank Sediment Using Microelectrodes." *Environmental Science & Technology*, 53:16:9352–9360.

Maestre, J.P., D.G. Wahman, and G.E. Speitel, Jr. 2013. "Monochloramine Cometabolism by *Nitrosomonas europaea* Under Drinking Water Conditions." *Water Research*, 47:13:4701–4709.

Maestre, J.P., D.G. Wahman, and G.E. Speitel, Jr. 2016. "Monochloramine Cometabolism by Mixed-culture Nitrifiers Under Drinking Water Conditions." *Environmental Science & Technology*, 50:12:6240–6248.

Mahmood, F., J.G. Pimblett, N.O. Grace, and W.M. Grayman. 2005. "Evaluation of Water Mixing Characteristics in Distribution System Storage Tanks." *Journal AWWA*, 97:3:74–88.

Mahmood, F., J. Pimplett, C. Hill, and Z. Chowdhury. 2009. "CFD Modeling and Spreadsheet Tool to Evaluate Mixing and Water Quality in Storage Tanks." In *2009 AWWA Annual Conference Proceedings*.

Margerum, D.W., L.M. Schurter, J. Hobson, and E.E. Moore. 1994. "Water Chlorination Chemistry: Nonmetal Redox Kinetics of Chloramine and Nitrite Ion." *Environmental Science & Technology*, 28:2:331–337.

McGuire, M.J., N.I. Lieu, and M.S. Pearthree. 1999. "Using Chlorite Ion to Control Nitrification." *Journal AWWA*, 91:10:52–61.

McGuire, M.J., X.Y. Wu, N.K. Blute, D. Askenaizer, and G. Qin. 2009. "Prevention of Nitrification Using Chlorite Ion: Results of a Demonstration Project in Glendale, Calif." *Journal AWWA*, 101:10:47–59.

Pintar, K.D.M., R.M. Slawson, and P.M. Huck. 2000. "Investigation of Conditions and Control Strategies Influencing Nitrification in a Chloraminated Bench-Scale Distribution System." In *Water Quality Technology Conference*: American Water Works Association.

Pope, P.G. 2006. *Haloacetic Acid Formation During Chloramination: Role of Environmental Conditions, Kinetics, and Haloamine Chemistry*. Doctoral dissertation. Department of Civil, Architectural, and Environmental Engineering, The University of Texas at Austin.

Pressman, J.G., W.H. Lee, P.L. Bishop, and D.G. Wahman. 2012. "Effect of Free Ammonia Concentration on Monochloramine Penetration Within a Nitrifying Biofilm and Its Effect on Activity, Viability, and Recovery." *Water Research*, 46:3:882–894.

Pryor, M., and W. Lovins. 2009. "Biostability in the Pinellas County Utilities Distribution System Nitrification Control." In *Florida Section of AWWA Fall Conference*.

Rungvetvuthivitaya, M., R.G. Song, M. Campbell, and C. Ray. 2014. "A Kinetic Study of Nitrification Inhibition in Water Distribution Systems Using Low Levels of Chlorite." *Journal of Water Supply Research and Technology-Aqua*, 63:6:497–506.

Sarker, D.C., A. Sathasivan, C.A. Joll, and A. Heitz. 2013. "Modelling Temperature Effects on Ammonia-oxidising Bacterial Biostability in Chloraminated Systems." *Science of the Total Environment*, 454:88–98.

Speitel, G.E., Jr, J.P. Maestre, and D.G. Wahman. 2014. *Monochloramine Cometabolism by Ammonia-Oxidizing Bacteria*. Denver: Water Research Foundation.

Skadsen, J. 1993. "Nitrification in a Distribution System." *Journal AWWA*, 85:7:95–103.

Srinivasan, S., and G.W. Harrington. 2007. "Biostability Analysis in Drinking Water Distribution Systems." *Water Research*, 41:10:2127–2138.

Valentine, R.L., K. Ozekin, and P.J. Vikesland. 1998. *Chloramine Decomposition in Distribution System and Model Waters*. Denver: AWWA Research Foundation and AWWA.

Vikesland, P.J., K. Ozekin, and R.L. Valentine. 2001. "Monochloramine Decay in Model and Distribution System Waters." *Water Research*, 35:7:1766–1776.

Vikesland, P.J., K. Ozekin, and R.L. Valentine. 1998. "Effect of Natural Organic Matter on Monochloramine Decomposition: Pathway Elucidation through the Use of Mass and Redox Balances." *Environmental Science & Technology*, 32:10:1409–1416.

Vikesland, P.J., and R.L. Valentine. 2000. "Reaction Pathways Involved in the Reduction of Monochloramine by Ferrous Iron." *Environmental Science & Technology*, 34:1:83–90.

Vikesland, P.J., and R.L. Valentine. 2002a. "Iron Oxide Surface-Catalyzed Oxidation of Ferrous Iron by Monochloramine: Implications of Oxide Type and Carbonate on Reactivity." *Environmental Science & Technology*, 36:3:512–519.

Vikesland, P.J., and R.L. Valentine. 2002b. "Modeling the Kinetics of Ferrous Iron Oxidation by Monochloramine." *Environmental Science & Technology*, 36:4:662–668.

Vokes, C.M. 2007. "Impact of Ozone and Biological Filtration on Water Quality Parameters in Arlington, Texas." *Ozone: Science & Engineering*, 29:4:261–271.

Wahman, D.G. 2018. "Web-Based Applications to Simulate Drinking Water Inorganic Chloramine Chemistry." *Journal AWWA*, 110:11: E43–E61.

Wahman, D.G., J.P. Maestre, and G.E. Speitel, Jr. 2016. "Monochloramine Cometabolism by Drinking Water Relevant Nitrifying Biofilm." *Journal AWWA*, 108:7:E362–E373.

Wahman, D.G., and G.E. Speitel, Jr. 2012. "The Relative Importance of Nitrite Oxidation by Hypochlorous Acid Under Chloramination Conditions." *Environmental Science & Technology*, 46:11:6056–6064.

Wahman, D.G., and Speitel, G.E., Jr. 2015. "Hydroxylamine Addition Impact to *Nitrosomonas europaea* Activity in the Presence of Monochloramine." *Water Research*, 68:719-730.

Wahman, D.G., G.E. Speitel, Jr., and M.V. Machavaram. 2014. "A Proposed Abiotic Reaction Scheme for Hydroxylamine and Monochloramine Under Chloramination Relevant Drinking Water Conditions." *Water Research*, 60:218–227.

White, G.C. 1999. *Handbook of Chlorination and Alternative Disinfectants*. 4th edition. New York: John Wiley & Sons, Inc.

Wilczak, A., M. Boozarpoor, A. Wong, J. Drago, L. Chau, and L. Leong. 2012. "To Switch or Not to Switch Periodically from Chloramine to Free Chlorine for Nitrification Prevention?" In *California-Nevada Section American Water Works Association Conference*. Santa Clara, CA.

Wilczak, A., D.R. Hokanson, R. Trussell, M. Boozarpour, and A.F. Degraca. 2010. "Water Conditioning for LCR Compliance and Control of Metals Release in San Francisco's Water System." *Journal AWWA*, 102:3:52–64.

Wilczak, A., L.L. Hoover, and H.H. Lai. 2003. "Effects of Treatment Changes on Chloramine Demand and Decay." *Journal AWWA*, 95:7:94–106.

Wilczak, A., L.L. Hoover, D.J. Song, H.H. Lai, S.B. Kachur, and C.W. Long. 1999. "Investigations of Oxidation, TOC Removal and Biological Filtration Effects on Chloramine Stability and DBP Formation at EBMUD, CA." In *Proceedings of the AWWA Water Quality Technology Conference*: American Water Works Association.

Wilczak, A., U.J. Smith, L.L. Hoover, and H.H. Lai. 2000. "Pilot and Full-Scale Evaluation of Chloramine Decay and HAA6 Formation in Ozonated/Biofiltered Surface Water." In *Proceeding of the AWWA Water Quality Technology Conference*: American Water Works Association.

Wong, A.R., M. Quady, C. Smith, A. Wilczak, and B. Carlomagno. 2005. "Monitoring Locations and Sampling Capabilities for Distribution System Reservoirs." In *2005 AWWA Annual Conference Proceedings*.

Zhang, H., and S.A. Andrews. 2013. "Effects of Pipe Materials, Orthophosphate, and Flow Conditions on Chloramine Decay and NDMA Formation in Modified Pipe Loops." *Journal of Water Supply Research and Technology-Aqua*, 62:2:107–119.

Zhang, Y., and M. Edwards. 2010." Nutrients and Metals Effects on Nitrification in Drinking Water Systems." *Journal AWWA*, 102:7:56–66.

Zhu, E., V. Monks, and R. Song. 2010. "Effectiveness of Early Application of Sodium Chlorite in Nitrification Prevention – a Demonstration-Scale Study." In *Proceedings of the AWWA Water Quality Technology Conference*: American Water Works Association.

This page intentionally blank.

Chapter 5

Microbiology of Nitrifying Microorganisms

Helen Y. Buse, John M. Regan, Hélène Baribeau

INTRODUCTION

This chapter presents the microbiology fundamentals of nitrifiers and nitrification with an emphasis on nitrifiers in drinking water and references to other environments as necessary. It is divided into four sections: taxonomy, morphology, and physiology of nitrifying microorganisms; isolation and enumeration techniques; growth conditions for nitrifiers; and their inactivation by various disinfectants. The discussion of the current knowledge about the biology of nitrifying microorganisms and their inactivation by the most common drinking water disinfectants presented in this comprehensive chapter can help water professionals better understand the nitrification-related phenomena that occur in drinking water distribution systems. Table 5-1 presents a summary of the information found in this chapter.

TAXONOMY, MORPHOLOGY, AND PHYSIOLOGY OF NITRIFYING MICROORGANISMS

Nitrification is commonly described as a two-step process performed by metabolically distinct groups of microorganisms: ammonia-oxidizing bacteria (AOB) and ammonia-oxidizing archaea (AOA) are responsible for the oxidation of ammonia (NH_3) to nitrite (NO_2^-), which is subsequently converted to nitrate (NO_3^-) by nitrite-oxidizing bacteria (NOB). When the nitrification process stops at the first step (ammonia oxidation to nitrite), it is often called "incomplete nitrification." When the first step is followed by the second step (nitrite oxidation to nitrate), the combined process is often called "complete nitrification."

Table 5-1 Key points from Chapter 5

Taxonomy, morphology, and physiology of nitrifying microorganisms	• Nitrification is carried out by various lithotrophic bacteria belonging to several classes of the Pseudomonadota (formerly Proteobacteria), heterotrophic bacteria and fungi, and archaea. • Among lithotrophic nitrifiers, ammonia-oxidizing microorganisms (AOM) are responsible for the oxidation of ammonia to nitrite and their genera have the prefix *Nitroso-*, nitrite-oxidizing bacteria (NOB) are responsible for the oxidation of nitrite to nitrate and their genera have the prefix *Nitro-*, and complete ammonia oxidizer are responsible for the oxidation of ammonia to nitrate and belong to the genus *Nitrospira*. • The first step, ammonia oxidation to nitrite, is often called incomplete nitrification, whereas the combination of both steps, oxidation of ammonia to nitrate, is called complete nitrification. • AOM and NOB are not phylogenetically related, and NOB appear more sensitive to environmental conditions, which may explain why incomplete nitrification can happen and nitrite is often present in water distribution systems experiencing nitrification. • Many nitrifiers are found in loose or compact cell aggregates (with slime layers). • Competition between heterotrophs and nitrifiers for dissolved oxygen, nitrogen, and space when growing attached to a surface is significant. Both groups of microorganisms can also benefit from interacting with each other.
Isolation and enumeration techniques	• Molecular methods for the detection, identification, and enumeration of nitrifiers have gained interest in the drinking water industry. However, they are not readily amenable to use by all water systems, resulting in the reliance on surrogate chemical indicators to detect nitrifier activity. Thus, nitrification in drinking water distribution systems is mainly monitored using indicator parameters such as nitrite, nitrate, ammonia, and chlorine residual. • Enumeration can be performed using a most probable number technique for direct nitrifier detection. This culturing technique requires an incubation period of 3–4 weeks at 25–30°C. Culturing techniques may not recover all target microorganisms, thus underestimating the organisms present. • Large serological diversity of nitrifiers and the coexistence of several serotypes in one ecosystem can limit the effectiveness of serological techniques.
Growth conditions	• Many factors affect the growth of nitrifiers, including substrate concentration, temperature, pH, oxygen concentration, alkalinity, light intensity, microbial community composition, and presence of inhibitory substances. • The optimum temperature for nitrifier growth ranges from 20–30°C (68–86°F), and the optimum pH ranges from 7.5 to 8.0. Each reduction in pH value of one unit decreases the concentration of free ammonia (NH_3), the substrate of AOM, by 1 order of magnitude. • Growth rates for nitrifiers are considerably less than those for heterotrophic bacteria, with doubling times for nitrifiers estimated at 11–50 h.
Inactivation of nitrifiers	• Inactivation of nitrifiers is affected by various parameters, including exposure time, water matrix, temperature, pH, turbidity, natural organic material concentration and composition, and disinfectant dosage, with the reported rates of nitrifier inactivation dependent on the enumeration method used. The ability of nitrifiers to grow in tight clusters of cells surrounded by a slime layer and to attach to surfaces, sediments, and other debris renders them more resistant to disinfection as compared to other drinking water bacteria. • Nitrifiers are much more sensitive to free chlorine than to chloramines. • Chlorite, the main byproduct of chlorine dioxide, is very effective at inactivating nitrifiers in a preventive context. • Nitrifiers can be partially inhibited by light. It also appears that nitrifiers have the ability to recover from inhibition during dark periods.

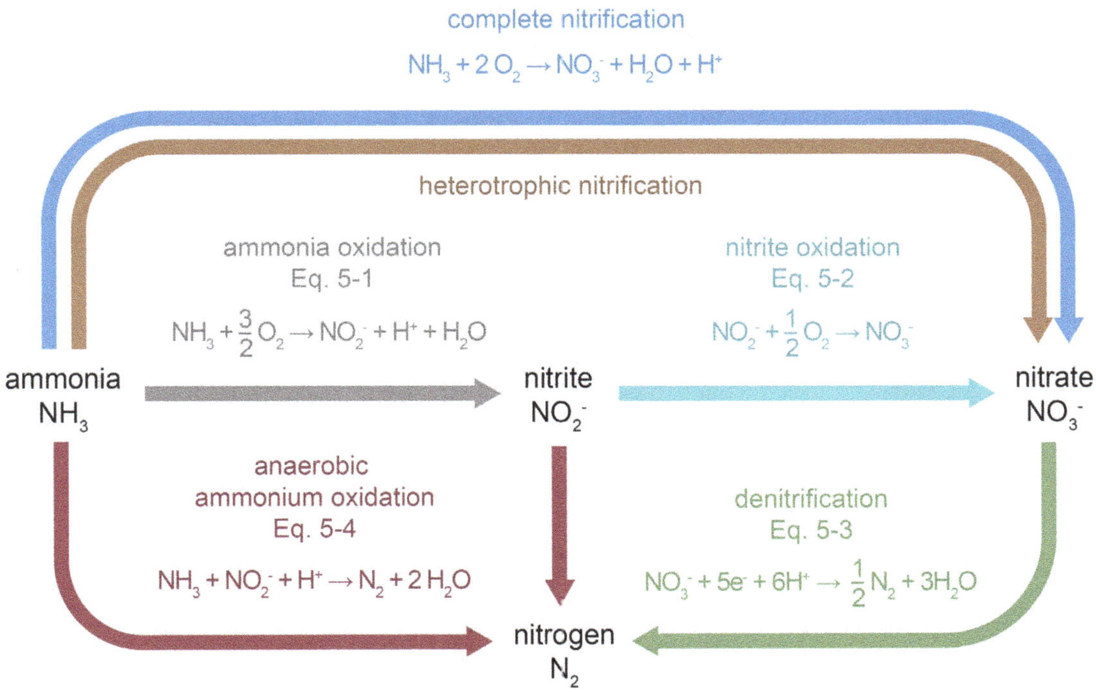

Figure 5-1 Microbially catalyzed nitrogen transformations of potential significance in drinking water distribution systems. For Eq. 5-3, the nitrate reduction half reaction can be coupled with organic electron donors for heterotrophic denitrification or H_2 for lithotrophic denitrification.

It was recently discovered that some bacteria (termed *comammox*, for complete ammonia oxidizer) can carry out both steps, ammonia and nitrite oxidation (Daims et al. 2015; van Kessel et al. 2015). Additionally, heterotrophic nitrification can occur when heterotrophics (i.e., microorganisms that use organic carbon), such as *Paraccocus pantotrophus* (formerly *Thiosphaera pantotropha*), couple nitrification and denitrification to convert ammonia into nitrogen gas without accumulation of intermediates such as nitrite and hydroxylamine (NH_2OH) (Rainey et al. 1999; Robertson et al. 1988). These reactions are depicted in Figure 5-1.

Most nitrifiers are aerobic lithoautotrophs, i.e., microorganisms that use molecular oxygen as the electron acceptor (aerobic); oxidize inorganic substrates (in this case, ammonia and/or nitrite) as the electron donor (lithotrophic); and fix carbon dioxide (CO_2) as a source of carbon for production of cell material (autotrophic). This chapter emphasizes the microbiology of these lithotrophic nitrifiers under aerobic conditions, with only brief acknowledgement of anoxic reactions they can perform when oxygen is limited as well as heterotrophic nitrification.

Ammonia-Oxidizing Microorganisms

Mechanism of ammonia oxidation. The first step of the nitrification process (i.e., aerobic oxidation of ammonia to nitrite, Eq. 5-1 in Figure 5-1) is the most extensively studied nitrification reaction in drinking water distribution systems. Ammonia rather than ammonium ion (NH_4^+) is the substrate of the energy-generating system (Suzuki et al. 1974), which is supported by the finding of Kleiner (1985) that cell membranes are highly permeable to NH_3 but not to NH_4^+. This suggests that pH of 7 and lower, where NH_4^+ predominates and NH_3 is less than 1%, may limit the growth of ammonia-oxidizing microorganisms (AOM), whereas higher pHs may promote growth of AOM. The process is actually a two-stage reaction, with the production of an intermediate, hydroxylamine that is then oxidized to nitrite (Prosser 1990). Though not shown in Eq. 5-1, small amounts of hydroxylamine intermediate can be released, and nitrous oxide (N_2O) and nitric oxide (NO) can also be produced under low oxygen conditions (Bock et al. 1992; Doetsch and Cook 1973). The catabolic reaction summarized in Eq. 5-1 yields much less energy than typical heterotrophic reactions, and these autotrophic microorganisms need to invest a considerable amount of energy into carbon fixation, so they have slower growth rates and lower cell yields as compared to aerobic heterotrophs.

Taxonomy of ammonia-oxidizing microorganisms. Based on morphologic distinction (i.e., difference in form and structure), AOB were originally separated into five genera: *Nitrosomonas*, *Nitrosococcus*, *Nitrosospira*, *Nitrosovibrio*, and *Nitrosolobus*. Since the emergence of molecular biology tools, the taxonomy of nitrifiers was reassessed using sequence similarities of their 16S rRNA gene (16S rDNA). Most lithotrophic AOB (and all AOB characterized in fresh water) currently belong to two classes of the Pseudomonadota (formerly Proteobacteria) (Oren and Garrity 2021) phylum (which is one of the major lineages in the Bacteria domain and consists of a large and diverse group of gram-negative bacteria):

- The *Betaproteobacteria* class comprises the genera *Nitrosomonas*, *Nitrosospira* (which formerly included strains of *Nitrosolobus*, *Nitrosovibrio*, and *Nitrosospira* itself), and *Nitrosococcus mobilis*.
- The *Gammaproteobacteria* class comprises only *Nitrosococcus* (*N. oceanus*, *N. watsonii*, and *N. halophilus*), which are marine AOB.

Their 16S rDNA sequence similarities are shown in Figure 5-2, which presents a phylogenetic tree (graphical representation of the similarities of DNA sequences). The lengths of the branches reflect how the organisms differ from one another in the particular gene of interest.

In the last two decades, the known diversity of lithotrophic AOM has expanded to include the archaea domain as well as NOB with the previously overlooked capability of also oxidizing ammonia (Daims et al. 2015; van Kessel et al. 2015). *Nitrosopumilus maritimus* is an AOM belonging to the *Archaea* domain and was isolated from a tropical marine aquarium tank (Könneke et al. 2005). This AOA is representative of the ubiquitous marine Group 1 of the *Crenarchaeota* kingdom and contains genes for all three subunits (*amoA*, *amoB*, and *amoC*) of ammonia monooxygenase (AMO), the key enzyme responsible for ammonia oxidation. Cultivation studies suggest that *N. maritimus* tends to have autotrophic characteristics (Könneke et al. 2005). *Nitrososphaera viennensis* is also an AOA that was isolated from soil (Tourna et al. 2011). These AOA are also shown in Figure 5-2.

Speciation of ammonia-oxidizing microorganisms in drinking water distribution systems. Historically, *Nitrosomonas* has been the most frequently mentioned genus associated with the first step of the nitrification process in drinking water, wastewater, and seawater, and *Nitrosomonas europaea* has been extensively studied as the model AOM (e.g., Chain et al. 2003). However, until recently, AOM were mostly identified by culturing techniques, which underestimates their levels and biases species recovery since AOB and AOA grow poorly in artificial growth medium (Belser and Mays 1982). Belser and Schmidt (1978a) observed that media selectivity and enumeration efficiency of the culturing technique used in the early days may have favored the recovery of *Nitrosomonas* over the other genera. Accordingly, results obtained from the introduction of molecular methods now suggest that *Nitrosospira* spp. are the most common AOB in soils and fresh water, not *Nitrosomonas* spp. as was previously thought due to culturing results. Moreover, our understanding of microorganisms involved in this part of the nitrogen cycle has only recently expanded to include AOA and comammox bacteria, so their potential presence and contribution to ammonia oxidation were completely overlooked in all environments.

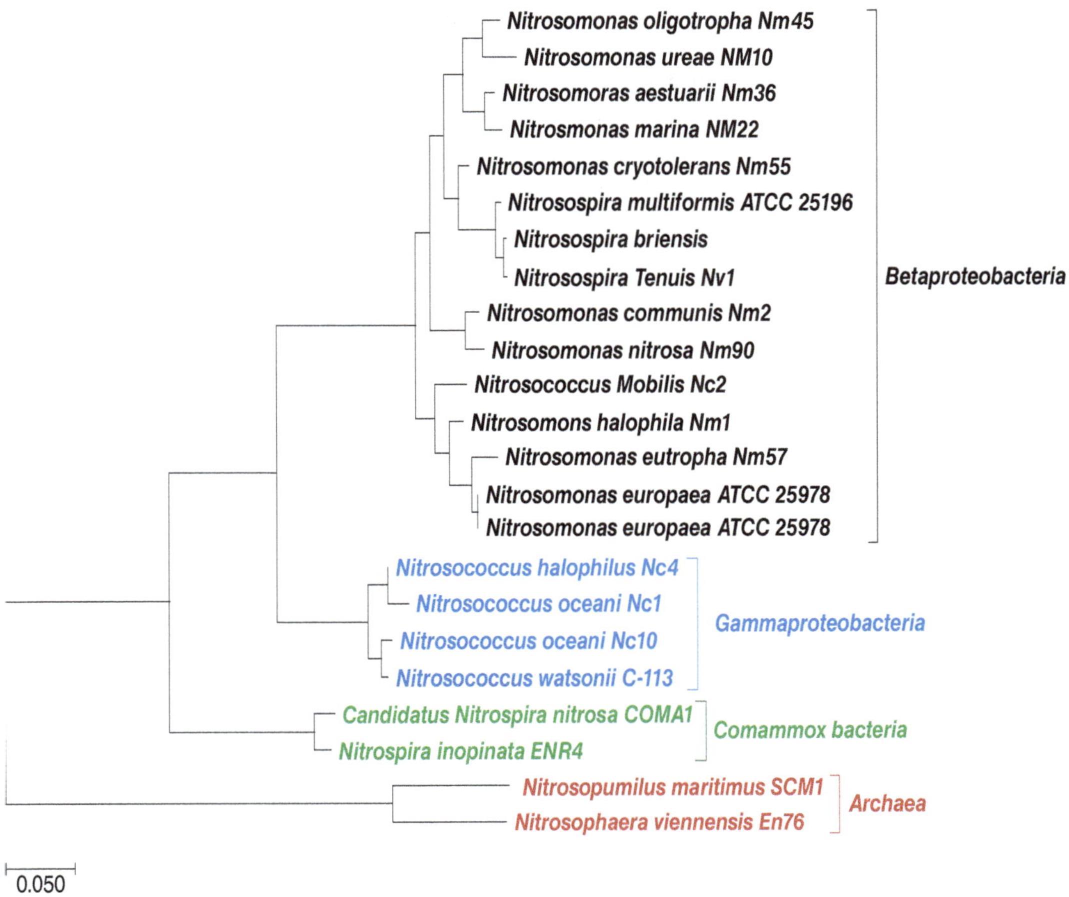

Figure 5-2 Phylogenetic tree of ammonia-oxidizing microorganisms based on multiple alignment of nearly full-length AOB, AOA, and comammox 16S rDNA sequences. AOB and AOA catalyze Eq. 5-1, and comammox catalyze both Eq. 5-1 and 5-2 in Figure 5-1. Scale bar represents a 5-percent sequence difference. The tree was generated using the neighbor-joining distance matrix method excluding positions with gaps.

AOA are found in various habitats, including hot/thermal springs, marine and fresh waters, soils, wastewater treatment systems, and drinking water distribution systems, where they may outnumber their counterpart, AOB (Waak et al. 2019; Li et al. 2018; Scott et al. 2015; You et al. 2009). Several studies indicate AOA can be found in chloraminated and chlorinated drinking water distribution systems and in systems operated with low residual (Wang et al. 2014; Waak 2019). They have also been found at various stages during groundwater treatment and granular activated carbon filtration (Kasuga et al. 2010; van der Wielen et al. 2009). Furthermore, AOA seem to be more sensitive to chlorination than chloramination (Wang et al. 2014; Kasuga et al. 2010); however, low concentrations of AOA in chlorinated systems could be due to the lack of sufficient ammonia substrate necessary for their growth (Scott et al. 2015). Collectively, this should be taken into consideration in this summary of the literature, as the combination of molecular detection and enumeration techniques and our more complete representation of nitrifying microorganisms may change our understanding of nitrifiers moving forward.

Eichler and colleagues (2006) studied the bacterial community dynamics of an entire chlorinated drinking water distribution system from source to tap using single-strand conformation polymorphism (SSCP) fingerprinting of 16S rRNA or its genes. This drinking water system was supplied by two surface water reservoirs, and treatment consisted of coagulation, flocculation, sand filtration, and chlorination prior to storage and distribution. They observed that the bacterial community remained rather constant after chlorination and up to the tap. The bacterial community also did not change substantially over several months. Eichler and colleagues (2006) observed the presence of *Betaproteobacteria* AOB, including the species *Nitrosospira briensis* (phylotypes 41 and 43) and *Nitrosomonas ureae* (phylotype 42). The AOB species *N. briensis* (phylotype 43) was abundant in all drinking water samples. In a study of groundwater with elevated dissolved nitrogen concentrations caused by mill-tailings contamination, 16S rDNA and *amoA* analyses showed that AOB diversity was dominated by *Nitrosomonas* species over *Nitrospira* species, with the latter detected in only 20 percent of groundwater samples (Ivanova et al. 2000).

When using other molecular techniques, Regan and colleagues (2003) observed that *Nitrosomonas* (and particularly strains from the *Nitrosomonas oligotropha* cluster) were ubiquitous in full-scale drinking water distribution systems receiving chloraminated water, and *Nitrosospira* constituted a negligible fraction of the AOB community. A subsequent study observed the exclusive detection of *N. oligotropha*-like AOB in chloraminated drinking water distribution systems (Regan et al. 2006). Similarly, Regan and colleagues (2002) observed the presence of *N. oligotropha* and *Nitrosomonas ureae* in pilot-scale chloraminated drinking water distribution systems, with a considerably smaller presence of *Nitrosospira*-like AOB. Regan and colleagues (2002, 2003) explained that the predominance of *N. oligotropha* over other *Nitrosomonas* species could be associated with the low concentrations of free ammonia in the distribution systems studied (<0.20 mg/L ammonia nitrogen). The reported half-saturation constant (or half-velocity coefficient, K_s, equivalent to the growth-limiting substrate concentration at half the maximum specific growth rate) for *N. oligotropha* strains is 0.42–1.05 mg/L ammonia nitrogen, which is approximately 1 order of magnitude lower than the K_s of other *Nitrosomonas* species.

Lipponen and colleagues (2004) examined biofilm nitrifiers in polyvinyl chloride (PVC) pipes connected to two Finnish drinking water distribution systems. Like Regan and colleagues (2002), Lipponen and colleagues predominantly observed *Nitrosomonas*-like AOB over other nitrifiers when searching for the *amoA*-like sequences of biofilm clones.

In bench-scale biofilm reactors fed with nitrifiers collected from nitrifying circulating bed reactors, Nogueira and colleagues (2002) used molecular methods to identify the AOB present. Results showed that the AOB population was most likely affiliated with the *Nitrosomonas europaea/eutropha* group. No AOB belonging to the *Nitrosospira* cluster were detected, and *Nitrosococcus mobilis* was also absent. Although no AOA were detected in groundwater-fed biofilters using metagenomic techniques (Gülay et al. 2019), another study found that AOA were present in drinking water from the groundwater treatment train and distribution system as assessed by quantitative polymerase chain reaction (PCR) and phylogenetic analysis of the *amoA* gene (van der Wielen 2009).

AOM diversity has also been extensively studied in other environments, including sewage effluents, nitrifying activated sludge, freshwater lakes, and soils. These characterizations are outside the scope of this manual.

Morphology of ammonia-oxidizing microorganisms. Nitrifiers present a wide variety of morphologic characteristics, as presented in Table 5-2. Figure 5-3A–F shows the diversity of morphologies of AOB and comammox isolated from different environments. Watson and colleagues (1989) mention that cells of *Nitrosomonas* are rich in cytochrome C-type pigment, which imparts a yellowish to reddish color to cell suspensions. This pigment has an absorption peak at 415 nm, which means that exposure of *Nitrosomonas* cells to light at that particular wavelength may inhibit these organisms (see nitrification control using ultraviolet [UV] light in Chapter 7). As for the AOA *N. maritimus*, it appears as straight rods with a diameter of 0.17–0.22 μm and a length of 0.5–0.9 μm (Könneke et al. 2005).

Table 5-2 Characteristics of various ammonia-oxidizing microorganisms

Characteristics	*Nitrosopumilus*	*Nitrosomonas*	*Nitrosococcus*	*Nitrosospira*	*Nitrosospira* Formerly *Nitrosolobus*	*Nitrosospira* Formerly *Nitrosovibrio*
Typical cell shape:	Straight rod	Straight rod	Coccus	Helical	Lobed	Curved rod
Cell size:						
Width (μm)	0.17–0.22	0.7–1.5	–	0.3–0.8	1.0–1.5	0.3–0.4
Length (μm)	0.5–0.9	0.5–2.4	–	1.0–8.0	1.0–2.5	1.1–3.0
Diameter (μm)	–	–	1.5–2.2	–	–	1.0–1.2*
Motility	–	D	+	D	+	+
Mol% G+C of DNA†	33–34	45–54	48–51	53–55	53–56	54

–: 90 percent or more of strains are negative
+: 90 percent or more of strains are positive
D: Different reactions observed in different taxa.
* Spherical form is present in the early stationary phase of growth.
† Percent ratio of (G+C)/(A+T+G+C), where G, C, A, and T represent the relative molar amounts of guanine, cytosine, adenine, and thymine, respectively, in a given organism. The value varies according to genus and species and provides a useful criterion in microbial taxonomy.
Adapted from Qin et al. 2017; Könneke et al. 2005; Holt et al. 2000; Watson et al. 1989; Ford 1980; Doetsch and Cook 1973

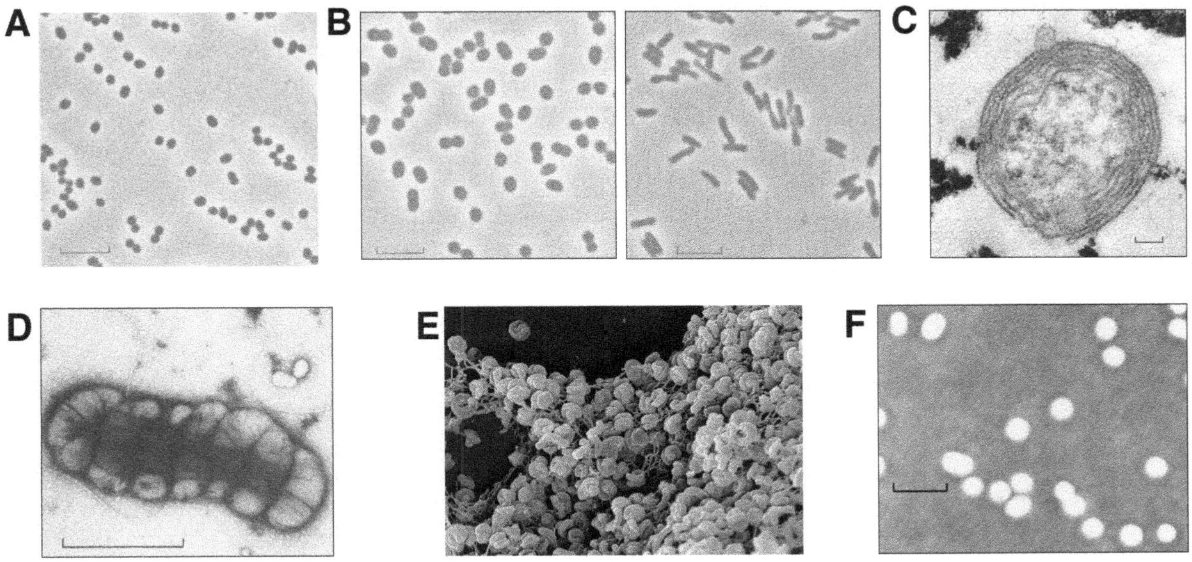

Figure 5-3 Microscopy images of nitrifying microorganisms. Modified phase-contrast images are shown for *Nitrosomonas europaea* ATCC 25978 (A), *Nitrosomonas* species terrestrial strains (B), and *Nitrosococcus mobilis* (F) with 5-μm scale bar. Transmission electron micrographs are shown for *Nitrosomonas* species isolated from a drinking water reservoir (C, 0.1-μm scale bar) and *Nitrosospira briensis* negatively stained cell (D, 1-μm scale bar). A cell aggregate of the comammox bacterium *Nitrospira inopinata* visualized by electron microscopy (E). Panels A, B, D, and F are reprinted with permission of Bergey's Manual Trust; panel C with permission of American Society for Microbiology, and panel E with permission from, and credit to, Anne Daebeler, Stefano Romano, and Holger Daims.

Nitrifier cells occur singly, in pairs, or, occasionally, in short chains. Many of the nitrifiers occur in either loose or compact cell aggregates (Figure 5-4). In the case of AOB, these aggregates occur in nature and in enrichment cultures and have been observed in drinking water distribution systems but not in pure cultures (Holt et al. 2000). Figure 5-4A shows loosely associated cells embedded in and surrounded by a soft slime layer

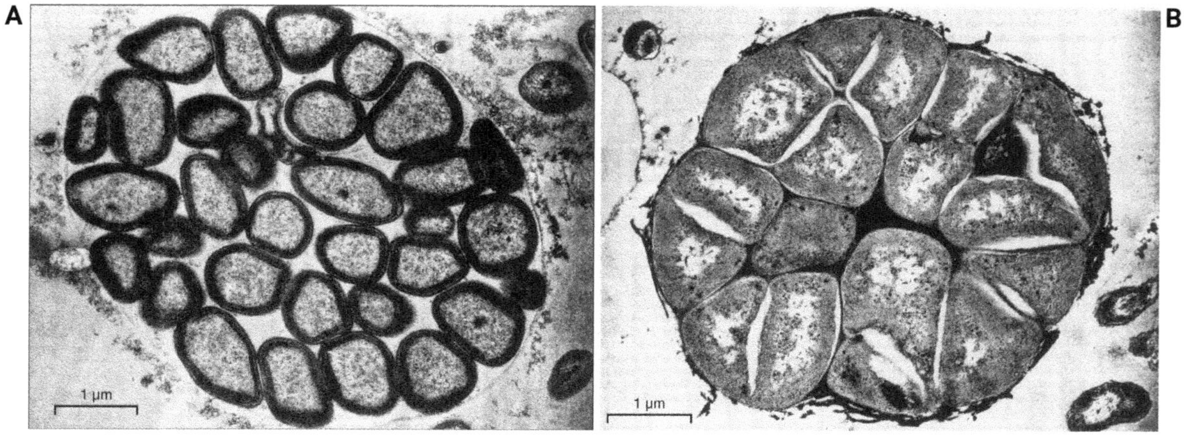

Source: Watson et al. 1981

Figure 5-4 Nitrifying bacterial aggregates of *Nitrosomonas europaea* cells. Loose (A) and compact (B) aggregates are shown here. Scale bar represents 1 µm.

of low electron density with no distortion or compression of cells. Conversely, closely packed and compact cell clusters are embedded in and surrounded by a firm slime layer of high electron density (Figure 5-4B) (Watson et al. 1989; Holt et al. 2000). These cells are tightly compressed and are often distorted, and the cytomembranes are frequently localized in the outermost surface of the cells. These slime layers protect nitrifiers from disinfectants (Stewart and Lieu 1997). As observed with other microorganisms, Wolfe and colleagues (1990) and Stewart and Lieu (1997) suggested that AOB may be more densely concentrated in drinking water distribution system biofilms than in the water column. The advantages of growth in cell aggregations are well known and may include protection against disinfectants, greater resistance against substrate limitations (Stein and Arp 1998; Keen and Prosser 1987) and greater resistance outside of optimum pH range (Tarre and Green 2004).

Nitrite-Oxidizing Microorganisms

Mechanism of nitrite oxidation. The second step of the nitrification process (aerobic nitrite oxidation to nitrate, Eq. 5-2 in Figure 5-1) is a single step without reaction intermediates (Bock et al. 1992; Prosser 1990; Doetsch and Cook 1973), and it does not produce protons that potentially lower the pH of the surrounding medium, like the first step of nitrification. According to Prosser (1990) and Bock and colleagues (1992), nitrite oxidation is a reversible process. As with the metabolism of AOM, the catabolic reaction of NOB yields much less energy than do typical heterotrophic reactions, and their autotrophic production of cells requires a considerable amount of energy, so they also have slower growth rates and lower cell yields as compared to aerobic heterotrophs.

Taxonomy of nitrite-oxidizing microorganisms. NOB are more phylogenetically diverse as compared to AOB and have been identified in various classes of the phylum Pseudomonadota, the phyla Nitrospirae and Nitrospinae, and most recently the phylum Chloroflexota (Daims et al. 2016). *Nitrobacter* is the most frequently mentioned genus associated with the second step of the nitrification process, although it is not necessarily the dominant genus in drinking water, and there are six other genera that can also oxidize nitrite to nitrate: *Nitrospina*, *Nitrococcus*, *Nitrospira*, *Ca.* Nitrotoga, *Nitrolancea*, and *Ca.* Nitromaritima. *Nitrobacter* forms a tight cluster within the *Alphaproteobacteria* class, *Nitrococcus mobilis* is a member of the *Gammaproteobacteria* class, *Ca.* Nitrotoga is in the *Betaproteobacteria* class, *Nitrospina* and *Ca.* Nitromaritima are in the phylum Nitrospinae and loosely related to the *Gammaproteobacteria* class, and *Nitrospira* constitute their own phylum *Nitrospirae* (Daims et al. 2016; Regan 2001). This diversity was further expanded with the discovery of *Nitrolancea* in the phylum Chloroflexota (Sorokin et al. 2012), and Kitzinger et al. (2018) even speculated on the possible existence of nitrite-oxidizing archaea based on metagenomic sequence analysis results. The genera *Nitrococcus*, *Nitrospira*, and *Nitrospina* are generally restricted to marine environments and grow optimally at high salt concentrations (Prosser 1990), although exceptions have been observed: one *Nitrospira* strain was isolated from soil by Bock and colleagues (1986), and *Nitrospira moscoviensis* was isolated from freshwater (Hovanec et al. 1998).

Figure 5-5 presents the phylogenetic relatedness of NOB based on 16S rDNA sequences and illustrates the high sequence similarity of *Nitrobacter* strains and the diversity represented by *Nitrospira* strains. Known comammox bacteria are also noted on this figure.

Speciation of nitrite-oxidizing bacteria in drinking water distribution systems and different environments. Although *Nitrobacter* is the NOB genus that is the most often cited in the literature, it may not be the dominant group in drinking water. Using molecular techniques, Regan and colleagues (2003) observed that *Nitrospira* (and particularly *Nitrospira moscoviensis*) were detected in most of the samples collected from full-scale chloraminated drinking water distribution systems, while *Nitrobacter* were detected in a few samples only. Similarly, Regan and colleagues (2002) observed that the NOB communities of pilot-scale drinking water distribution systems were composed primarily of *Nitrospira*, although *Nitrobacter* was detected in some samples. Regan and colleagues (2003, 2002) explained that the low-nitrite environment may have selected for *Nitrospira*, which has an estimated K_s of 0.14 mg/L NO_2^--N, approximately 1–2 orders of magnitude lower than the K_s for *Nitrobacter* species. Moreover, Waak and colleagues (2019) analyzed microbial communities in a chloraminated system and found lower biomass, species richness, and *Nitrospira* sequences and higher abundances of *Nitrosomonas*, *Mycobacterium*, and *Desulfovibrio* sequences in contrast to a drinking water distribution system operated with no disinfectant residual. Although chloramination may reduce biomass, these results indicated that the potential selection of microorganisms that might be associated with human disease (e.g., *Mycobacterium*) and pipe corrosion (e.g., *Desulfovibrio*) warrants further investigation (Waak et al. 2019).

Eichler and colleagues (2006) studied the bacterial community dynamics of an entire chlorinated drinking water distribution system from source to tap using SSCP fingerprinting of 16S rRNA or its genes. They observed that the bacterial community remained rather constant after chlorination and up to the tap. The bacterial community also did not change substantially over several months. The NOB species *Nitrospira moscoviensis* (phylotypes 61 and 96) were abundant in all drinking water samples, whereas members of the phylum *Nitrospira* (phylotypes 58, 61, 76, and 96) occurred only after chlorination (Eichler et al., 2006) (see Figure 5-5).

Work by Daims and colleagues (2006) and Maixner and colleagues (2006) suggested that *Nitrospira* can be separated into two sublineages, and that Sublineage II grows more effectively in environments with low nitrite concentrations than does Sublineage I. This suggests that some subgroups of *Nitrospira* can tolerate and may prefer low nitrite environments, such as those offered by drinking water systems.

Nitrite oxidizers have been observed and studied in many different environments, such as fresh water and activated sludge within an industrial wastewater treatment plant; however, much of that work is beyond the scope of this manual.

Morphology of nitrite-oxidizing bacteria. Nitrifiers present a wide variety of characteristics, as presented in Table 5-3. Figure 5-3E shows an example of an NOB, specifically a recently characterized comammox bacterium. Cell suspensions of *Nitrobacter* and *Nitrococcus* show characteristic absorption peaks at 420, 440, 550, 587, and 600 nm (Watson et al. 1989). *Nitrobacter* membranes impart a brownish color that is typical of NOB.

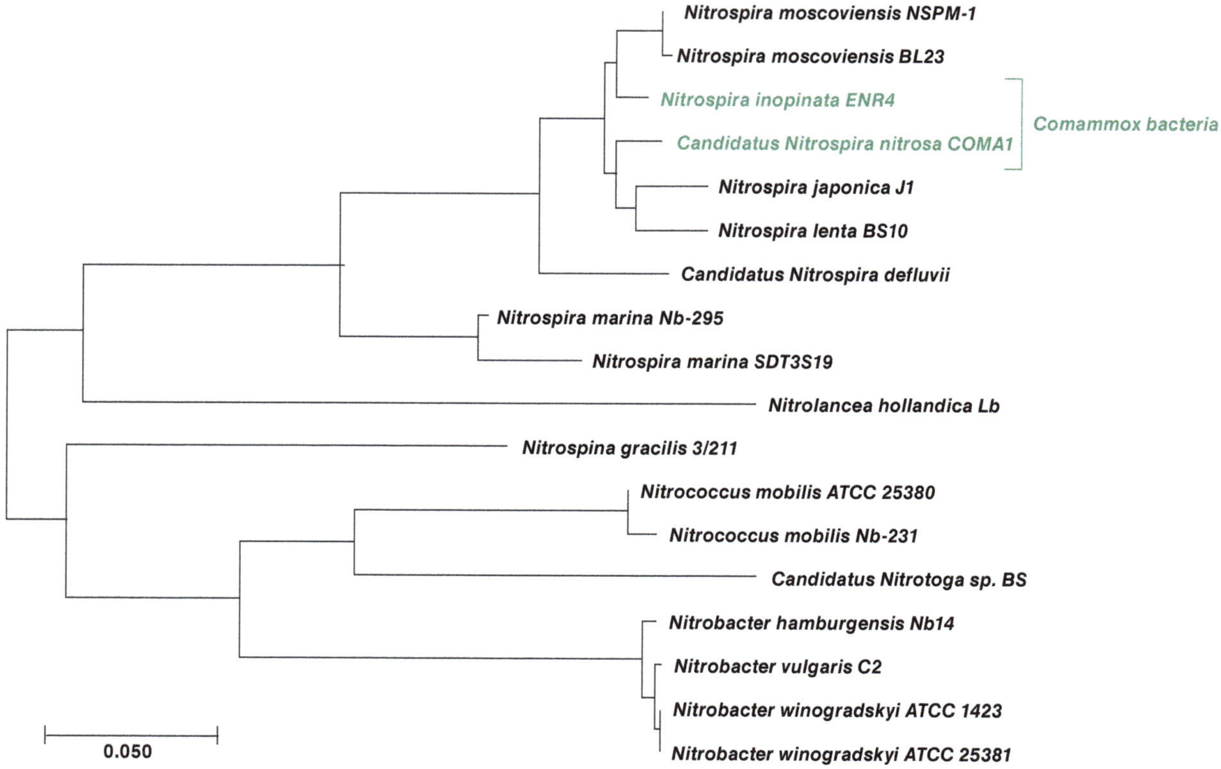

Figure 5-5 Phylogenetic tree of nitrite-oxidizing bacteria (NOB) based on a multiple alignment of NOB 16S rDNA sequences, with NOB capable of complete ammonia oxidation indicated in green. Scale bar represents a 5-percent sequence difference.

Table 5-3 Characteristics of various nitrite-oxidizing bacteria

Characteristics	*Nitrobacter*	*Nitrococcus*	*Nitrolancea*	*Ca.* Nitromaritima	*Nitrospina*	*Nitrospira*	*Ca.* Nitrotoga
Typical cell shape:	Straight rod	Coccus	Straight rod	ND	Straight rod	Helical	Curved rod
Cell size:							
Width (µm)	0.5–0.8	NA	1.0-1.2	ND	0.3–0.4	0.3–0.4	0.5
Length (µm)	1.0–2.0	NA	2.0-4.0	ND	1.7–6.6	0.8–1.0	1.0
Diameter (µm)	NA	1.5–1.8	NA	ND	NA	NA	NA
Mol% G+C of DNA*	60–62	51	63	ND	58	50	50

NA, not applicable; ND, no data

* Percent ratio of (G+C)/(A+T+G+C), where G, C, A, and T represent the relative molar amounts of guanine, cytosine, adenine, and thymine, respectively, in a given organism. The values vary according to genus and species and provide a useful criterion in microbial taxonomy.

Adapted from Kitzinger et al. 2018; Sorokin et al. 2012; Holt et al. 2000; Watson et al. 1989; Ford 1980; Doetsch and Cook 1973

Autotrophic and Heterotrophic Nitrification

The emphasis of nitrification research in chloraminated drinking water distribution systems has been focused on lithoautotrophic populations, and the relative contributions of autotrophic and heterotrophic nitrification in drinking water distribution systems are poorly understood (Verstraete and Focht 1977, Killham 1986). Currently, there is insufficient information regarding these processes in drinking water, but the microbiology of autotrophic and heterotrophic nitrification has been more extensively studied in wastewater and other environments. Therefore, findings from these other contexts will be discussed briefly in this section, as they may also pertain to as yet uncharacterized reactions occurring in drinking water systems.

As previously mentioned, most nitrifiers are obligate lithoautotrophs. It has, however, been shown that certain AOB like *Nitrosomonas europaea* can metabolize some organic compounds to a limited extent while in the presence of their inorganic substrate ammonia (Hommes et al. 2003; Verstraete and Focht 1977), which may classify them as lithoorganotrophs. Also, some *Nitrobacter* species are mixotrophic microorganisms, i.e., microorganisms that can obtain energy by the metabolism of both inorganic and organic substrates (Holt et al. 2000; Watson et al. 1989; Ford 1980). However, most strains of *Nitrobacter* cells grow much slower heterotrophically, and heterotrophic growth is much less efficient than autotrophic growth in the presence of their inorganic substrate nitrite (Watson et al. 1989; Ford 1980).

Of interest to the drinking water industry, AOB can also biodegrade trihalomethanes (THMs) through an enzymatic process known as *cometabolism* (Wahman et al. 2005, 2006). Cometabolism refers to the biodegradation of a chemical through nonspecific enzymatic catalysis without any significant benefit to the microorganism (Dalton and Stirling 1982). Although the byproducts of THM biodegradation by AOB are highly reactive and degrade almost immediately after they are produced, they are also toxic to the AOB (Wahman et al. 2005). This process is further discussed below.

While lithotrophic nitrifying microorganisms are thought to be the most important nitrifiers in nature, various species of prokaryote and eukaryote heterotrophic (i.e., microorganisms that use organic carbon sources) nitrifiers can also oxidize ammonia (Bock et al. 1992; Watson et al. 1989; Killham 1986; Focht and Verstraete 1977; Verstraete and Alexander 1973, 1986). This process of heterotrophic nitrification (where ammonia is converted to nitrate, as illustrated in Figure 5-1) occurs at a much slower rate than lithotrophic nitrification and is not directly associated with energy generation, as the growth of the known heterotrophic nitrifiers is completely dependent on the oxidation of organic substrates (Bock et al. 1992; Focht and Verstraete 1977). One possible explanation for heterotrophic nitrification is to produce nitrate that can then be used as an electron acceptor in denitrification (Eq. 5-3) (Castignetti and Hollocher 1984). Nonetheless, heterotrophic nitrification may be greater than autotrophic nitrification in some environments, such as forest soils (Schimel et al. 1984). Heterotrophic nitrifying organisms produce gaseous nitrogen oxides, and the final product of heterotrophic nitrification can sometimes be nitrate but is often nitrite. This abundance of nitrite may explain why, in many environments, the number of NOB is much higher than the number of AOB.

Microorganisms capable of heterotrophic nitrification include the bacteria *Alcaligenes faecalis*, numerous *Pseudomonas* species, and *Paraccocus pantotrophus*, fungi, and even some algae. Methane-oxidizing bacteria are also able to oxidize ammonia to nitrite, although they cannot grow from this reaction (Bock et al. 1992). There are some similarities between the ammonia and the methane monooxygenase enzymes of the nitrifiers and the methanotrophs, respectively (Bédard and Knowles 1989); therefore, caution must be taken when interpreting *amoA* PCR results because methanotrophs can generate false positives. Because the occurrence and significance of heterotrophic nitrification in drinking water have not been established, the rest of this chapter focuses on lithotrophic nitrification.

Interactions Between Nitrifiers and Other Microbial Community Members

In drinking water distribution systems, nitrifying microorganisms are part of a mixed culture ecology in which heterotrophic microorganisms participate and typically dominate, depending on substrate availabilities. Ecological interactions within these communities can be significant in terms of overall microbial water quality beyond nitrification. Several authors have reported on the competition between heterotrophs and nitrifiers for dissolved oxygen, nitrogen, and space, and there are also potential benefits for each group that may result from these interactions. These considerations are summarized below.

- Heterotrophs produce organic compounds that stimulate some nitrifiers (as described above for heterotrophic or mixotrophic growth of nitrifiers).
- Heterotrophs biodegrade organic compounds that are inhibitory to the nitrifiers.
- Heterotrophs produce extracellular polymers that improve aggregation of both heterotrophs and nitrifiers into flocs and biofilms.
- The faster-growing heterotrophs protect the slower-growing nitrifiers from detachment by predominating in the outer layer of multispecies biofilms.
- Autotrophic nitrifiers produce and release soluble microbial products (SMPs) that augment the heterotroph substrate supply and contribute to chloramine decay, which SMPs may be particularly important in low organic contexts such as drinking water distribution systems.
- In the atypical situation of dissolved oxygen being depleted in a drinking water distribution system, the nitrite and nitrate products of nitrification can serve as electron acceptors for other microorganisms.

When conducting chemostat experiments with an NOB *Nitrobacter* sp. and the AOB *N. europaea*, Rittmann and colleagues (1994) demonstrated that both nitrifiers produce SMP that can support heterotrophic bacteria. The first evidence was the presence of significant concentrations of soluble chemical oxygen demand in the chemostat effluent, even though the influent was free of organic compounds. The second evidence was the maintenance of a small population of heterotrophic bacteria, apparently through utilization of the nitrifier-produced SMP. Herath and colleagues (2018) demonstrated the contribution of SMPs to chloramine decay in laboratory scale reactors fed with treated drinking water supplemented with different concentrations of organic carbon. Studies by Barker and Stuckey (1999) identified SMP as humic and fulvic acids, polysaccharides, proteins, amino acids, nucleic acids, and structural cell components.

Nogueira and colleagues (2002) explained that microbial competition within a mixed-culture biofilm may result in stratification of the biofilm structure, with the faster-growing heterotrophs predominantly in the outer layers and the slow-growing nitrifiers staying deeper inside the biofilm. This stratification may create a disadvantage to the nitrifiers when the bulk liquid oxygen concentration is low. However, the heterotrophic layer on the outside part of the biofilm may present an advantage to the nitrifiers by protecting them from detachment (Nogueira et al. 2002) and disinfectant. However, a study from Schramm and colleagues (1996) of nitrifying biofilms from trickling filters of an aquaculture water recirculation system suggested the opposite, i.e., nitrification was restricted to a narrow zone of 50 μm on the very top of the film. *Nitrosomonas* and *Nitrobacter* were mainly detected in the upper oxic zone of the biofilm, with substantially lower concentrations in the anoxic layers and occasionally at the bottom of the biofilm. In a subsequent study involving a nitrifying fluidized bed reactor operated with low ammonia concentrations to resemble a natural freshwater habitat, Schramm and colleagues (1998) demonstrated the occurrence of complete nitrification in the outer 125 μm of the aggregates (Figure 5-6).

Kindaichi and colleagues (2004) studied interactions between nitrifying and heterotrophic bacteria in a carbon-limited nitrifying biofilm fed only with ammonium ion as an energy source. In this study, autotrophic nitrifying biofilms were collected from primary settling tank effluent from a wastewater treatment plant and then cultured with synthetic medium in a biofilm reactor. Since no external organic carbon was added in the experiment, SMP produced by nitrifiers provided the sole organic substrates for heterotrophic bacteria. Molecular methods (microautoradiography–fluorescence in situ hybridization [FISH]) were used to determine the community structure and

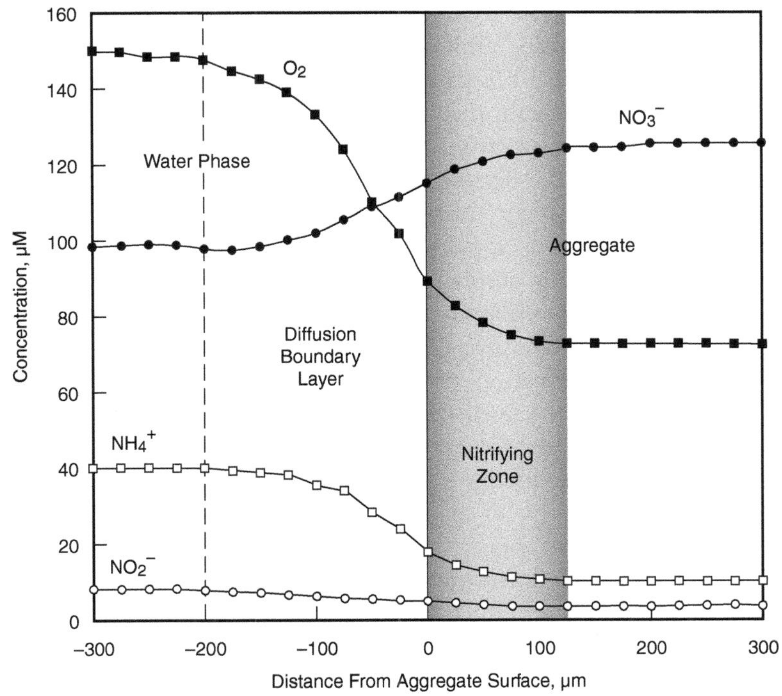

Adapted from Schramm et al. 1998

Figure 5-6 Typical microprofiles of oxygen, ammonium ion, nitrite, and nitrate concentrations in nitrifying aggregates. The gray area marks the zone with nitrifying activity; negative distance indicates water phase, and positive distance represents the biofilm.

spatial organization in the biofilm. Results indicated that the biofilm was composed of 50 percent nitrifying bacteria (22 percent AOB and 28 percent *Nitrospira* phylum [NOB]) and 50 percent heterotrophic bacteria. Results also indicated that a pair of nitrifiers (AOB and NOB) supported a heterotrophic bacterium via production of SMP. The heterotrophic community was composed of bacteria that were physiologically and metabolically diverse and to some extent metabolically redundant, which ensured the stability of the ecosystem as a biofilm. Different groups of heterotrophs were responsible for mineralizing different low-and high-molecular-weight organic compounds produced or released by nitrifiers. Moreover, the FISH results suggested that primary degraders directly colonize or occur around the nitrifying clusters that excrete SMP. These primary degraders recruit other species to form metabolically structured and functionally integrated biofilm communities that ensure maximum utilization of SMP produced by nitrifiers and prevent buildup of metabolites or waste materials of nitrifiers to significant levels. Similar information in drinking water systems is lacking, although the relatively low concentrations of organic carbon found in drinking water suggest that a similar response may be observed.

Zhang and Edwards (2009) observed that in the absence of disinfectant and in natural waters with measurable concentrations of organic material, nitrification did not increase suspended and fixed heterotrophic plate counts (HPCs) because the concentrations of organic compounds produced by nitrifiers were insignificant (nitrification of 2 mg/L ammonia-N can produce 20 μg/L C of soluble organic carbon) compared to the organic carbon concentrations of natural waters. The differences in fixed HPCs in the absence and presence of nitrification were less than an order of magnitude.

Modeling work conducted by Zhang and colleagues (2010) suggested that organic carbon concentrations need to be much higher than in drinking water conditions (i.e., greater than 10 mg/L C) to significantly decrease nitrification by allowing heterotrophs to outcompete nitrifiers for dissolved oxygen (modeling was conducted at high ammonia concentration, i.e., 2 mg/L ammonia nitrogen). Subsequent work by the same group (Zhang and Edwards 2010; Zhang et al. 2010) showed that nitrifiers may be outcompeted by heterotrophs for phosphorus, however.

Denitrification. Denitrifying bacteria thrive in practically all habitats: e.g., soil, fresh water, polluted habitats, oceans, plants, animals, and humans (Zumft 1992). Although it has not been extensively studied nor identified as a major phenomenon in drinking water distribution system, some evidence suggests denitrification may happen (Eq. 5-3 of Figure 5-1). Oligotrophic aerobic denitrifiers, identified as *Zoogloea* sp. and *Acinetobacter* sp., were isolated from different drinking water reservoirs and were able to efficiently remove nitrate at temperatures of 10–30°C (or 50–86°F) and at pH 7–8 (Huang et al. 2015; Su et al. 2015). Denitrification, potentially mediated by zinc release from galvanized iron pipes, was also observed under simulated drinking water conditions, but the possibility that zinc inhibited nitrification could not be ruled out (Zhang et al. 2010) as high levels of zinc have been previously shown to repress the nitrification process and promote denitrification in soil (Lv et al. 2022). Collectively, it is likely that denitrification is a marginal process that may occur only in portions of the distribution system where there is a low oxygen concentration and other conditions conducive to this process develop; thus, only the fundamentals and microbiology of denitrification are presented in this chapter.

In the scientific literature, denitrification refers to the dissimilatory reduction of one or both of the ionic nitrogen oxides (nitrate and nitrite) to the gaseous oxides (nitric oxide [NO] and nitrous oxide [N_2O]), which may themselves be further reduced to dinitrogen (N_2). (Eq. 5-3 shows this complete reduction to N_2.) The nitrogen oxides act as terminal electron acceptors, and the process can be referred to as anaerobic respiration. Most denitrifiers are facultative anaerobic bacteria, preferentially using oxygen, if it is present, over nitrate. The oxygen concentration at which denitrification stops has been reported to be 0.2 mg/L in pure cultures (Grau et al. 1982) and 0.3–1.5 mg/L in activated sludge systems (Henze 1991). Similar values in drinking water systems are not available in the literature. Some species such as *Paracoccus denitrificans* can denitrify under aerobic conditions (Zumft 1992), while others can denitrify under partial oxygen concentrations. An organic electron donor is generally used, although some denitrifiers can grow lithoautotrophically using hydrogen gas (H_2) or reduced sulfur compounds as the electron donor and CO_2 as the carbon source (Knowles 1982).

The microorganisms involved in denitrification are distinctly different from those involved in the nitrification process. They are also biochemically and taxonomically very diverse. Denitrifiers are predominantly Gram-negative Bacteria, but several Archaea belonging to the genus *Haloarcula* have also been identified (Zumft 1992). Zumft (1992) surveyed known denitrifiers and report an extensive list of 130 species within more than 50 genera. Half of the species identified by this author are members of only three genera: *Pseudomonas* (28 members), *Neisseria* (13 members), and *Bacillus* (12 members).

Previous work indicates that AOB can also carry out denitrification by using nitrite as a terminal electron acceptor, but simultaneous oxidation of ammonia is required as a source of electrons. Poth and Focht (1985) suggest that the purpose of such behavior is to conserve oxygen for use by the AMO; reduce production of nitrite, which may accumulate to toxic levels; and decrease competition for oxygen by NOB by denying them their source of substrate. Similarly, *Nitrobacter*—but not *Nitrospira*—can reduce nitrate to nitrite in the presence of an electron donor such as NADH or methylviologen/benzylviologen (Sundermeyer-Klinger et al. 1984; Tanaka et al. 1983).

Anaerobic ammonium oxidation Another metabolism that could potentially be supported by nitrification under low dissolved oxygen conditions is anaerobic ammonia oxidation (anammox, Eq. 5-4 of Figure 5-1). While not reported in drinking water contexts, it has been extensively studied in the wastewater treatment context as an emerging approach for nitrogen removal. Anammox bacteria use ammonia as their electron donor and nitrite as their electron acceptor, and they are also autotrophic, so they have the potential to complement AOM when dissolved oxygen is very low. Whether this occurs in microenvironments within drinking water distribution systems is presently unreported.

ISOLATION AND ENUMERATION OF NITRIFYING MICROORGANISMS

Methods for nitrifying microorganism detection and enumeration are available, although many of them are complex, time consuming, and/or require sophisticated equipment or training. Although quantitative and rapid molecular methods for nitrifier detection, identification, and enumeration are now available, they are not readily amenable for use by all water systems. Thus, most water suppliers continue to rely on surrogate chemical indicators to detect, identify, and/or enumerate nitrifiers. As discussed in Chapter 6, these indicator parameters include decrease in monochloramine residual, decrease in ammonia concentration, and increase in nitrite and possibly nitrate concentrations. However, relying on indicator parameters presents several drawbacks.

- The indicators are the result of nitrification and occur once nitrification is already well underway; they do not predict the onset of nitrification.
- The trends are system dependent, and some of them may be observed during nitrification episodes in some systems and not in others.
- The trends in nitrification indicators may be due to factors other than nitrification.

Outside of indicator parameters, the most common methods for nitrifier detection and identification are presented in the following sections and are separated by type of technique: culturing techniques that rely on growth of microorganisms followed by enumeration, and nonculturing techniques that include microscopy and molecular methods.

When identifying and enumerating nitrifiers or determining inactivation kinetics, method selection is very important as studies have shown different microorganism recovery when different methods are used (Wahman et al. 2009, 2010; Oldenburg et al. 2002; Belser and Schmidt 1978a). Oldenburg et al. (2002) suggested that nonculturing techniques may be more appropriate to describe inactivation rates in water distribution systems. Their results showed inactivation rates that were three orders of magnitude higher with a culture-dependent most probable number (MPN) technique than with a culture-independent bacterial-viability based method. Using drinking water distribution system samplings, Regan and colleagues (2006) also came to the same conclusion, where a molecular method (real-time PCR) resulted in significantly higher AOB concentrations as compared to a culturing technique (MPN cultivation).

Also of importance is the need to assess microorganisms' viability and activity, as only viable cells are likely to alter their surrounding environment in the case of nitrifiers or pose a threat in the case of pathogens. Other than culturability, viability can be assessed by detection of metabolic activity, RNA transcripts that tend to degrade rapidly after cell death, or an intact membrane. This is particularly important when molecular and some serological techniques are used, since culture-based methods implicitly indicate viability. Approaches to characterize and assess viability are presented in Nocker and Camper (2009), and a method to evaluate nitrifiers' activity was proposed by Völsch and colleagues (1990).

Cell-membrane fatty-acid composition has also been used to identify microorganisms, relying on the fact that the fatty acids present in cell membranes of the same species or strains are consistent when grown under similar conditions. The method can also be used to indicate the organism viability, microbial activity, nutritional/physiological status, level of stress, and others. This method is however not described in this manual as it is seldom used, and contradictory results have been reported in the literature with respect to nitrifiers, which may be explained by the fact that several factors influence lipid composition in microorganisms including but not limited to temperature, pH, quantity and quality of carbon and nitrogen nutrients, oxygen levels, and presence of trace metals.

Culturing Techniques

The traditional approach for identifying bacteria involves culturing the bacteria on solid growth media, isolating individual colonies that develop, and evaluating attributes such as morphology, staining properties, and nutritional requirements of the colonies. In the case of nitrifiers, however, the energy yields from ammonia and nitrite oxidations are low and the energy required for autotrophic growth is high, leading to small biomass yields and low maximum specific growth rates. Growth of visible colonies on solid media takes an extended time, and elimination of heterotrophic contaminants is difficult because of their higher growth rate and the presence of trace organics in agar solidifying agents. Thus, conventional microbiology involving culture on/in nutrient media is difficult, and recovery efficiencies are low: 0.1–1 percent for environmental samples (Wolfe and Lieu 2001). Also, it is recognized that culturing methods considerably underestimate organisms present because of the inability of any medium to support all microorganisms, the presence of viable but nonculturable bacteria, the limited sample size that can be processed, possible cell aggregation in response to stress conditions (see Morphology of AOB), and the lack of immediate results (the incubation period for organisms such as nitrifiers is 3–4 weeks). Despite their limitations, culturing techniques have provided a substantial amount of important information throughout the years (e.g., see Table 5-5).

Many media types and associated methods have been developed, and a number of those are presented in Appendix A Table 5-1 for AOB. Usually, the same medium is used both for AOB and NOB, and only the detection method changes after incubation (nitrite for AOB, nitrate for NOB). Nonetheless, several kinds of media have been used for NOB isolation, and one of them is presented in Appendix A Table A-2. Of the several media that are available, Belser and Schmidt (1978a) reported that the Soriano and Walker (1968) medium was superior to others for enrichment of AOB from soil for the following reasons: it requires the shortest incubation time for MPN enumeration; it is convenient to prepare; it includes an indicator; and precipitate does not form, which facilitates isolation. A literature review also shows that this medium has been used more extensively than any others.

For AOB, incubations are usually conducted in darkness, at 25–30°C typically for 3–4 weeks, although an incubation period of up to 4 months has been used in the literature. Enumeration can be performed by serial dilutions using an MPN technique based on the production of nitrite, which is determined colorimetrically using sulfanilic acid and N,N-dimethyl-α-naphtylamine. Lipponen and colleagues (2002) have used microtiter plates, a solution of diphenylamine in sulfuric acid, and a microtiter plate reader to read the absorbance at a wavelength of 630 nm, indicating the presence of nitrite or nitrate. The presence of nitrite was measured using Griess–Ilosvay reagent solution (Alexander and Clark 1965) to the wells and measuring the absorbance at a wavelength of 540 nm using the microplate reader. Elbanna and colleagues (2012) have also utilized the MPN technique where enumeration was based on pH changes resulting from acid production during nitrification, which was determined colorimetrically via pH indicators.

French et al. (2012) cultured both AOA and AOB from lake sediments in a mineral salts (MS) medium containing NaCl, KCl, $CaCl_2$, $MgSO_4$, and trace elements. AOA were enriched from these sediment samples by culturing in the MS medium supplemented with NH_4^+ and incubated in the dark at 27°C. Recently, a novel method for culturing and obtaining single colonies of AOA on agar medium was demonstrated by immobilizing cells in the gelling agent gellan gum then overlaying with synthetic freshwater medium (Klein et al. 2022). Nitrite released by AOA cells in the solid layer is able to diffuse into the overlaying liquid layer, which can be periodically replaced with fresh medium, thus eliminating the toxic accumulating of nitrite and allowing AOA cells and colonies to form in the solid layer.

For the identification of the comammox *Nitrospira inopinata* (Daims et al. 2015), biofilms were sampled from metal pipe walls containing hot water raised from an oil exploration well and incubated at 46°C in the enrichment medium described in Appendix A Table A-3. After addition of antibiotics and increasing the ammonium concentration, an enrichment culture was obtained from which *Nitrospira inopinata* was isolated (Daims et al. 2015).

Several studies have observed that results differ when different culture media are used, which is due to media selectivity and enumeration efficiency. When enumerating 34 AOB isolates from various environments using different ammonium sulfate ($(NH_4)_2SO_4$) concentrations, Suwa and colleagues (1994) observed that MPN estimates varied depending on the isolates and growth conditions. Belser and Schmidt (1978a) examined several media and observed that each medium was selective to particular strains, and the media differed greatly in incubation times required to reach maximum counts. Wahman and colleagues (2010) observed different disinfection kinetics of *Nitrosomonas europaea* when the media composition changed (namely the phosphate and ionic strength), based on enumeration performed using two nonculturing techniques, i.e., the Live/Dead® *Bac*Light™ Bacterial Viability and Counting Kit (Fisher Scientific, Waltham, Ma.) and propidium monoazide quantitative PCR.

HACH Company (Loveland, Colo.) offers biological activity reaction test (BART™) detectors as easy diagnostic tools to help identify the presence and activity of various bacteria. The BART for Nitrifying Bacteria (N-BART) detects the production of nitrite in water samples. The test offers the following advantages: the procedure does not require specialized training, incubation is performed at room temperature, presence/absence of AOB can be observed in 5 days, and results are easy to interpret. This incubation time is much shorter in comparison to the 3–4 weeks for the MPN technique as mentioned above yet still may not capture a severe nitrification event in real time. The presence of AOB can be observed by the formation of a pink coloration with an order of magnitude concentration. The sensitivity of the test is estimated at 1,000 colony-forming units per mL. As a culture-base method, the N-BART is subject to the same limitations as mentioned above.

Microscopy- and Molecular-Based Techniques

Microscopy-based detection methods and serological techniques (i.e., involving serum or plasma, such as immunofluorescent antibody-based techniques) have been used to study the species composition of nitrifiers, relationships between groups of bacteria, and inactivation kinetics (Wahman et al. 2010, 2009; Oldenburg et al. 2002; Völsch et al. 1990; Belser and Schmidt 1978b). These techniques eliminate the lengthy incubation period that is necessary when using culturing techniques. In the case of serological techniques, however, they require initial isolation of pure cultures to raise antibodies for subsequent detection assays and may require confirmation for identification (Holt et al. 2000; Stephen et al. 1996). Also, the large serological diversity of nitrifiers and the coexistence of several serotypes in one ecosystem seriously limit the efficiency of serological techniques for environmental samples by requiring the use of several antibodies to cover every genus of AOB (Belser and Schmidt 1978b).

Other microscopy-based techniques involve epifluorescent microscopy using various fluorochromes or dyes, such as acridine orange, 4,6-diamidino-2-phenylindole (DAPI), or Live/Dead *Bac*Light. These dyes are not specific to nitrifiers but rather detect all microorganisms present by indiscriminately staining nucleic acids. Therefore, they are only suitable for studying pure cultures of nitrifiers (e.g., for growth and inactivation kinetics of pure cultures) or coupled with another microscopic technique that allows the specific labelling of target cells (e.g., FISH, discussed below).

Molecular techniques have eliminated several of the limitations observed with culturing, serological, or microscopy techniques. The 16S rRNA gene is carried by all bacterial and archaeal species, and the precise sequence of a given 16S rRNA gene is indicative of the species from which it was derived. It is therefore often the target of choice for analysis of complex or unknown bacterial communities. Alternatively, genes related to a specific function of interest, such as ammonia oxidation, can also be targeted in molecular analyses. While molecular techniques do not require culturing, they do typically require information about gene sequences of interest to allow the design or selection of primers or probes that complement the targeted gene(s), and they are limited by database representation of gene sequences; otherwise, our knowledge of phylogenetic and metabolic diversity can bias the results of molecular approaches. This shortcoming is illustrated by the recent discoveries of several microbial groups involved in nitrification that had gone undetected in earlier molecular studies of nitrifying communities, but newer metagenomic techniques now allow us to identify previously undetected microorganisms and pathways. Target nucleic acid amplification techniques, such as PCR, are often needed for detection of low copy number bacterial DNA, such as observed in low-biomass environmental samples. Specific microorganisms can also be detected by genetic probe hybridization with or without prior nucleic acid amplification. When hybridization follows nucleic acid amplification, it is used as a confirmation step in microorganism identifications.

A large variety of molecular methods are available for microorganism detection, identification, and community analysis, and a complete coverage of these techniques is beyond the scope of this manual. Much of the early molecular-based work in chloraminated distribution systems involved the use of denaturing gradient gel electrophoresis, terminal restriction fragment length polymorphism, cloning and sequencing, and FISH. The former techniques in that list are community screening approaches that enabled the detection and identification of prominent populations and significant changes in community composition. However, they have given way to more advanced next generation sequencing (NGS) methods that have much higher resolution and have become considerably more affordable in recent years, particularly in light of the massive increase in data provided by NGS approaches. FISH is a whole-cell technique that involves the fluorescent labeling of the rRNA of target organisms. It can be coupled with other compatible labeling techniques (e.g., DAPI staining), and labeled cells can be visualized microscopically or analyzed through flow cytometry. Although FISH is often used as a tool to evaluate microbial activity and has been used in numerous studies on nitrification, it is not as effective with many AOB because their rRNA contents are rather independent of their nitrifying activity, i.e., starvation condition (Zhang et al. 2010; Bollmann et al. 2005; Wagner et al. 1995). It is also expected that nitrifying bacteria in drinking water samples may contain low rRNA content due to the low-substrate conditions of this particular environment, resulting in low cell fluorescent intensity (Regan et al. 2006). To overcome this issue, Regan and colleagues (2006) used dual probe labeling and tyramide signal amplification.

NGS and other meta-omics techniques (e.g., metagenomics, metatranscriptomics, metaproteomics) are commonly used to correlate changes between microbial community structure and water quality parameters. These techniques have also been used to elucidate microbially mediated processes in drinking water distribution systems with the overall goal being to identify improved operating conditions that support a stable, resilient, and pathogen-free microbiome and thus, safe drinking water (Zhang and Liu 2019; Tan et al. 2015). Using metagenomic techniques, comammox *Nitrospira* and *Nitrosomonas* were found to be the main drivers of ammonia oxidation in groundwater-fed biofilters, *Nitrospira* bacteria were the only nitrite oxidizers identified, and no AOA were detected in those systems (Gülay et al. 2019).

Pinto and colleagues (2016) identified a comammox *Nitrospira*-like organism from biologically active filters in a drinking water treatment plant. The genome constructed from this microorganism contained the nitrite oxidoreductase alpha subunit (*nxrA*) gene required for nitrite oxidation and the genes required for ammonia oxidation (ammonia monooxygenase [*amoA, amoB, amoC*] and hydroxylamine oxidoreductase [*hao*]), indicating full comammox potential (Pinto et al. 2016). Collectively, these studies highlight the utility of meta-omic approaches to further our understanding of microbially mediated nitrification processes in various parts of the drinking water distribution system.

Quantitative PCR methods—such as competitive PCR, quantitative real-time PCR, digital droplet PCR, and digital PCR—are routinely used for the culture-independent enumeration of microorganisms. Competitive PCR quantification includes a competitive sequence that serves as an internal control. Among others, it has been used in groundwaters (Ivanova et al. 2000), wastewaters (Dionisi et al. 2002), and activated sludge (Bjerrum et al. 2002). Quantitative real-time PCR uses a fluorescently labeled probe or a DNA-binding fluorochrome and measures fluorescence emitted continuously during the amplification reaction. Calculations of target gene abundance are based on the initial exponential phase of the PCR (Heid et al. 1996). This method has been applied to nitrifier enumeration in soils (Okano et al. 2004; Hermansson and Lindgren 2001), activated sludge (Araki et al. 2004), chloraminated drinking water distribution systems (Zhang et al. 2010; Regan et al. 2004, 2006), and disinfection kinetics (Wahman et al. 2009, 2010). Regan and colleagues (2007) developed a method to subdivide AOB present into *Nitrosomonas oligotropha*, non-*oligotropha* species of *Nitrosomonas*, and *Nitrosospira* from comparative *amoA* sequence analysis of the *Betaproteobacteria* subclass of AOB. Notably, Wahman et al. (2009) coupled propidium monoazide (PMA) staining to quantitative PCR to eliminate the amplification of DNA from membrane-compromised cells, thereby adding a viability assay to their nitrifier quantitation.

For these molecular methods, primers and probes specific to nitrifiers have been developed and target genes encoding the key enzymes required for ammonia oxidation (e.g.*amoA, amoB, amoC, hao*; involved in Eq 5-1 in Figure 5-1) and nitrite oxidation (*nxrA*; involved in Eq. 5-2 in Figure 5-1), as well as highly conserved regions within the 16S and 23S rRNA gene regions (Appendix Table A-4). AMO is a membrane-bound protein that starts the nitrification process by oxidizing ammonia into hydroxylamine (NH_2OH), after which the HAO, located in the periplasm, oxidizes the NH_2OH into nitrite (Arp et al. 2002). In the case of AOA and AOB, the *amoA* gene is often used as a target gene (Meinhardt et al. 2015; Francis et al. 2005). Additionally, the *amoA* and *hao* genes can be used to distinguish between the clade A (COM-A) and clade B (COM-B) comammox *Nitrospira* (Yu et al. 2018; Daims et al. 2015, 2016; van Kessel et al. 2015). For AOA, the biomarkers that can be used include the *amoA* gene, tetraether lipids that are specific for *Crenarchaeota*, and soprenoid glycerol dialkyl glycerol tetraethers (GDGTs) that are membrane lipids unique of A*rchaea* (You et al. 2009). Results suggested that there is a good correlation between the amount of GDGT (or crenarchaeol alone) and the abundance of Archaeal *amoA* gene copies.

Yu and colleagues (2018) demonstrated that there was a higher diversity of COM-A members than β-AOB members in enriched cultures of tidal river sediments, with COM-A *amoA* genes making up approximately 0.4–5.6% of the total *amoA* genes. A similar trend was observed in assembled genomes of drinking water samples in which AOB *amoA* sequences were three-fold higher than comammox *amoA* sequences (Wang et al. 2017). However, in groundwater-fed rapid sand filters, 52–81% of *amoA* sequences were attributed to comammox organisms (Palomo et al. 2016), indicating that populations of AOA, AOB, and comammox can vary significantly depending on the type of environment. Furthermore, monitoring AOB, AOA, *Nitrospira,* and *Nitrobacter* during chlorine maintenance, or burn, period revealed that AOB may play an important role in nitrification in distributed water, *Nitrospira* was highly associated with nitrite oxidation, and their reduction is temporary as regrowth occurs within a few months after this burn period (Wang et al. 2014, Starke et al. 2013).

Examples of PCR primers and hybridization probes are presented in Appendix A, Tables A-4 and A-5, respectively. Unfortunately, a sequence specific and unique to all AOB or all NOB does not exist. Thus, a combination of different primers is necessary to identify nitrifiers.

GROWTH CONDITIONS FOR NITRIFYING MICROORGANISMS

This section presents the general growth conditions of nitrifiers. Substantial species-level differences have been observed, and the phenotypic diversity of AOB, AOA, and NOB is also still not well understood. Therefore, this section presents some general principles useful for understanding the influence of several environmental parameters on nitrifier growth, and Table 5-4 presents reported ranges of growth characteristics for representative genera of nitrifiers, but comprehensive coverage of growth kinetics for nitrifiers encountered in drinking water distribution systems is not provided because those values are generally unknown and population/context specific. As with most lithoautotrophs, nitrifiers grow slowly and have low cell yields (Table 5-4) relative to heterotrophic bacteria (Grady et al. 1999) due to the low amount of energy obtained from ammonia or nitrite oxidation (Prosser 1990) and the need to invest approximately 80 percent of this energy into carbon dioxide (CO_2) fixation (Kelly 1978). The growth rate of AOB can be faster than that of NOB under some conditions, which may contribute to the observed accumulation of nitrite in some environments such as in drinking water distribution systems, where nitrite concentrations reaching 1.0 mg/L NO_2^--N have been reported (Harrington et al. 2002; McGuire et al. 1999; Odell et al. 1996; Wilczak et al. 1996; Skadsen 1993). Jung et al. (2022) evaluated the ammonia-oxidation kinetic properties of 12 AOA strains, isolated from marine, soil, sediment, and hot springs, and representing the four currently described AOA phylogenetic lineages. Their results supported the hypothesis that AOA are generally adapted to lower ammonia substrate concentrations compared to AOB. However, their study revealed that AOA can have a broad range of substrate affinities, which can be impacted by temperature, pH, and cellular morphology (Jung et al. 2022).

Table 5-4 Growth characteristics of nitrifying bacteria (Grady et al. 1999; Ford 1980; Painter 1970)

Characteristics	*Nitrosomonas* (AOB)	*Nitrobacter* (NOB)
Generation time (h)	8–36	12–59
Observed yield (g cell/g N oxidized)	0.04–0.13	0.02–0.07
Maximum specific growth rate (day^{-1})	0.33–2.2* 0.34–2.21†	0.14–1.39
Half-saturation concentration (mg/L as ammonia nitrogen)	1.0 at 20°C (68°F)*, † 10 at 30°C (86°F)* 0.06 to 5.6 (sum of NH_3 and NH_4^+)†	1.3–8.4

* Ford(1980)

† Grady et al. (1999)

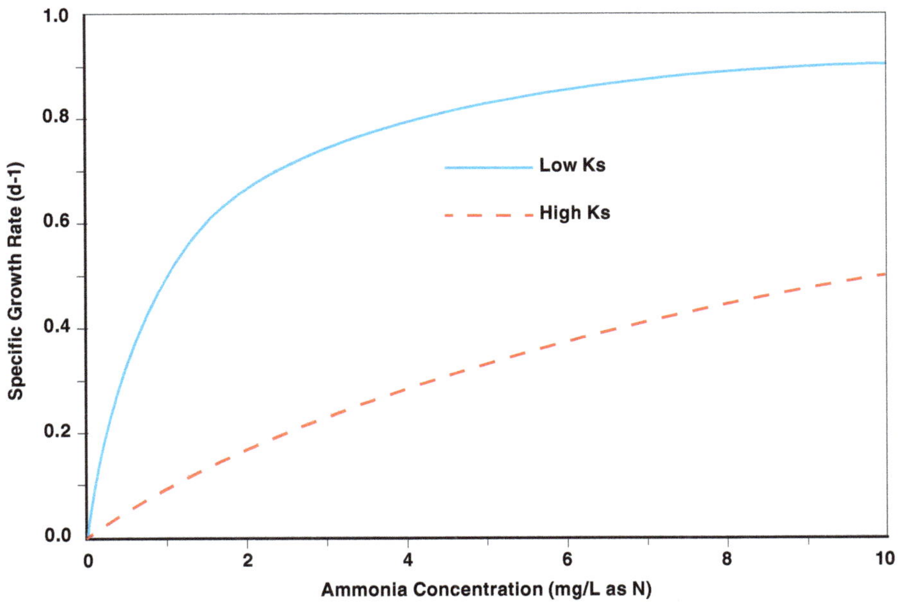

Figure 5-7 Generalized plot of Monod kinetics, showing the comparative specific growth rates of two ammonia oxidizers with the same maximum specific growth rate (μ_{max} = 1 d^{-1}) and different affinities for ammonia (low K_s = 1 mg/L as N, high K_s = 10 mg/L as N).

The growth kinetics of nitrifying microorganisms can be generally modeled using the empirical Monod equation, which relates growth rate to the concentration of a rate-limiting substrate as follows:

$$\mu = \mu_{max} \frac{S}{K_S + S} \qquad \text{(Eq 5-5)}$$

where μ is the specific growth rate (day^{-1}), μ_{max} is the maximum specific growth rate (day^{-1}), S is the rate-limiting substrate concentration (mg/L), and K_s is the half-saturation concentration (mg/L). Typically, ammonia or nitrite is the rate-limiting substrate of interest for nitrifiers, but it can be oxygen in some environments (discussed in following sections). This relationship between growth rate and ammonia concentration is illustrated in Figure 5-7. The features of Monod kinetics are defined by the two parameters μ_{max} and K_s, where μ_{max} describes the maximum growth rate that is approached when there is sufficient ammonia or nitrite and K_s is the ammonia or nitrite concentration at which the growth rate is one-half μ_{max}. A bacterium with a low K_s has a high affinity for the rate-limiting substrate and may outcompete other bacteria at low substrate concentrations, whereas a bacterium with a high μ_{max} is most competitive at higher substrate concentrations. In the context of chloraminated drinking water distribution systems, where there would typically be very low concentrations of ammonia and nitrite, AOM and NOB with low K_S values would have a kinetic advantage (neglecting potential differences in other growth and inactivation factors). In a water-reuse scenario with relatively higher ammonia concentrations, low K_S may be less of a determining factor.

Many other factors besides substrate concentration also affect the growth rate of nitrifiers, including temperature, pH, light intensity, oxygen concentration, and microbial community composition. AOB generally grow optimally at 25–30C and a pH of 7.5–8.0, with *Nitrosospira* having a slightly broader temperature range (Holt et al. 2000; Watson et al. 1989). NOB growth optima are typically at 20–30C and a pH of 7–8, though some *Nitrobacter* extend these ranges from 5–37C and pH 6.5–8.5 (Holt et al. 2000; Watson et al. 1989). The critical factors are detailed in the following sections.

Growth Substrates

Inorganic nitrogen. Most strains of AOB and NOB grow optimally at ammonia and nitrite substrate concentrations of 28–140 mg/L and 28–420 mg/L ammonia nitrogen, respectively (Watson et al. 1989). According to Prosser (1990), liquid cultures of nitrifiers (similar to nitrifiers that would be suspended in the bulk fluid of a distribution system) require minimum ammonia concentrations on the order of 0.05–2 mg/L ammonia nitrogen and nitrite concentrations on the order of 0.05–0.2 mg/L NO_2^--N. These latter ranges of concentrations are similar to the conditions found in drinking water distribution systems. The specific growth rates for AOB and NOB can decrease significantly at low substrate concentrations, such as in drinking water distribution systems. AOA have also been shown to grow at ammonia concentrations that are about 100-fold lower than levels required by AOB, with growth inhibition of AOA at ammonia levels conducive for AOB growth (Madigan et al. 2019).

As mentioned previously, there is diversity in the AOB and NOB with respect to growth attributes, and some strains thrive at higher substrate concentrations (Prinčič et al. 1998; Suwa et al. 1994) while others thrive at lower concentrations and may actually be inhibited at higher concentrations (Suwa et al. 1994). These differences in kinetic parameters affect strain selection in a given environment, and drinking water seems to favor the enrichment of nitrifiers with a high affinity for low concentrations of substrate. This consideration also contributes to the poor and variable recovery of different nitrifiers by standard MPN media. Cai and colleagues (2018) utilized a culture-dependent approach and observed a synergism between *Nitrosomonas*-like AOB and *Nitrobacter*-like NOB under oligotrophic conditions at 25°C. Cultures were heavily influenced by either NH_4^+ (0.9 mg/L), NO_2^- (maximum of 16 mg/L), or both, as well as bicarbonate, HCO_3^- (18 mg/L), concentrations; although growth competitiveness was observed when both AOB and NOB were present, robust nitrification occurred (Cai et al. 2018).

Nitrifiers are subject to both substrate (ammonia and nitrite) and product (nitrite and nitrate) inhibition; if the concentration of either the substrate or the product is too high, the rate of nitrification decreases (Focht and Verstraete 1977). While the results of some studies are reported in this manual, they all suggest that inhibitory concentrations of substrate and product are much higher than those found in drinking water systems.

Dissolved oxygen. Nitrification is an aerobic process with molecular oxygen required for respiration by both AOM and NOB. The theoretical oxygen demand for oxidation of ammonia to nitrite is 3.43 g O_2 per g of ammonia-N oxidized (per Eq 5-1 of Figure 5-1), and the subsequent demand for nitrite oxidation to nitrate is 1.14 g O_2 per g nitrite-N oxidized (per Eq 5-2 of Figure 5-1). Thus, the total theoretical oxygen requirement is 4.57 g O_2 to oxidize ammonia to nitrate. Actual oxygen demand is slightly lower than this (Grady et al.1999; Ford 1980) due to the fact that some of the reduced nitrogen is assimilated as cellular material. In any case, approximately 4 mg/L of oxygen are consumed during complete nitrification of 1 mg/L of ammonia nitrogen in a chloraminated water system. This may explain why several authors report that dissolved oxygen concentrations below 0.5–2.5 mg/L O_2 may inhibit nitrification (2.0 mg/L is recognized in practice) (USEPA 1993). These data also suggest that oxygen requirements for nitrifiers are fulfilled in most drinking water systems. If dissolved oxygen is low, such as at the base of a thick biofilm or in sediment of a poorly mixed tank, the Monod equation can be expanded to account for rate limitations due to both the electron donor (ammonia or nitrite) and the electron acceptor (dissolved oxygen). This is called multiplicative- or dual-Monod kinetics and is shown in Eq 5-6:

$$\mu = \mu_{max}\left(\frac{S_1}{K_{S1}+S_1}\right)\left(\frac{S_2}{K_{S2}+S_2}\right) \qquad \text{(Eq 5-6)}$$

where S_1 and K_{S1} refer to one of the rate-limiting substrates (e.g., ammonia) and S_2 and K_{S2} refer to the other rate-limiting substrate (in this context, dissolved oxygen). For typical dissolved oxygen concentrations encountered in drinking water distribution systems, S_2 would be sufficient such that the second parenthetical term would be close to 1, and the equation would just reduce to Eq 5-5.

In addition to these growth-rate effects, earlier observations indicated that the oxidation of ammonia by *Nitrosomonas* under low oxygen concentrations is accompanied by production of N_2O or NO (Watson et al. 1989). Goreau and colleagues (1980) observed enhanced yield of N_2O and lower production of NO_2^- at low oxygen concentrations. As the oxygen concentration decreased from 7–0.18 mg/L, the production of N_2O increased from 0.002 to nearly 0.1 mol N_2O per mol NO_2^- at the expense of the production of NO_2^-. The growth rate varied only slightly (approximately 30 percent) over the entire oxygen range. It is now understood that some AOB have an alternate denitrification metabolism involving nitrite, nitric oxide, and nitrous oxide reduction that is used at low oxygen concentrations (Schmidt et al. 2004). It is unknown at present if AOA are capable of utilizing an alternate denitrification pathway; however, AOA are able to survive at lower nutrient and ammonia levels than are needed for AOB growth (Madigan et al. 2019). It should be noted that denitrification by traditional denitrifying microorganisms presents an alternate reaction at low oxygen concentrations, with the reduction of nitrate and nitrite sequentially to NO, N_2O, and N_2 (Eq 5-3 in Figure 5-1). This is not a known reaction in distribution systems but can be deliberately induced in biological treatment systems.

Organic matter. Most nitrifiers are obligate lithotrophs. As such, nitrifiers oxidize inorganic substrates, such as ammonia and nitrite, as sole sources of energy and do not require organic carbon for energy. Some strains of AOB can grow mixotrophically, incorporating organic compounds (pyruvate, formate, acetate, glucose, peptone). This may exhibit various results in AOB, from no effect to stimulation to inhibition (Focht and Verstraete 1977). In any case, organotrophic growth of lithotrophic AOB (use of organic carbon for both carbon and energy sources) has never been observed (Prosser 1990; Watson et al. 1989); this is not the case, however, for some NOB, as explained below.

Some *Nitrobacter* species are facultative lithotrophs, as they can also use organic substrates as an energy and carbon source (Holt et al. 2000; Watson et al. 1989; Ford 1980). As such, the incorporation of organic carbon (acetate, casein hydrolysate, glycerol, pyruvate) may increase NOB growth rates and cell yields (Prosser 1990; Watson et al. 1989; Focht and Verstraete 1977). However, an inorganic source is the preferred medium, and organotrophic growth of *Nitrobacter* species is slower than growth on nitrite and slower than the growth of other heterotrophs. Various other groups of organisms, including heterotrophic bacteria, fungi, and algae, can also be responsible for nitrification, although at a much slower rate than lithotrophic nitrifying bacteria. These other nitrifying microorganisms require organic substrates, but discussion of these organisms is outside the intent of this manual.

Several authors suggest that removal of natural organic matter (NOM) may prevent nitrification by enhancing the biological stability of the water. However, because AOM generally use CO_2 as a carbon source, NOM is not the source of biological instability from the AOM perspective (Harrington et al. 2002, 2003). Instead, ammonia is the source of biological instability, and NOM enhances this instability by reacting with chloramines to release ammonia, which promotes AOM growth. A discussion on chloramine demand and decay is presented in Chapter 4.

Phosphorus. Earlier work suggested that nitrification is favored by the presence of calcium and phosphate (Doetsch and Cook 1973). This was confirmed by van der Aa and colleagues (2000), who observed that low-temperature water should contain at least 10 µg/L PO_4^{3-} as -P to maintain sufficient nitrification in a rapid sand filter at a plant with an ammonia-impacted source water. These authors dosed phosphoric acid (30 µg/L PO_4^{3-} as -P) in the drinking water treatment plant in the winter to stimulate growth and re-established ammonia removal within the plant. A laboratory study on the effects of pipe material, inorganic carbon, phosphorus, copper, and zinc on nitrification showed nitrifier inhibition at <5 μg/L PO_4^{3-} as -P in PVC pipe sections (Zhang and Edwards 2010).

Pipe material effects on phosphate availability included the liberation of phosphate from new soft copper pipe sections and the removal of soluble phosphate from lead pipe corrosion. Water matrix effects were also reported, with high alkalinity levels reducing phosphate uptake associated with lead pipe corrosion. These results suggest that phosphorus limitation may inhibit the growth of nitrifiers in some drinking water systems, and there are complex matrix and pipe interactions that can affect phosphate bioavailability even in systems with added phosphate (Zhang and Edwards 2010).

These results indicate that the two methods typically used to control lead and copper corrosion—i.e., pH and alkalinity adjustment—or the use of an orthophosphate-base corrosion inhibitor may help sustain nitrification in phosphorus-limited distribution systems. The effect of a zinc orthophosphate corrosion inhibitor on nitrification has not been documented in detail, but it is possible that the inhibitory effect of zinc may counteract the effect of phosphate.

Temperature

The optimum growth temperature for nitrifiers has been reported in the range of 20–30°C, although the optimum temperature for nitrite oxidation may be lower. Generally, the optimum growth temperature reported in the literature is 30°C for AOB, 37°C for AOA, and 28°C for NOB (Wu et al. 2013; Holt et al. 2000). At 5–30°C, temperature affects nitrifier growth according to an Arrhenius relation (i.e., biological activity doubles with every 10°C water temperature increase) (Wong-Chong and Loehr 1975). This is consistent with the findings of a laboratory study in which AOB were incubated in dechlorinated treated drinking water (Kirmeyer et al. 1995). Results showed that AOB grew at all water temperatures tested (10, 15, and 25°C), although growth took 3–4 weeks longer at 10°C than at the other temperatures. In the presence of chloramine, nitrification occurred earlier at 25°C than at 15°C (Odell et al. 1996).

There have also been studies on the effects of water temperature on nitrifier inactivation by chloramine, which show the combined temperature-chloramine effects on nitrifier growth and inactivation. Lieu and colleagues (1993) conducted extensive inactivation studies using AOB cultures inoculated in drinking water samples in the presence of different chloramine concentrations (1.7, 2.0, and 2.5 mg/L) and chlorine-to- ammonia-N weight ratios (3:1, 4:1, and 5:1). (Note: this study did not differentiate monochloramine and dichloramine or free ammonia concentrations.) Results show that all chloramine conditions tested prevented the growth of AOB (<0.2 MPN/mL) at 10°C and AOB did not recover. At 15 and 25°C, AOB were able to recover at different rates following exposure to 1.7 and 2.0 mg/L chloramines but not following exposure to 2.5 mg/L chloramines. The rates of both inactivation and regrowth were faster at 25°C than at 15°C (Lieu et al. 1993). Contact time and chloramine dosage controlled the growth of AOB at 15°C, and the chlorine-to-ammonia-N ratio did not have a significant effect. At 25°C, however, nitrification was controlled by chloramine dosage, contact time, and chlorine-to-ammonia-N ratio.

Using bench-scale distribution system simulators, Pintar and colleagues (2000) showed that nitrification occurred at 12°C at a chloramine residual of 0.05–0.6 mg/L and a chlorine-to-ammonia-N weight ratio of 3:1. When the bench-scale system temperature was decreased to 6°C, nitrite production was also observed following a period of acclimation, showing that AOB are capable of metabolic activity at low temperatures. Pintar and colleagues (2000) observed that at lower temperature, a lag or acclimation period occurs prior to nitrite production.

Quinlan (1980) examined the effects of temperature and substrate concentrations on the rate of ammonia and nitrite oxidation by mixed and pure cultures of AOB and NOB. Results showed that the maximum rate of substrate oxidation increased with temperature and substrate concentration. Thus, the determination of optimum temperature for AOB and NOB growth needs to consider substrate concentrations as well. These results also suggest that for substrate concentrations in the range of 0.5–10 mg/L ammonia nitrogen, the optimum temperature for nitrite production (i.e., oxidation of ammonia to nitrite) exceeds the optimum temperature for nitrite consumption (i.e., oxidation of nitrite to nitrate) (Quinlan 1980). As a result, thermal enrichment of a nitrifying system should generally favor the production of nitrite over nitrate, with accumulation of nitrite.

Finally, it is important to distinguish between short-term temperature dependence (i.e., shock temperature response) and long-term temperature dependence (Christensen and Harremoës 1978). The short-term temperature dependence reflects the immediate ability of microorganisms to change reaction rates. The long-term temperature dependence is a combination of adaptation by organisms originally present and selection of organisms favored by the new temperature conditions.

pH

Nitrifying bacteria are very sensitive to pH (Figures 5-8 and 5-9). Generally, the optimum pH reported in the literature is 7.5–8.0 for AOB and 7.6–7.8 for NOB. For AOA, while acidophilic growth and metabolic activity have been detected at pH 2.5–5.5 in acidic soils (Hatzenpichler 2012; Stahl and de la Torre 2012), the isolate *Nitrososphaera viennensis* and three freshwater AOA enrichment cultures showed optimal growth at pH values of 7–7.5 (French et al. 2012; Tourna et al. 2011). Growth was observed within a pH range of approximately 2 pH units (pH 7.5–9) and was inhibited at the acidic pH of 6 (French et al. 2012). The maximum specific growth rate of *Nitrosomonas europaea* is observed at pH 8.0, but a 54 percent reduction in growth rate is observed at pH 7, and growth stops at pH 6.5 (Keen and Prosser 1987). The effect of pH on AOB can be explained partially by the equilibrium of NH_4^+ and NH_3, which possesses a pK_a value (i.e., the pH at which there is a 1:1 molar ratio of NH_4^+ and NH_3) of 9.26 at 20°C. As previously presented, NH_3, not NH_4^+, is the substrate for ammonia oxidation, and each reduction in pH value of one unit decreases the concentration of free ammonia by one order of magnitude. This contributes to the sharp decrease in specific growth rates observed with decreasing pH values within the pH range where AOB can survive. This also implies that the optimum pH value for growth depends on ammonia concentration.

The work of Quinlan (1984) supports this interdependency of optimum pH and substrate concentration. Figure 5-8 describes the pH dependence of the rate of ammonia oxidation by *Nitrosomonas europaea* and shows a decrease in optimum pH (i.e., the pH that produces the maximum rate of oxidation) with increasing total ammonia concentration (Quinlan 1984). Even though the shift in optimum pH with increasing ammonia concentration is relatively small (approximately 0.8 pH units for 1–50 mg/L total ammonia as N), the consequences for nitrification control are important. For each mole of ammonia-N oxidized, 2 moles of hydrogen ion are produced. Thus, as ammonia is consumed, the pH should decrease, and the growth rate of nitrifiers should decrease. However, this tendency is contrary to the shift in optimum pH; as ammonia concentrations decrease, the optimum pH increases. Consequently, a constant set point for pH will not control nitrification. Instead, the set point must be varied to optimize the rate of ammonia oxidation (Quinlan 1984) or to prevent it when attempting to prevent or control nitrification. In drinking water systems, where the change in ammonia concentration is very small, a decreased nitrifier growth rate at lower pH would likely occur.

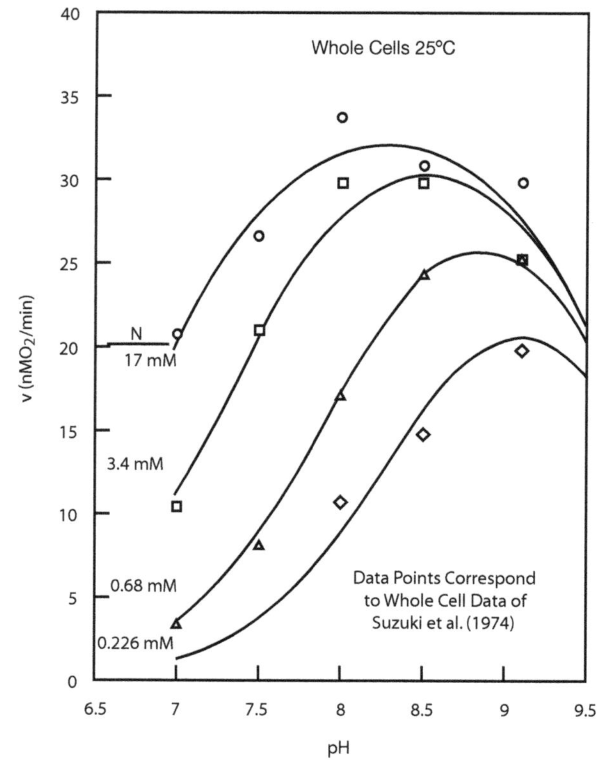

Source: Quinlan 1984

v, reaction velocity rate

Figure 5-8 The pH and ammonia concentration dependences of the ammonia oxidation rate for *Nitrosomonas europaea*.

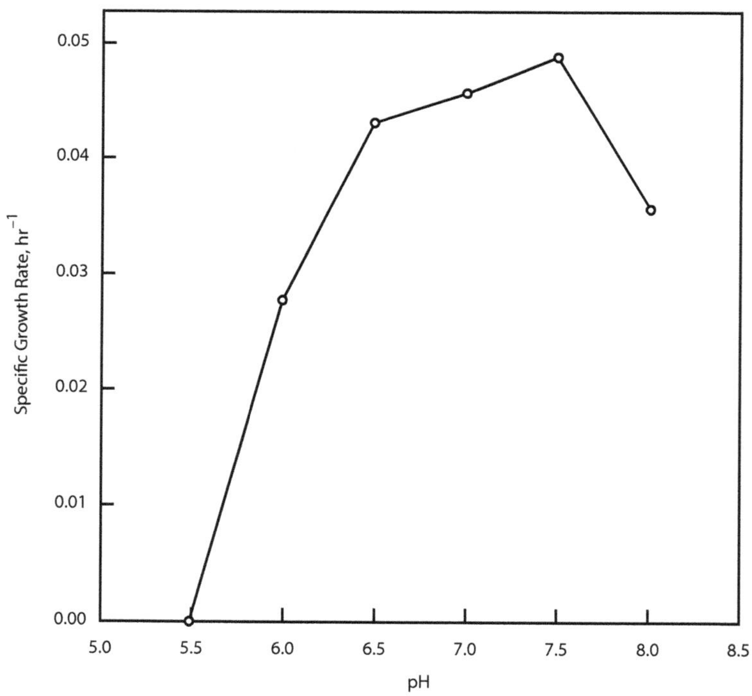

Source: Prosser 1990

Figure 5-9 Effect of pH on maximum specific growth rate of *Nitrobacter* species.

Despite the neutrophilic (i.e., optimum near neutral pH) growth of pure cultures of AOB and NOB, nitrification did occur in enrichment cultures within a pH range of 6.5 to nearly 9.0 in soil at pH values below 4.0. Prosser (1990) reports four possible explanations for this phenomenon: the existence of acidophilic strains of AOB and NOB, surface growth, the existence of microenvironments of neutral pH value, and heterotrophic nitrification. An additional explanation may include long detention times that favor the slow nitrification process. Also, the slime layer in which nitrifiers are embedded may protect them against unfavorable conditions. Killham (1987) reported that cells of nitrifiers attached to glass beads were able to nitrify under acidic conditions in which unattached cells did not survive. Some of these microenvironments may exist in drinking water distribution systems, such as in biofilms on pipes or sediment, that allow the growth of nitrifiers where the bulk water conditions would suggest otherwise. Although drinking water systems contain levels of substrates that are much lower than those used in the enrichment cultures to draw these conclusions, some of the explanations previously presented could also apply to drinking water systems.

As previously explained for the temperature dependence, it is also appropriate to distinguish between short-term and long-term pH effects (Christensen and Harremoës 1978). Most investigations have focused on short-term pH changes, but the long-term acclimation of nitrifiers to nonoptimal pH conditions may allow their survival and growth under conditions that would not permit their development otherwise (USEPA 1993).

Alkalinity

Alkalinity can have direct effects on nitrifier growth, as these bacteria generally grow autotrophically, i.e., using carbon dioxide as a carbon source. It can also have several indirect effects. As mentioned above, the first step of nitrification produces protons and thereby reduces alkalinity, and this may lead to a significant decrease in pH (DiGiano et al. 2002). Approximately 7 mg alkalinity as $CaCO_3$ is destroyed per mg NH_3-N oxidized (Ford 1980). However, in a water quality survey of ten drinking water distribution systems, Wilczak and colleagues (1996) did not observe any correlation between alkalinity and the occurrence of nitrification. The authors explained this observation by the fact that in drinking water, only small changes in alkalinity would result from nitrification, which may not have been detected during the survey (Wilczak et. al. 1996). Other indirect effects of alkalinity on nitrifier growth are the reduction of toxic effects from metals, particularly copper and zinc, and the associated increase of available phosphate to support biomass growth (Zhang and Edwards 2010).

Inhibitory Substances

Ford (1980) and USEPA (1993) reported that nitrifiers are extremely sensitive microorganisms that may experience severe upsets in the presence of very small concentrations of inhibitory substances. These inhibitory substances include heavy metals, cyanides, halogenated compounds (e.g., perchlorate, chloroform), phenols, mercaptans, and thiourea (Ford 1980). Watson and colleagues (1989) also reported that the oxidation of ammonia to nitrite is sensitive to inhibition by many compounds at low concentrations (such as copper-binding agents, acetylene), whereas the oxidation of NH_2OH (an intermediate compound in the oxidation of ammonia to nitrite) to nitrite is unaffected or much less susceptible. Metal ions such as copper, nickel, chromium, and zinc can also inhibit nitrification (Sarker et al. 2018; Zhang and Edwards 2010; Ford 1980). The complete or substantial inhibition of oxygen uptake by *Nitrosomonas* occurs at 0.01 M of iron, aluminum, copper, zinc, lead, and manganese and at 0.0002 M of nickel (Painter 1970), which can play a role in nitrifier inhibition with different pipe materials (Zhang and Edwards 2010).

THMs have also been shown to indirectly inhibit AOB. These compounds can be cometabolized by AOB (Wahman et al. 2005), producing reactive THM-derived cometabolism byproducts that inactivate AOB (Speitel et al. 2005). Brominated THM species appear to be more toxic to AOB than their chlorinated counterparts. The THM concentrations necessary to both prevent and halt nitrification are a function of the nitrifying community composition (Speitel et al. 2010; Wahman et al. 2006). As expected, inactivation of AOB through byproduct toxicity from THM cometabolism depends on both THM concentration and exposure time. As a result, lower THM concentrations are necessary to delay the onset of nitrification as compared to concentrations necessary to inactivate a nitrifying biofilm (Speitel et al., 2010). In all cases, however, the THM concentrations necessary to prevent nitrification exceed the regulatory limit of 80 μg/L for the sum of the four THMs. Nevertheless, this mechanism of AOB inhibition may be contributing to the overall considerations regarding episodic nitrification susceptibility of distributed water (Speitel et al. 2010).

INACTIVATION BY DISINFECTANTS

The ability of selected chemical disinfectants and light to inactivate nitrifiers is presented in this section. Throughout this discussion, readers need to be aware of the following observations:

1. Reported microbial inactivation rates are measured or calculated values that are affected by the method used to enumerate microorganism viability (survival after exposure to disinfectants). This is particularly important in the case of nitrifiers. For example, when conducting bench-scale inactivation experiments of *Nitrosomonas europaea* with chloramines, Harrington and colleagues (2003) observed that the culturing technique (MPN method) showed significantly higher rates of inactivation (approximately three orders of magnitude) than the culture-independent *Bac*Light staining/microscopy method that assays membrane integrity as a proxy for viability. Also, the *Bac*Light method appeared to yield inactivation rates that were more consistent with the rates required in a distribution system undergoing nitrification (Harrington et al. 2003; Oldenburg et al. 2002). Similarly, Wahman (2009) performed bench-scale chloramine inactivation tests on *N. europaea* using *Bac*Light and propidium monoazide quantitative, real-time polymerase chain reaction (PMA-qPCR). They reported a better parameter estimation of *N. europaea* monochloramine disinfection using the delayed Chick-Watson model, and while both assays gave similar lag coefficients, the PMA-qPCR analysis of viability gave a lower (i.e., more conservative) inactivation rate constant. Such differences in viability counts are frequently reported for different analytical techniques.
2. Disinfectant stability is affected by various parameters, including incubation time, water matrix, temperature, pH, turbidity, NOM concentration and composition, and disinfectant dosage (Chapter 4). In turn, disinfectant stability can affect apparent (calculated or measured) microbial inactivation rates, depending on the capability of measuring or compensating for disinfectant degradation during a given study.
3. Nitrifying bacteria can grow in tight clusters of cells surrounded by a slime layer (e.g., Figure 5-4), whereby they also tend to attach to surfaces, sediments, and other debris. These nitrifier biofilms may be very resistant to disinfection. The metabolic status of the microorganisms, which is linked to their growth conditions, may also be responsible for the formation of cell aggregates. The formation of aggregates and biofilms also complicates the enumeration of viable cell density.

Inactivation by Chloramine

Based on reported concentrations (disinfectant concentration, C, multiplied by disinfectant contact time, T), the inactivation of 99 percent of planktonic AOB at a typical chloramine residual of 1.5–2 mg/L as Cl_2 should be achieved in less than 30 min (Regan 2001), which is much shorter than the average time that chloraminated water stays in reservoirs and distribution systems. Nonetheless, nitrification episodes are common in these systems, even in the presence of high chloramine residuals. There are several possible explanations for the observed growth and persistence of nitrifiers in chloraminated systems, such as the association of nitrifiers with biofilms and tuberculation or sediment accumulation imparting some degree of protection from the disinfectant.

Limitations of the MPN detection methods may also overpredict the disinfectant efficacy due to poor recovery of cells in the growth medium and the underestimation of cell density when nitrifiers are aggregated. It is also possible that the presence of the disinfectant causes selection for or adaptation of strains with increased resistance to chloramine (Cunliffe 1991; Wolfe 1990). (See additional discussion in the following sections.)

Only a few inactivation studies on *Nitrosomonas* or *Nitrospira* (prevalent AOB and NOB genera, respectively, in drinking water distribution systems) have been reported. Wolfe and Lieu (2001) suggested that NOB were more susceptible to disinfection by chloramine than AOB because the oxidation of nitrite to nitrate was not observed to occur when the chloramine residual was lower than that required to inhibit ammonia oxidation.

Inactivation experiments and results are summarized in Table 5-5. Findings from Wolfe et al. (1990) indicate that an AOB isolate grown in chlorine-neutralized tap water was approximately 11 times more resistant to chloramines than when grown in culture medium. Cunliffe (1991) reported the monochloramine resistance of a mixed population of nitrifiers collected from a drinking water reservoir at doses up to 5.2 mg/L Cl_2 (no distinction was made between AOB and NOB) (Figure 5-10). As shown in Table 5-5, the CT_{99} value measured was much higher in the Cunliffe (1991) study (760 mg • min/L Cl_2) than that obtained by Wolfe and colleagues (1990) (3–33 mg•min/L Cl_2). Differences in experimental design and nitrifier speciation could explain the disparity between the various sets of laboratory results. As an example, several researchers have shown that bacteria grown in tap water may have higher disinfectant resistance than bacteria grown in culturing media (Wolfe and Olson 1985; Ridgway and Olson 1982).

Lieu and colleagues (1993) conducted extensive inactivation studies using AOB cultures inoculated in drinking water samples. The results of this study were reported in the previous section on temperature effects. This study reported AOB regrowth occurred in 77 percent of the samples following exposure to monochloramine at 1.7 mg/LCl_2, regardless of the chlorine-to-ammonia-N ratio or contact time. When AOB were exposed to 2.0 mg/L monochloramine, regrowth appeared in 67 percent of the samples, and recovery depended on both the chlorine-to-ammonia-N ratio and contact time. Regrowth did not occur in any of the samples exposed to 2.5 mg/L monochloramine as Cl_2, except when the chlorine-to-ammonia-N ratio was 3:1 and the contact time only 2 days, representing 26 percent of the samples.

In bench-scale experiments, Harrington and colleagues (2003) observed that the rate of *Nitrosomonas europaea* inactivation by monochloramine decreased with increasing pH from 7.0 to 9.0. They suggested that where nitrification is better controlled by chloramines at lower pH, inactivation was due to dichloramine, because the dichloramine fraction of total chlorine decreases as pH increases from 7 to 9 (Figure 5-11; chloramine speciation and the breakpoint curve are described in Chapters 6 and 8). However, the rate of ammonia release from chloramine decay decreases with increasing pH (Harrington et al. 2003; Valentine et al. 1998). As a result, the amount of substrate available for AOB growth decreases with

Table 5-5 Concentration (CT)$_{99}$ values* for ammonia-oxidizing bacteria (AOB) inactivation by chloramine (adapted from Harrington et al. 2003)

CT_{99} (mg•min/L Cl_2)	pH	Temperature °C (°F)	Enumeration Method	Chlorine-to-Ammonia-N Ratio (mg Cl_2:mg NH_3-N)	Type of Culture (Pure or Mixed)	Reference
19,000	9.0	20 (68)	*Bac*Light (cell membrane integrity as an indicator of viability)	4:1	Pure	Harrington et al. (2003)
9,800	8.0	15 (59)	MPN–Soriano and Walker media (incubation at 28°C for 23 days)	3:1; 4:1; and 5:1	Pure	Lieu et al. (1993)
3,900	8.0	20 (68)	*Bac*Light (cell membrane integrity as an indicator of viability)	4:1	Pure	Harrington et al. (2003)
1,900	7.0	20 (68)	*Bac*Light (cell membrane integrity as an indicator of viability)	4:1	Pure	Harrington et al. (2003)
760	8.0	30 (86)	MPN–Skinner and Walker media (incubation at 30°C for 28 days)	(not specified)	Mixed	Cunliffe (1991)
33	8.2	23 (73)	MPN–Soriano and Walker media (incubation at 28°C for 21 days)	3:1	Pure	Wolfe et al. (1990)‡
22	9.0	20 (68)	MPN–Soriano and Walker media (incubation at 28°C for 21 days)	4:1	Pure	Harrington et al. (2003)
14	8.6	20 (68)	MPN–Soriano and Walker media (incubation at 28°C for 21 days)	4:1	Pure	Harrington et al. (2003)
5.0–9.3	8.0	21 (70)	MPN–Soriano and Walker media (incubation at 26°C for 25–30 days)	5:1	Mixed	Baribeau et al. (2001)
3	8.2	23 (73)	MPN–Soriano and Walker media (incubation at 28°C for 21 days)	3:1	Pure	Wolfe et al. (1990)†
0.39	7.0	21 (70)	MPN–Soriano and Walker media (incubation at 26°C for 25–30 days)	5:1	Mixed	Baribeau et al. (2001)

* CT_{99} values represent the time (min) required to inactivate 99 percent of the microorganisms with a concentration of 1 mg/L Cl_2.
† Inactivation experiments conducted on an AOB isolated from a treated drinking water reservoir and grown in modified Soriano and Walker medium.
‡ Inactivation experiments were conducted identically to [2], except the AOB isolate was grown in dechlorinated tap water with ammonium sulfate ($(NH_4)_2SO_4$) used as a dechlorinating agent.
MPN: most probable number.

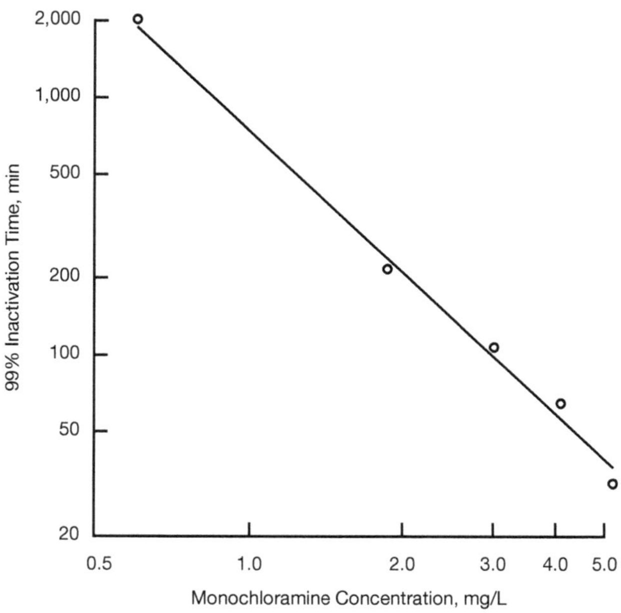

Source: Cunliffe 1991

Figure 5-10 Inactivation of mixed populations of nitrifying bacteria by monochloramine.

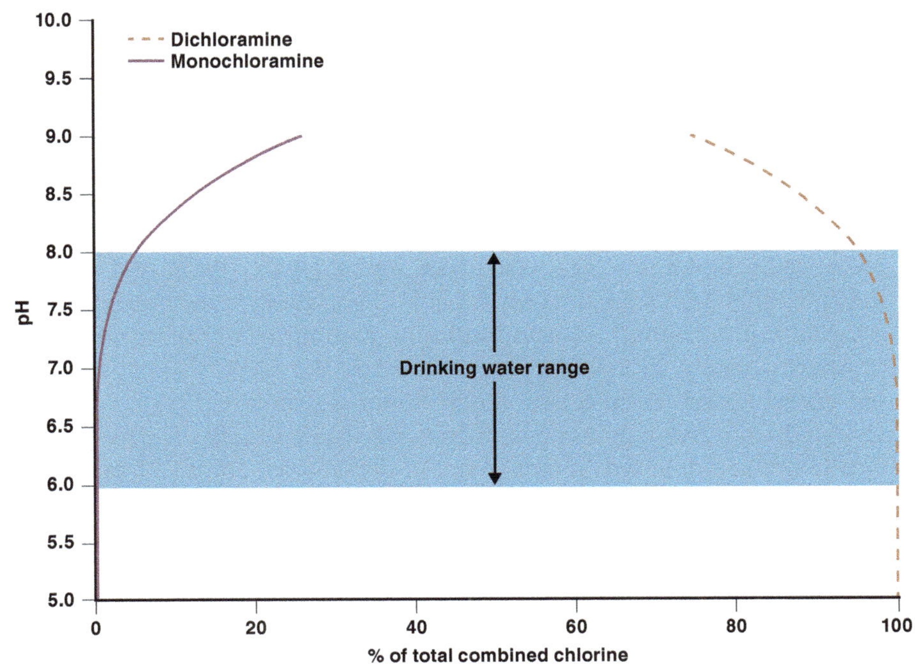

Source: Wahman and Pressman 2014

Figure 5-11 Distribution of mono- and dichloramine as a function of pH

increasing pH, suggesting that increasing pH may control nitrification. Moreover, AOB growth drops off at pH values above 9, so operating at a higher distributed water pH may help prevent AOB growth (Hill and Webster 2008). Therefore, these pH effects are opposite each other, and their net effect on nitrification control would depend on whether inactivation or substrate limitation was the dominant consideration. Harrington and colleagues (2003) stated that if ammonia is present at a concentration where the rate of AOB growth exceeds the rate of AOB inactivation, then net AOB growth can theoretically occur in the presence of a chloramine residual. This phenomenon of AOB growth in the presence of chloramine residual is indeed typically seen in bulk waters at the onset of a nitrification episode, especially at total chlorine residuals below 1 mg/L Cl_2.

Kirmeyer and colleagues (1995) suggested that the apparent resistance of established nitrifier populations to high chloramine doses may be due to a higher rate of chloramine degradation and ammonia release, resulting in additional substrate for AOB and/or mass transfer limitation of the disinfectant in clusters, biofilms, and sediments, which harbor nitrifiers. Regan (2001) also hypothesized mechanisms that may explain the additional protection offered by biofilms: the development of a chloramine concentration profile within the biofilm due to mass transport limitation of the disinfectant within the biofilm matrix, allowing a layer of microorganisms within the biofilm that is not exposed to disinfectant; a gradient of growth physiologies within the biofilm producing a gradient of chloramine resistance; and biofilm growth inducing the expression of genes that result in an increased resistance to chloramines that is not expressed in suspended-growth cultures. Lee and colleagues (2011) demonstrated that, compared to free chlorine, monochloramine is more effective at penetrating biofilms. After 30 min of 2.7 mg/L disinfectant exposure, 0.2 mg/L of chlorine (8%) and 0.4 mg/L of monochloramine (15%) penetrated the same biofilm depths of 50 μm (disinfectant concentrations are expressed as mg/L Cl_2). However, the increased penetration did not correlate with greater inactivation of biofilm microorganisms (Lee et al. 2011).

Interestingly, the chloramine concentration profile would also be accompanied by a free ammonia concentration profile. The reaction of chloramine at the surface of the biofilm could release ammonia for nitrifiers that are located inside the biofilm. A microelectrode-based study of chlorine and chloramine penetration into biofilms did not support the first hypothesis, as chloramine was observed to fully penetrate the biofilm (Lee et al. 2011). The other hypotheses have not been tested, but Ridgway and Olson (1982) have demonstrated the selection or adaptation of more resistant bacterial strains to chloramines.

Nitrification has also been observed in drinking water distribution systems in the presence of chloramine residuals exceeding 2 mg/L Cl_2. Wolfe and colleagues (1990, 1988) indicated that AOB were not only capable of surviving in the presence of monochloramine residual at 1.2–1.5 mg/L Cl_2 and a chlorine-to-ammonia-N ratio of 3:1 (water temperature of approximately 25°C and pH of about 8.2) but were also capable of growing in the presence of these disinfectant concentrations. Cunliffe (1991) examined 1,184 samples collected from five chloraminated drinking water supplies for the presence of nitrifiers. Nitrifiers were detected in 64 percent of all samples and in 21 percent of samples with more than 5.0 mg/L monochloramine as Cl_2. In a water quality survey of ten drinking water distribution systems, Wilczak and colleagues (1996) observed that even at high chloramine concentrations in finished water (3–6 mg/L Cl_2), increases in nitrite and nitrate concentrations above 50 μg/L N occurred in the distribution systems. Harrington et al. (2002), Odell et al. (1996), and Kirmeyer et al. (1995) observed that a chloramine residual above 2.0 mg/L Cl_2 may prevent the onset of nitrification. However, once a population of nitrifiers was established, a similar chloramine concentration did not inhibit the growth of nitrifiers.

When conducting bench-scale inactivation experiments using chloramines and a culture-independent enumeration technique (*Bac*Light) for AOB, Harrington and colleagues (2003) concluded that inactivation was faster at pH 7 than at pH 9. Figure 5-12 illustrates the clear relationship between inactivation and pH. Similar trends were observed when using a culture-dependent enumeration method (MPN), although the values of inactivation rate constant were different, as explained at the beginning of this section.

Inactivation by Free Chlorine

In drinking water, nitrification occurs in chloraminated distribution systems. Some chloraminated systems, however, practice periodic free chlorination to limit nitrification (Chapters 4 and 8). Free chlorine is a stronger oxidant than monochloramine and generally, a stronger disinfectant to inactivate microorganisms. In drinking water, free chlorine is found as hypochlorous acid (HOCl) and hypochlorite ion (OCl^-), with a pK_a of 7.4 at 20°C. Below pH 7.4, free chlorine is predominantly in the form of HOCl, which is considered a stronger disinfectant than is OCl^-.

Wolfe and colleagues (1990) reported that free chlorine was 13 times more effective than monochloramine for inactivating suspended AOB. These authors measured a 99-percent inactivation of an AOB culture after 2–3 min of exposure to free chlorine at Cl_2 at 23°C, and pH 8.2. Skadsen (1993) reported that the only effective method against the severe nitrification episodes observed in their drinking water distribution system was the intermittent use of free chlorine as final disinfectant. A residual of 0.7–1.8 mg/L Cl_2 controlled and prevented further nitrification. This being said, intermittent use of free chlorine has shown limited success, as presented in Chapter 7 and in Alfredo (2021).

Lee and colleagues (2018, 2011) directly compared biofilm penetration of chlorine and monochloramine, using microelectrodes to measure both disinfectant and dissolved oxygen and *Bac*Light to assess cell viability. They reported a 170-times higher penetration of monochloramine as compared to free chlorine. This was supported by the dissolved oxygen concentration data, which showed nearly complete penetration in the monochloramine-

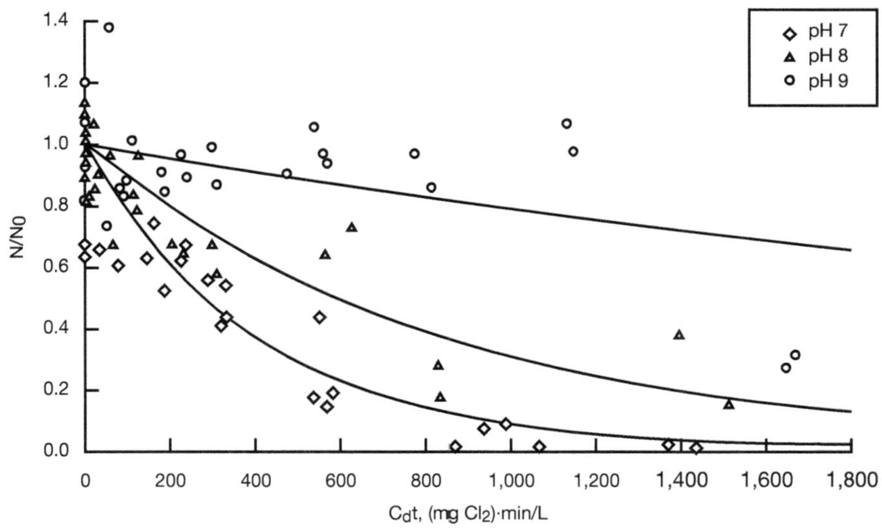

Source: Harrington et al. 2003

Figure 5-12 Data and fitted regression plots for *BacLight*-based *Nitrosomonas europaea* inactivation experiments using the Chick–Watson model (n = 1). C_d is the concentration of disinfectant; N, concentration of live microorganisms at time t; N_0, initial concentration of live microorganisms.

exposed biofilm (i.e., oxygen penetration indicated limited respiration as cells were inactivated by monochloramine throughout the biofilm). The *Bac*Light data showed that chlorine was more effective at disrupting the membranes of cells at the biofilm surface, resulting in significant decreases in biofilm thickness and increased sloughing of biofilm material, while the monochloramine-exposed biofilm had a small fraction of viable cells that were calculated to correspond to a 2–4-log inactivation and an absence of biofilm sloughing.

Inactivation by Chlorine Dioxide

Although chlorine dioxide may oxidize nitrite to nitrate (Narkis and Kott 1992), chlorite and chlorate (the main byproducts of chlorine dioxide oxidation) may block the microbial oxidation of ammonia to nitrite and nitrite to nitrate. Belser and Mays (1980) observed that 1.06 g/L sodium chlorate ($NaClO_3$) inhibits the oxidation of NO_2^- to NO_3^- by *Nitrobacter* spp. but does not affect the oxidation of NH_4^+ to NO_2^- by *Nitrosomonas europaea*. Hynes and Knowles (1983) demonstrated that the oxidation of NH_4^+ by *N. europaea* was strongly inhibited by sodium chlorite ($NaClO_2$), and the oxidation of NO_2^- by *Nitrobacter winogradskyi* was inhibited by both chlorate and chlorite. *N. winogradskyi* reduced chlorate to chlorite under both aerobic and anaerobic conditions. In mixed *N. europaea-N. winogradskyi* cell suspensions, the oxidation of both NH_4^+ and NO_2^- was inhibited in the presence of 835 mg/L ClO_3^- after a 2-h lag period. Hynes and Knowles (1983) concluded that in mixed culture, NH_4^+ oxidation is inhibited by chlorite produced by reduction of chlorate by the NOB.

Using laboratory experiments with a mixed culture of AOB collected from a drinking water distribution system, McGuire and colleagues (1999) examined the effect of chlorite in AOB inactivation. Results showed a 2–3-log inactivation of AOB after 30 min of contact time to 1 mg/L chlorite. Even at a low chlorite concentration (0.05 mg/L), AOB inactivation of 4–5 log was observed after 24 h of contact time. These trends were confirmed during field investigations in five drinking water distribution systems. Results showed more stable chloramine residual and ammonia concentration and thus less nitrification in the systems in which chlorite was present (McGuire et al. 1999). Subsequent studies have shown mixed results (Chapter 7), with evidence of chlorite not being effective by itself at stopping nitrification that is already taking place (McGuire et al. 2009).

McGuire and colleagues (1999) proposed the following mechanism to explain the effect of chlorite on AOB. The oxidation of ammonia has been shown to take place on the cytoplasmic membrane of AOB (Murray and Watson 1965). In acidic conditions, chlorite will react to produce chlorine dioxide according to the following equation (McGuire et al. 1999):

$$5NaClO_2 + 5HCl \rightarrow 4ClO_2 + 5NaCl + HCl + 2H_2O \quad \text{(Eq 5-7)}$$

McGuire and colleagues (1999) suggested that small, localized concentrations of chlorine dioxide inside the cell could affect AOB in several ways, including alteration of the cell membrane permeability, impairment of the cell's enzyme and protein functions, and damage to the nucleic acids (Stewart and Olson 1996).

The efficacy of using chlorine dioxide or chlorite to limit nitrification in full-scale distribution systems is discussed in Chapter 7.

Inactivation by Light

Although the effect of light is minimal in drinking water distribution systems, it can be significant in water treatment plants. Growth of *Nitrosomonas* can be inhibited by sunlight (380–415 nm) and UV light (Wolfe and Lieu 2001; Watson et al. 1989; Alleman et al. 1987; Hooper and Terry 1974) but does not seem to be fully inactivated or limited by them (Odell et al. 1996). Olson (1981) observed 50 percent inhibition of AOB, and NOB at light intensities approximately three orders of magnitude lower than the intensity of full sunlight. In 1962, Schön and Engel (Prosser 1990) and, in 1965, Bock (Prosser 1990) found that strains of *Nitrobacter* were much more sensitive to light than strains of *Nitrosomonas* and suggested that photooxidation of cytochrome C was the mechanism of inhibition by light. Yoshioka and Saijo (1984) also observed differences in sensitivity between AOB and NOB, with the latter being more sensitive during short-term experiments.

The ability to recover from inhibition during dark periods is also important. Hooper and Terry (1974) and other authors reported recovery of AOB from photoinhibition within hours of exposure. Wolfe and Lieu (2001) reported that nitrifiers have an excision repair mechanism for DNA; therefore, low levels of nitrifiers may be recovered from partially shaded environments. Yoshioka and Saijo (1984) also observed that an extended exposure to light in the absence of ammonia extended the recovery period significantly.

Alleman and colleagues (1987) observed that *Nitrosomonas* cells maintained under aerobic conditions without an external ammonium source were fully inhibited within 10 min under fluorescent and indirect natural light. However, these cells were protected against this inhibition during active respiration periods and anoxic exposure conditions. *Nitrosomonas* had recovered by 40 percent of their respiration rate after a "dark contact period" of approximately 2½–3 h (Alleman et al. 1987). According to Prosser (1990), the extent of photoinhibition and length of recovery period depend on the intensity of the illumination, the attenuation properties of the water, and the circulation of the water in the case of natural environments.

As an application of the inhibitory effect of light, the Los Angeles Department of Water and Power retrofitted some covered storage reservoirs with the addition of fluorescent blacklights to prevent the onset of nitrification (White and Adams 2010). Results have shown negligible nitrite production in reservoirs with the fluorescent lighting but with a significant increase in nitrite concentrations in control reservoirs that did not have the lights. Thus, fluorescent lighting may be a strategy for limiting nitrifier growth in stored water.

KEY TAKEAWAYS

Nitrification is now known to be carried out by three distinct groups of nitrifiers—AOM (which includes AOA and AOB), NOB, and comammox. Microorganisms in these three groups are distinguished "operationally" by using a different substrate—ammonia or nitrite. AOM are responsible for the oxidation of ammonia to nitrite, NOB are responsible for the oxidation of nitrite to nitrate, and comammox are responsible for the oxidation of ammonia to nitrate. Within each group, nitrifiers belong to several classes of the Pseudomonadota phylum, the Thaumarchaeota phylum, and the phyla Nitrospirae, Nitrospinae, and Chloroflexota: AOM genera have the prefix Nitroso-; NOB genera have the prefix Nitro-; and comammox belong to the genus *Nitrospira*.

In drinking water distribution systems, AOM are of major operational concern because their activity produces nitrite, which contributes to the decomposition of monochloramine. NOB require the presence of nitrite, which is produced by AOM, and NOB activity does not in itself contribute to the instability of water quality because they convert nitrite to stable, fully oxidized nitrate.

AOMs and NOB growing in drinking water distribution systems are selected for their affinity to the low levels of substrates present and are selected to withstand environmental stresses. To do so, they can grow in clumps (cell aggregates) and are primarily concentrated in protective layers of the biofilm and sediments.

Because nitrifiers belong to diverse groups, serological and molecular methods cannot capture them all in a single assay. Their slow growth rate indicates that conventional culturing techniques require long incubation periods (3–4 weeks). Culturing techniques are also not species specific. As a result, the detection, identification, and enumeration of nitrifiers are complex and limited by available analytical methods. This may explain why information is still lacking regarding the presence and speciation of nitrifiers in the environment, particularly in drinking water; however, the prevalent use of molecular-based methods (e.g., quantitative PCR, metagenomic, whole-genome sequencing) will help fill these knowledge gaps.

Limiting the growth of nitrifiers in chloraminated distribution systems will depend on many factors that can be summarized as: environmental factors (distribution system operation and maintenance practices; presence of disinfectants; presence of biofilms and deposits that can protect nitrifiers against disinfectants; high water turnover rate, which would reduce nitrifier growth and proliferation in the bulk water) and growth rate factors (water temperature, pH, substrates, microelements, etc.). Factors that affect the growth and inactivation of nitrifiers are at times consistent with each other, but at other times they are in opposition.

Examples of growth factors that are consistent with each other include

- Nitrifier specific growth rate decreases with decreasing substrate concentration (ammonia for AOB or nitrite for NOB). This is consistent with observations suggesting that nitrification is limited at low ammonia residuals in drinking water distribution systems.
- The maximum specific growth rate of AOB increases with increasing pH (up to pH ~8.0), which is within the pH range where AOB can survive.

Example of growth factors that are opposed to each other include

- Under drinking water distribution conditions, an increase in water temperature increases the growth of nitrifiers but also increases inactivation kinetics. As a result, there is a certain temperature specific to each water where net inactivation is maximum.

- The ammonia oxidation reaction indicates that pH decreases as ammonia is consumed. However, optimum pH for nitrifier growth increases as ammonia concentration decreases. Consequently, the set point to optimize the rate of ammonia oxidation varies. Typically, however, a pH decrease due to nitrification in drinking water would not be substantial enough to get far away from optimum growth conditions.

Examples of inactivation factors that are opposed to each other include

- Under drinking water distribution conditions, an increase in water temperature increases the growth of nitrifiers but also increases inactivation kinetics. As mentioned above, there is a certain temperature specific to each water where net inactivation is maximum.
- The rate of nitrifier inactivation by monochloramine decreases with increasing pH (under typical drinking water conditions), suggesting that monochloramines may better control nitrification at lower pH. However, the rate of ammonia released from chloramine decay decreases with increasing pH. These opposing trends and some practical observations suggest that increasing pH may better control nitrification.
- Higher chloramine concentrations improve nitrifier disinfection efficacy (limiting nitrification) but also provide higher concentrations of ammonia following chloramine decay (enhancing nitrification). Therefore, increasing chloramine residual entering the distribution system is effective only if chloramine is stable and a high residual reaches the affected areas.

When examining nitrifier inactivation by disinfectants, considerations should include the measured inactivation rates that depend on the method used to enumerate microorganism viability; nitrifier growth in biofilms, which confers resistance to disinfectants; and bacteria grown in water carrying a disinfectant residual may have higher disinfectant resistance than bacteria grown in culturing media without a disinfectant residual.

REFERENCES AND BIBLIOGRAPHY

Alexander, M., and F.E. Clark. 1965. "Nitrifying Bacteria." In: *Methods of Soil Analysis*. Part 2. C.A. Black, ed. Madison, WI.: American Society of Agronomy Publishers. 1477–1483.

Alfredo, K. 2021. "The 'Burn': Water Quality and Microbiological Impacts Related to Limited Free Chlorine Disinfection Periods in a Chloramine System." *Water Research*, 197:117044.

Alleman, J.E., V. Keramida, and L. Pantea-Kiser. 1987. "Light-induced *Nitrosomonas* inhibition." *Water Research*, 21:4:499–501.

Alzerreca, J.J., J.M. Norton, and M.G. Klotz. 1999. "The *amo* Operon in Marine, Ammonia-Oxidizing Gamma-Proteobacteria." *FEMS Microbiology Letters*, 180:1:21–29.

Araki, N., T. Yamaguchi, S. Yamazaki, and H. Harada. 2004. "Quantification of amoA Gene Abundance and Their *amoA* mRNA Levels in Activated Sludge by Real-time PCR." *Water Science & Technology*, 50:8:1–8.

Arp, D.J., L.A. Sayavedra-Soto, and N.G. Hommes. 2002. "Molecular Biology and Biochemistry of Ammonia Oxidation by *Nitrosomonas europaea.*" *Archives of Microbiology,* 178:4:250–255.

Baribeau, H., C.A. Kinner, R. De Leon, and P.A. Rochelle. 2001. "Nitrification in Steel and Concrete Drinking Water Reservoirs: A Case Study." AWWA Annual Conference, Washington, D.C., June 17–21.

Barker, D.J., and D.C. Stuckey. 1999. "A Review of Soluble Microbial Products (SMP) in Wastewater Treatment Systems." *Water Research,* 33:14:3063–3082.

Bédard, C., and R. Knowles. 1989. "Physiology, Biochemistry, and Specific Inhibitors of CH_4, NH_4^+, and CO Oxidation by Methanotrophs and Nitrifiers." *Microbiological Reviews,* 53:1:68–84.

Belser, L.W., and E.L. Schmidt. 1978a. "Diversity of the Ammonia-Oxidizing Nitrifier Population of a Soil." *Applied and Environmental Microbiology,* 36:4:584–588.

Belser, L.W., and E.L. Schmidt. 1978b. "Serological Diversity Within a Terrestrial Ammonia-Oxidizing Population." *Applied and Environmental Microbiology,* 36:4:589–593.

Belser, L.W., and E.L. Mays. 1982. "Use of nitrifier activity measurements to estimate the efficiency of viable nitrifier counts in soils and sediments." *Applied and Environmental Microbiology,* 43:4:945–948.

Bjerrum, L., T. Kjaer, and N.B. Ramsing. 2002. "Enumerating Ammonia-oxidizing Bacteria in Environmental Samples Using Competitive PCR." *Journal of Microbiological Methods,* 51:2:227–239.

Bock, E., H.P. Koops, B. Ahlers, and H. Harms. 1992. "Oxidation of Inorganic Nitrogen Compounds as Energy Source." In: *The Prokaryotes,* 2nd ed. A. Balows, H.G. Trüper, M. Dworkin, W. Harder, and K.H. Schleifer, eds. New York: Springer-Verlag.

Bock, E., H.P. Koops, and H. Harms. 1986. "Cell Biology of Nitrifying Bacteria." In: *Nitrification.* J.I. Prosser, ed. Oxford, U.K.: IRL Press.

Bollmann, A., I. Schmidt, A.M. Saunders, and M.H. Nicolaisen. 2005. "Influence of Starvation on Potential Ammonia-Oxidizing Activity and *amoA* and mRNA of *Nitrosospira briensis.*" *Applied and Environmental Microbiology,* 71:3:12761–282.

Caffrey, J.M., N. Bano, K. Kalanetra, and J.T. Hollibaugh. 2007. "Ammonia Oxidation and Ammonia-Oxidizing Bacteria and Archaea from Estuaries with Differing Histories of Hypoxia." *ISME Journal,* 1:7:660–662.

Cai, M., S.K. Ng, C.K. Lim, H. Lu, Y. Jia, and P.K.H. Lee. 2018. "Physiological and Metagenomic Characterizations of the Synergistic Relationships between Ammonia- and Nitrite-Oxidizing Bacteria in Freshwater Nitrification." *Frontiers in Microbiology,* 9:280.

Castignetti, D., and T.C. Hollocher. 1984. "Heterotrophic Nitrification Among Denitrifiers." *Applied and Environmental Microbiology,* 47:4:620–623.

Chain, P., J. Lamerdin, F. Larimer, W. Regala, V. Lao, M. Land, L. Hauser, A. Hooper, M. Klotz, J. Norton, L. Sayavedra-Soto, D. Arciero, N. Hommes, M. Whittaker, and D. Arp. 2003. "Complete Genome Sequence of the Ammonia-Oxidizing Bacterium and Obligate Chemolithoautotroph *Nitrosomonas europaea*." *Journal of Bacteriology*, 185:9:2759–2773.

Christensen, M.H., and P. Harremoës. 1977. "Biological Denitrification of Sewage: A Literature Review." *Progress in Water Technology*, 8:4–5:509–555.

Christensen, M.H., and P. Harremoës. 1978. "Nitrification and Denitrification in Wastewater Treatment." In: *Water Pollution Microbiology*, Vol. 2. R. Mitchell, ed. New York: John Wiley & Sons.

Connell, G.F. 1996. *The Chlorination/Chloramination Handbook*. Denver: AWWA.

Cunliffe, D.A. 1991. "Bacterial Nitrification in Chloraminated Water Supplies." *Applied and Environmental Microbiology*, 57:11:3399–3402.

Daims, H., F. Maixner, S. Lücker, K. Stoecker, K. Hace, and M. Wagner. 2006. "Ecophysiology and Niche Differentiation of *Nitrospira*-like Bacteria, the Key Nitrite Oxidizers in Wastewater Treatment Plants." *Water Science and Technology*, 54:1:21–27.

Daims, H., E.V. Lebedeva, P. Pjevac, P. Han, C. Herbold, M. Albertsen, N. Jehmlich, M. Palatinszky, J. Vierheilig, A. Bulaev, R. H. Kirkegaard, M. von Bergen, T. Rattei, B. Bendinger, P.H. Nielsen, and M. Wagner. 2015. "Complete nitrification by *Nitrospira* bacteria." *Nature*, 528:7583:504–509.

Daims, H., S. Lücker, and M. Wagner. 2016. "A New Perspective on Microbes Formerly Known as Nitrite-Oxidizing Bacteria." *Trends in Microbiology*, 24:9:699–712.

Dalton, H., and D.I. Stirling. 1982. "Cometabolism." *Philosophical Transactions of the Royal Society B, Biological Science*, 297:1088:481–496.

DiGiano, F.A., W. Zhang, A. Travaglia, D.E. Francisco, and M. Wood. 2002. *Occurrence of Bacterial Regrowth and Nitrification in the Raleigh Distribution System and Development of an EPANET Model for Future Assessments*. Raleigh, NC: Water Resources Research Institute of the University of North Carolina.

Dionisi, H.M., G. Harms, A. Layton, S. Hawkins, I. Gregory, A. Meyers, K. Robinson, and G. Sayler, 2002. Abstract Q-320 of the *102*nd *General Meeting of the American Society of Microbiol.*, 433,

Doetsch, R.N., and T.M. Cook. 1973. *Introduction to Bacteria and Their Ecology*. Baltimore: University Park Press.

Eichler, S., R. Christen, C. Höltje, P. Westphal, J. Bötel, I. Brettar, A. Mehling, and M.G. Höfle. 2006. "Composition and Dynamics of Bacterial Communities of a Drinking Water Supply System as Assessed by RNA- and DNA-Based16S rRNA Gene Fingerprinting." *Applied and Environmental Microbiology*, 72:3:1858–1872.

Elbanna, K., R.M. El-Shahawy, and K.M. Atalla. 2012. "A New Simple Method for the Enumeration of Nitrifying Bacteria in Different Environments." *Plant, Soil and Environment,* 58:1:49–53.

Feben, D. 1935. "Nitrifying Bacteria in Water Supplies." *Journal AWWA,* 27:4:439–447.

Feng, T.H. 1966. "Behavior of Organic Chloramines in Disinfection." *Journal Water Pollution Control Federation,* 38:614–628.

Focht, D.D., and W. Verstraete. 1977. "Biochemical Ecology of Nitrification and Denitrification." In: *Advances in Microbial Ecology,* Vol. 1. (M. Alexander, editor). New York: Plenum Press.

Ford, D.L. 1980. "Troubleshooting and Control of Nitrification—Wastewater Treatment Systems." *American Institute of Chemical Engineering—Symposium Series,* 209:77:159–170.

Francis, C.A., K.J. Roberts, J M. Beman, A.E. Santoro, and B.B. Oakley. 2005. "Ubiquity and Diversity of Ammonia-Oxidizing Archaea in Water Columns and Sediments of the Ocean." *Proceedings of the National Academy of Sciences of the United States of America,* 102:41:14683–14688.

French, E., J.A. Kozlowski, M. Mukherjee, G. Bullerjahn, and A. Bollmann. 2012. "Ecophysiological Characterization of Ammonia-Oxidizing Archaea and Bacteria from Freshwater." *Applied and Environmental Microbiology,* 78:16:5773–5780.

Glover, H.E. 1985. "The Relationship Between Inorganic Nitrogen Oxidation and Organic Carbon Production in Batch and Chemostat Cultures of Marine Nitrifying Bacteria." *Archives of Microbiology,* 142:45–50.

Goreau, T.J., W.A. Kaplan, S.C. Wofsy, M.B. McElroy, F.W. Valois, and S.W. Watson. 1980. "Production of NO_2 and N_2O by Nitrifying Bacteria at Reduced Concentrations of Oxygen." *Applied and Environmental Microbiology,* 40:3:526–532.

Grady, C.P.L., Jr., G.T. Daigger, and H.C. Lim. 1999. *Biological Wastewater Treatment,* 2nd ed. New York: Marcel Dekker.

Graham, D.W., C.W. Knapp, E.S. Van Vleck, K. Bloor, T.B. Lane, and C.E. Graham. 2007. "Experimental Demonstration of Chaotic Instability in Biological Nitrification." *ISME Journal,* 1:5:385–393.

Grau, P., P.M. Sutton, M. Henze, S. Elmaleh, C.P.L. Grady Jr., W. Gujer, and J. Koller. 1982. "Recommended Notation for Use in the Description of Biological Wastewater Treatment Processes." *Water Research,* 16:1501.

Gunderson, K. 1966. "The Growth and Respiration of *Nitrosocystis oceanus* at Different Partial Pressures of Oxygen." *Journal of General Microbiology,* 42:3:387–396.

Gülay, A., S.J. Fowler, K. Tatari, B. Thamdrup, H.-J. Albrechtsen, W.A. Al-Soud, S.J. Sørensen, and B.F. Smets. 2019. "DNA- and RNA-SIP Reveal *Nitrospira* spp. as Key Drivers of Nitrification in Groundwater-fed Biofilters." *mBio,* 10:6:e01870–19.

Harms, G., A.C. Layton, H.M. Dionisi, I.R. Gregory, V.M. Garrett, S.A. Hawkins, K.G. Robinson, and G.S. Sayler. 2003. "Real-time PCR Quantification of Nitrifying Bacteria in a Municipal Wastewater Treatment Plant." *Environ Sci Technol,* 37:2:343–351.

Harrington, G.W., D.R. Noguera, C.C. Bone, A.I. Kandou, P.S. Oldenburg, J.M. Regan, and D. Van Hoven. 2003. *Ammonia from Chloramine Decay: Effects on Distribution System Nitrification.* Denver, Colo.: Awwa Research Foundation.

Harrington, G.W., D.R. Noguera, A.I. Kandou, and D.J. Vanhoven. 2002. "Pilot-Scale Evaluation of Nitrification Control Strategies." *Journal AWWA,* 94:11:78–89.

Hatzenpichler, R. 2012. "Diversity, Physiology, and Niche Differentiation of Ammonia-Oxidizing Archaea." *Applied and Environmental Microbiology,* 78:21:7501–7510.

Hawkins, S.A., K.G. Robinson, A.C. Layton, and G.S. Saylor. 2008. "Response of *Nitrobacter* spp Ribosomal Gene and Transcript Abundance Following Nitrite Starvation and Exposure to Mechanistically Distinct Inhibitors." *Environmental Science and Technology,* 42:3:901–907.

Heid, C.A., J. Stevens, K.J. Livak, and P.M. Williams. 1996. "Real Time Quantitative PCR." *Genome Research,* 6:10:986–994.

Helder, W., and R.T.P. de Vries. 1983. "Netherlands Jour." *Sea Res.,* 17:1.

Henze, M. 1991. "Methods for Wastewater and Biomass Characterization." In: *Proc. of the IAWPRC Workshop on Interactions of Wastewater, Biomass and Reactor Configurations in Biological Treatment Plants.* Copenhagen, Denmark, August.

Herath, B.S., A. Torres, and A. Sathasivan. 2018. "Effects of Feed Water NOM Variation on Chloramine Demand From Chloramine-decaying Soluble Microbial Products During Rechloramination." *Chemosphere,* 212:744-754.

Hermansson, A., and P.E. Lindgren. 2001. "Quantification of Ammonia-Oxidizing Bacteria in Arable Soil by Real-Time PCR." *Applied and Environmental Microbiology,* 67:2:972–976.

Hill, D.R., and T. Webster. 2008. "High pH Inhibits Nitrifying Bacteria." *OpFlow,* 34:7:20–22.

Holt, J.G., N.R. Krieg, P.H.A. Sneath, J.T. Staley, and S.T. Williams. 2000. "Aerobic Chemolithotrophic Bacteria and Associated Organisms." In: *Bergey's Manual of Determinative Bacteriology,* 9th ed. W.R. Hensyl, ed. Philadelphia, Pa.: Lippincott Williams & Wilkins.

Hommes, N.G., L.A. Sayavedra-Soto, and D.J. Arp. 2003. "Chemolithoorganotrophic Growth of *Nitrosomonas europaea* on Fructose." *Journal of Bacteriology,* 185:23:6809–6814.

Hooper, A.B., and K.R. Terry. 1974. "Photoinactivation of Ammonia Oxidation in *Nitrosomonas.*" *Journal of Bacteriology,* 119:3:899–906.

Hovanec, T.A., L.T. Taylor, A. Blakis, and E.F. Delong. 1998. "*Nitrospira*-Like Bacteria Associated With Nitrite Oxidation in Freshwater Aquaria." *Applied and Environmental Microbiology*, 64:1:258–264.

Huang, T.L., S.L. Zhou, H.H. Zhang, S.Y. Bai, X.X. He, and X. Yang. 2015. "Nitrogen Removal Characteristics of a Newly Isolated Indigenous Aerobic Denitrifier From Oligotrophic Drinking Water Reservoir *Zoogloea* sp. N299." *International Journal of Molecular Science*, 16:5:10038–10060.

Hynes, R.K., and R. Knowles. 1983. "Inhibition of Chemoautotrophic Nitrification by Sodium Chlorate and Sodium Chlorite: A Reexamination." *Applied and Environmental Microbiology*, 45:4:1178–1182.

Ike, N.R., R.L. Wolfe, and E.G. Means. 1988. "Nitrifying Bacteria in a Chloraminated Drinking Water System." *Water Science and Technology*, 20:11–12:441–444.

Isaac, R.A., and J.C. Morris. 1983. "Rates of Transfer of Active Chlorine Between Nitrogenous Substrates." In: *Water Chlorination: Environmental Impact and Health Effects*, Vol. 3. R.L. Jolley et al., eds. Ann Arbor, Mich.: Ann Arbor Science Publishers.

Ivanova, I.A., J.R. Stephen, Y.-J. Chang, J. Bruggemann, P.E. Long, J.P. McKinley, G.A. Kowalchuk, D.C. White, and S.J. Macnaughton. 2000. "A Survey of 16S rRNA and amoA Genes Related to Autotrophic Ammonia-oxidizing Bacteria of the Beta-Subdivision of the Class *Proteobacteria* in Contaminated Groundwater." *Canadian Journal of Microbiology*, 46:11:1012–1020.

Jones, G.L., and A.R. Paskins. 1982. "Influence of High Partial Pressure of Carbon Dioxide and/or Oxygen on Nitrification." *Journal of Chemical Technology and Biotechnology*, 32:1:213–223.

Jones, R.D., and R.Y. Morita. 1983. "Methane Oxidation by *Nitrosococcus oceanus* and *Nitrosomonas europaea*." *Applied and Environmental Microbiology*, 45:2:401–410.

Jung, M. Y., C.J. Sedlacek, K.D. Kits, A. J. Mueller, S. Rhee, L. Hink, G. W. Nicol, B. Bayer, L. Lehtovirta-Morley, C. Wright, J. R. de la Torre, C. W. Herbold, P. Pjevac, H. Daims, and M. Wagner. 2022. "Ammonia-oxidizing Archaea Possess a Wide Range of Cellular Ammonia Affinities." *ISME Journal*,16:1:272–283.

Juretschko, S., G. Timmermann, M. Schmid, K.H. Schleifer, A. Pommerening-Röser, H.P. Koops, and M. Wagner. 1998. "Combined Molecular and Conventional Analyses of Nitrifying Bacterium Diversity in Activated Sludge: *Nitrosococcus mobilis* and *Nitrospira*-Like Bacteria as Dominant Populations." *Applied and Environmental Microbiology*, 64:8:3042–3051.

Kane, M.D., L.K. Poulsen, and D.A. Stahl. 1993. "Monitoring the Enrichment and Isolation of Sulfate-Reducing Bacteria by Using Oligonucleotide Hybridization Probes Designed from Environmentally Derived 16S rRNA Sequences." *Applied and Environmental Microbiology*, 59:3:682–686.

Kasuga, I., H. Nakagaki, F. Kurisu and H. Furumai. 2010. "Predominance of Ammonia-Oxidizing Archaea on Granular Activated Carbon Used in a Full-Scale Advanced Drinking Water Treatment Plant." *Water Research*, 44:17:5039–5049.

Keen, G.A., and J.I. Prosser. 1987a. "Interrelationship Between pH and Surface Growth of *Nitrobacter*." *Soil Biology and Biochemistry*, 19:6:665–672.

Keen, G.A., and J.I. Prosser. 1987b. "Steady-State and Transient Growth of Autotrophic Nitrifying Bacteria." *Archives of Microbiology*, 147:1:73–79.

Kelly, D.P. 1978. *Companion to Microbiology*. A.T. Bull, and P.M. Meadow, eds. London: Longman.

Killham, K. 1986. "Heterotrophic Nitrification." In: *Nitrification*. J.I. Prosser, ed. Washington, D.C.: IRL Press.

Killham, K. 1987. "A New Perfusion System for the Measurement and Characterization of Potential Rates of Soil Nitrification." *Plant and Soil*, 97:267–272.

Kindaichi, T., T. Ito, and S. Okabe. 2004. "Ecophysiology Interaction Between Nitrifying Bacteria and Heterotrophic Bacteria in Autotrophic Nitrifying Biofilms as Determined by Microautoradiography-Fluorescence in Situ Hybridization." *Applied and Environmental Microbiology*, 70:3:1641–1650.

Kirmeyer, G.J., L.H. Odell, J.G. Jacangelo, A. Wilczak, and R.L. Wolfe. 1995. *Nitrification Occurrence and Control in Chloraminated Water Systems*. Denver, Colo.: AWWA Research Foundation and American Water Works Association.

Kitzinger, K., H. Koch, S. Lücker, C.J. Sedlacek, C. Herbold, J. Schwarz, A. Daebeler, A.J. Mueller, M. Lukumbuzya, S. Romano, N. Leisch, S.M. Karst, R. Kirkegaard, M. Albertsen, P. H. Nielsen, M. Wagner, and H. Daims. 2018. "Characterization of the First '*Candidatus* Nitrotoga' Isolate Reveals Metabolic Versatility and Separate Evolution of Widespread Nitrite-Oxidizing Bacteria." *mBio*, 9:4:e01186–18.

Klein, T., L. Poghosyan, J.E. Barclay, J.C. Murrell, M.I. Hutchings, and L.E. Lehtovirta-Morley. 2022. "Cultivation of Ammonia-oxidising Archaea on Solid Medium." *FEMS Microbiology Letters*, 369:1:fnac029.

Kleiner, D. 1985. "Bacterial Ammonium Transport." *FEMS Microbiology Reviews*, 1:2:87–100.

Knowles, R. 1982. "Denitrification." *Microbiological Reviews*, 46:1:43–70.

Könneke, M., A.E. Bernhard, J.R. de la Torre, C.B. Walker, J.B. Waterbury, and D.A. Stahl. 2005. "Isolation of an Autotrophic Ammonia-Oxidizing Marine Archaeon." *Nature*, 437:7058:543–546.

Kowalchuk, G.A., J.R. Stephen, W. De Boer, J.I. Prosser, T.M. Embley, and J.W. Woldendorp. 1997. "Analysis of Ammonia-Oxidizing Bacteria of the β Subdivision of the Class Proteobacteria in Coastal Sand Dunes by Denaturing Gradient Gel Electrophoresis and Sequencing of PCR-Amplified 16S Ribosomal DNA Fragments." *Applied and Environmental Microbiology*, 63:4:1489–1497.

Lee, W.H., D.G. Wahman, P.L. Bishop, and J.G. Pressman. 2011. "Free Chlorine and Monochloramine Application to Nitrifying Biofilm: Comparison of Biofilm Penetration, Activity, and Viability." *Environmental Science & Technology*, 45:4:1412–1419.

Lee, W.H., J.G. Pressman, and D.G. Wahman. 2018. "Three-dimensional Free Chlorine and Monochloramine Biofilm Penetration: Correlating Penetration with Biofilm Activity and Viability." *Environmental Science & Technology,* 52:4:1889–1898.

Leininger, S., T. Urich, M. Schloter, L. Schwark, J. Qi, G. W. Nicol, J.I. Prosser, S.C. Schuster, and C. Schleper. 2006. "*Archaea* Predominate Among Ammonia-Oxidizing Prokaryotes in Soils." *Nature,* 442:7104:806–809

Li, M., G. Wei, W. Shi, Z. Sun, H. Li, X. Wang, and Z. Gao. 2018. "Distinct Distribution Patterns of Ammonia-Oxidizing Archaea and Bacteria in Sediment and Water Column of the Yellow River Estuary." *Scientific Reports,* 8:1584.

Lieu, I.L., R.L. Wolfe, and E.G. Means III. 1993. "Optimizing Chloramine Disinfection for the Control of Nitrification." *Journal AWWA,* 85:2:84–90.

Lipponen, M.T.T., P.J. Martikainen, R.E. Vasara, K. Servomaa, O. Zacheus, and M.H. Kontro. 2004. "Occurrence of Nitrifiers and Diversity of Ammonia-Oxidizing Bacteria in Developing Drinking Water Biofilms." *Water Research,* 38:20:4424–4434.

Lipponen, M.T.T., M.H. Suutari, and P.J. Martikainen. 2002. "Occurrence of Nitrifying Bacteria and Nitrification in Finnish Drinking Water Distribution Systems." *Water Research,* 36:17:4319–4329.

Lv, H., C. Ji, J. Ding, L. Yu, and H. Cai. "High Levels of Zinc Affect Nitrogen and Phosphorus Transformation in Rice Rhizosphere Soil by Modifying Microbial Communities." *Plants (Basel),* 11:17:2271.

Madigan, M.T., J.M. Martinko, K.S. Bender, D.H. Buckley, and D.A. Stahl. 2019. *Brock Biology of Microorganisms.* 15th ed. Boston: Pearson.

Maixner, F., D.R. Noguera, B. Anneser, K. Stoecker, G. Wegl, M. Wagner, and H. Daims. 2006. "Nitrite Concentration Influences the Population Structure of *Nitrospira*-Like Bacteria." *Environmental Microbiology,* 8:8:1487–1495.

Matulewich, V.A., P.F. Strom, and M.S. Finstein. 1975. "Length of Incubation for Enumerating Nitrifying Bacteria Present in Various Environments." *Applied Microbiology,* 29:2:265–268.

McCaig, A.E., T.M. Embley, and J.I. Prosser. 1994. "Molecular Analysis of Enrichment Cultures of Marine Ammonia Oxidisers." *FEMS Microbiology Letters,* 120:3:363–368.

McCarty, P.L. 1964. "Thermodynamics of Biological Synthesis and Growth." In: *Proc. of the 11*th *International Conference on Water Pollution Research.*

McGuire, M.J., N.I. Lieu, and M.S. Pearthree. 1999. "Using Chlorite Ion to Control Nitrification." *Journal AWWA,* 91:10:52–61.

McGuire, M.J., X. Wu, N.K. Blute, D. Askenazier, and G. Qin. 2009. "Prevention of Nitrification Using Chlorite Ion: Results of a Demonstration Project in Glendale, Calif." *Journal AWWA,* 101:10:47–59.

McTavish, H., J.A. Fuchs, and A.B. Hooper. 1993. "Sequence of the Gene Coding for Ammonia Monooxygenase in *Nitrosomonas europaea*." *Journal of Bacteriology*, 175:8:2436–2444.

Meinhardt, K.A., A. Bertagnolli, M.W. Pannu, S.E. Strand, S.L. Brown, and D.A. Stahl. 2015. "Evaluation of Revised Polymerase Chain Reaction Primers for More Inclusive Quantification of Ammonia-oxidizing Archaea and Bacteria." *Environmental Microbiology Reports*, 7:2:354–363.

Mobarry, B.K., M. Wagner, V. Urbain, B.E. Rittmann, and D.A. Stahl. 1996. "Phylogenetic Probes for Analyzing Abundance and Spatial Organization of Nitrifying Bacteria." *Applied Environmental Microbiology*, 62:6:2156– 2162.

Murray, R.G.E., and S.W. Watson. 1965. "Structure of *Nitrocystis oceanus* and Comparison With *Nitrosomonas* and *Nitrobacter*." *Journal of Bacteriology*, 89:6: 1594–1609.

Narkis, N., and Y. Kott. 1992. "Comparison Between Chlorine Dioxide and Chlorine for Use as a Disinfectant of Wastewater Effluents." *Water Science & Technology*, 26:7–8:1483–1492.

Nocker A., and A.K. Camper. 2009. "Novel Approaches Toward Preferential Detection of Viable Cells Using Nucleic Acid Amplication Techniques." *FEMS Microbiology Letters*, 291:2:137–142.

Nogueira, R., L.F. Melo, U. Purkhold, S. Wuertz, and M. Wagner. 2002. "Nitrifying and Heterotrophic Population Dynamics in Biofilm Reactors: Effects of Hydraulic Retention Time and the Presence of Organic Carbon." *Water Research*, 36:2:469–481.

O'Neil, J.G., and J.F. Wilkinson. 1977. "Oxidation of Ammonia by Methane-Oxidizing Bacteria and the Effect of Ammonia on Methane Oxidation." *Journal of General Microbiology*, 100:407–412.

Odell, L.H., G.J. Kirmeyer, A. Wilczak, J.G. Jacangelo, J.P. Marcinko, and R.L. Wolfe. 1996. "Controlling Nitrification in Chloraminated Systems." *Journal AWWA*, 88:7:86–98.

Okano, Y., K.R. Hristova, C.M. Leutenegger, L.E. Jackson, R.F. Denison, B. Gebreyesus, D. Lebauer, and K.M. Scow. 2004. "Application of Real-Time PCR to Study Effects of Ammonium on Population Size of Ammonia-Oxidizing Bacteria in Soil." *Applied and Environmental Microbiology*, 70:2:1008–1016.

Oldenburg, P.S., J.M. Regan, G.W. Harrington, and D.R. Noguera. 2002. "Kinetics of *Nitrosomonas europaea* Inactivation by Chloramine." *Journal AWWA*, 94:10:100–110.

Olson, R.J. 1981. "Differential Photoinhibition of Marine Nitrifying Bacteria—a Possible Mechanism for the Formation of the Primary Nitrite Maximum." *Journal of Marine Research*, 39:2:227–238.

Oren, A., and G.M. Garrity. 2021. "Valid Publication of the Names of Forty-two Phyla of Prokaryotes." *International Journal of Systematic and Evolutionary Microbiology*, 71:10: doi: 10.1099.

Painter, H.A. 1970. "A Review of Literature on Inorganic Nitrogen Metabolism in Microorganisms." *Water Research*, 4:6:393–450.

Painter, H.A. 1986. "Nitrification in the Treatment of Sewage and Wastewater." In: *Nitrification*. J.I. Prosser, editor. Oxford: IRL Press:185–211.

Palomo, A., S.J. Fowler, A. Gülay, S. Rasmussen, T. Sicheritz-Ponten, and B.F. Smets. 2016. "Metagenomic Analysis of Rapid Gravity Sand Filter Microbial Communities Suggests Novel Physiology of *Nitrospira* spp." *ISME Journal*, 10:11:2569–2581.

Pintar, K., R.M. Slawson, and P.M. Huck. 2000. "Investigation of Conditions and Control Strategies Influencing Nitrification in a Chloraminated Bench-Scale Distribution System." In: *Proc. of the AWWA Water Quality Technology Conference*. Denver: AWWA.

Pinto, A.J., D.N. Marcus, U.Z. Ijaz, Q.M. Bautista-de Lose Santos, G.J. Dick, and L. Raskin. 2016. "Metagenomic Evidence for the Presence of Comammox Nitrospira-like Bacteria in a Drinking Water System." *mSphere*, 1:1:e00054-15.

Pjevac, P., C. Schauberger, L. Poghosyan, C.W. Herbold, M.A.H.J. van Kessel, A. Daebeler, M. Steinberger, M.S.M. Jetten, S. Lücker, M. Wagner, and H. Daims. 2017. "AmoA-targeted Polymerase Chain Reaction Primers for the Specific Detection and Quantification of Comammox *Nitrospira* in the Environment." *Frontiers in Microbiology*, 4:8:1508.

Pommerening-Röser, A., G. Rath, and H.P. Koops. 1996. "Phylogenetic Diversity Within the Genus *Nitrosomonas*." *Systematic and Applied Microbiology*, 19:3:344–351.

Poth, M., and D.D. Focht. 1985. "15N Kinetic Analysis of N2O Production by *Nitrosomonas europaea*: An Examination of Nitrifier Denitrification." *Applied and Environmental Microbiology*, 49:5:1134–1141.

Prinčič, A., I. Mahne, F. Megušar, E.A. Paul, and J.M. Tiedje. 1998. "Effects of pH and Oxygen and Ammonium Concentrations on the Community Structure of Nitrifying Bacteria from Wastewater." *Applied and Environmental Microbiology*, 64:10:3584–3590.

Prosser, J.I. 1990. "Autotrophic Nitrification in Bacteria." In: *Advances in Microbial Physiology*, Vol. 30. A.H. Rose and D.H. Tempest, editors. New York: Academic Press:125–181.

Purkhold, U., A. Pommerening-Röser, S. Juretschko, M.C. Schmid, H.P. Koops, and M. Wagner. 2000. "Phylogeny of All Recognized Species of Ammonia Oxidizers Based on Comparative 16S rRNA and amoA Sequence Analysis: Implications for Molecular Diversity Surveys." *Appl. Environ. Microbiol.*, 66:12:5368–5382.

Qin, W., K.R. Heal, R. Ramdasi, J.N. Kobelt, W. Martens-Habbena, A.D. Bertagnolli, S.A. Amin, C.B. Walker, H. Urakawa, M. Könneke, A.H. Devol, J.W. Moffett, E.V. Armbrust, G.J. Jensen, A.E. Ingalls, and D.A. Stahl. 2017. "*Nitrosopumilus maritimus* gen. nov., sp. nov., *Nitrosopumilus cobalaminigenes* sp. nov., *Nitrosopumilus oxyclinae* sp. nov., and *Nitrosopumilus ureiphilus* sp. nov., Four Marine Ammonia-oxidizing Archaea of the Phylum Thaumarchaeota." *International Journal of Systematic and Evolutionary Microbiology*, 67:12:5067–5079.

Quinlan, A.V. 1980. "The Thermal Sensitivity of Nitrification as a Function of the Nitrogen Substrate." *Water Research*, 14:10:1501–1507.

Quinlan, A.V. 1984. "Prediction of the Optimum pH for Ammonia-N Oxidation by *Nitrosomonas europaea* in Well-Aerated Natural and Domestic-Waste Waters." *Water Research*, 18:5:561–566.

Rainey, F.A., D.P. Kelly, E. Stackebrandt, J. Burghardt, A. Hiraishi, Y. Katayama, and A.P. Wood. 1999. "A Re-evaluation of the Taxonomy of *Paracoccus denitrificans* and a Proposal for the Combination *Paracoccus pantotrophus* Comb. Nov." *International Journal of Systematic Bacteriology*, 49:Pt 2:645–651.

Regan, J.M. 2001. *Microbial Ecology of Nitrification in Chloraminated Drinking Water Distribution Systems*. Doctoral dissertation, University of Wisconsin, Madison, Wisc.

Regan, J.M., A.Y. Cho, S. Kim, and C.D. Smith. 2004. "Monitoring Nitrification in Chloraminated Systems Using Molecular Detection Strategies." In: *Proc. of the AWWA Water Quality Technology Conference*. Denver: AWWA.

Regan, J.M., A.Y. Cho, S. Kim, and C.D. Smith. 2006. *Monitoring Ammonia-Oxidizing Bacteria in Chloraminated Distribution Systems*. Draft Final Report. Denver: Awwa Research Foundation.

Regan, J.M., G.W. Harrington, H. Baribeau, R. De Leon, and D.R. Noguera. 2003. "Diversity of Nitrifying Bacteria in Full-Scale Chloraminated Distribution Systems." *Water Research*, 37:1:197–205.

Regan, J.M., G.W. Harrington, and D.R. Noguera. 2002. "Ammonia- and Nitrite-Oxidizing Bacterial Communities in a Pilot-Scale Chloraminated Drinking Water Distribution System." *Applied and Environmental Microbiology*, 68:1:73–81.

Ridgway, H.F., and B.H. Olson. 1982. "Chlorine Resistance Patterns of Bacteria from Two Drinking Water Distribution Systems." *Applied and Environmental Microbiology*, 44:4:972–987.

Rittmann, B.E., J.M. Regan, and D.A. Stahl. 1994. "Nitrification as a Source of Soluble Organic Substrate in Biological Treatment." *Water Science & Technology*, 30:6:1–8.

Robertson, L.A., E.W.J. van Niel, R.A.M. Torremans, and J.G. Kuenen. 1988. "Simultaneous Nitrification and Denitrification in Aerobic Chemostat Cultures of *Thiosphaera pantotropha*." *Applied and Environmental Microbiology*, 54:11:2812–2818.

Rotthauwe, J.H., K.P. Witzel, and W. Liesack. 1997. "The Ammonia Monooxygenase Structural Gene amoA as a Functional Marker: Molecular Fine-Scale Analysis of Natural Ammonia-Oxidizing Populations." *Applied and Environmental Microbiology*, 63:12:4704– 4712.

Sarker, D.C., C.M. Patel, A. Heitz, and A.H.M.F. Anwar. 2018. "Evaluation of Zinc and Copper for Co-inhibition of Nitrification in Mild Nitrified Drinking Water." *Journal of Environmental Chemical Engineering*, 6:2:2939–2943.

Schimel, J.P., M.K. Firestone, and K.S. Kill-ham. 1984. "Identification of Heterotrophic Nitrification in a Sierran Forest Soil." *Applied and Environmental Microbiology*, 48:4:802–806.

Schmidt, I., R.J. van Spanning, and M.S. Jetten. 2004. "Denitrification and Ammonia Oxidation by *Nitrosomonas europaea* Wild-Type, and NirK- and NorB-Deficient Mutants." *Microbiology*, 150: Pt 12:4107–14.

Schramm, A., D. de Beer, M. Wagner, and R. Amann. 1998. "Identification and Activities in Situ of Nitrosospira and Nitrospira spp. as Dominant Populations in a Nitrifying Fluidized Bed Reactor." *Applied and Environmental Microbiology*, 64:9:3480–3485.

Schramm, A., L.H. Larsen, N.P. Revsbech, N.B. Ramsing, R. Amann, and K.H. Schleifer. 1996. "Structure and Function of a Nitrifying Biofilm as Determined by in Situ Hybridization and the Use of Microelectrodes." *Applied and Environmental Microbiology*, 62:12:4641–4647.

Scott, D., M.I. Van Dyke, W.B. Anderson, and P.M. Huck. 2015. "Influence of Water Quality on Nitrifier Regrowth in Two Full-Scale Drinking Water Distribution Systems." *Canadian Journal of Microbiology*, 61:12:965–976.

Skadsen, J. 1993. "Nitrification in a Distribution System." *Journal AWWA*, 85:7:95–103.

Soriano, S., and N. Walker. 1968. "Isolation of Ammonia-Oxidizing Autotrophic Bacteria." *Journal of Applied Bacteriology*, 31:4:493–497.

Sorokin, D.Y., S. Lücker, D. Vejmelkova, N.A. Kostrikina, R. Kleerebezem, W.I. Rijpstra, J.S. Damsté, D. Le Paslier, G. Muyzer, M. Wagner, M.C. van Loosdrecht, and H. Daims. 2012. "Nitrification Expanded: Discovery, Physiology, and Genomics of a Nitrite–Oxidizing Bacterium from the Phylum Chloroflexi." *ISME Journal*, 6:12:2245–2256.

Speitel Jr., G.E., B.M. Bayer, and R. Kannappan, 2010. *Significance of Trihalomethanes in Preventing Distribution System Nitrification in Chloraminated Waters*. Water Research Foundation, Denver, CO.

Speitel Jr., G.E., L.E. Katz, D.G. Wahman, and J.L. Fairey, 2005. *Cometabolism of Trihalomethanes in Nitrifying Biofilters*. Water Research Foundation, Denver, CO.

Stahl, D.A., and J.R. de la Torre. 2012. "Physiology and Diversity of Ammonia-Oxidizing Archaea." *Annual Review of Microbiology*, 66:1:83–101.

Starke, J.A., S. Yilmaz, A. Gorzalski, D. R. Noguera, and G. W. Harrington. 2013. Characterizing the Microbial Community Responsible for Nitrification. Web Report 4165. Water Research Foundation: Denver, CO.

Stein, L.Y. and D. Arp. 1998. "Ammonium Limitations Results in the Loss of Ammonia-Oxidizing Activity in *Nitrosomonas europaea*." *Applied and Environmental Microbiology*, 64:4:1514–1521.

Stephen, J.R., G.A. Kowalchuk, M.A.V. Bruns, A.E. McCaig, C.J. Phillips, and T.M. Embley. 1998. "Analysis of β-Subgroup Proteobacterial Ammonia Oxidizer Populations in Soil by Denaturing Gradient Gel Electrophoresis Analysis and Hierarchical Phylogenetic Probing." *Applied and Environmental Microbiology*, 64:8:2958–2965.

Stephen, J.R., A.E. McCaig, Z. Smith, J.I. Prosser, and T.M. Embley. 1996. "Molecular Diversity of Soil and Marine 16S rRNA Gene Sequences Related to β-Subgroup Ammonia-Oxidizing Bacteria." *Applied and Environmental Microbiology*, 62:11:4147–4154.

Stewart, M.H., and B.H. Olson. 1996. "Bacterial Resistance to Potable Water Disinfectants." In: *Modeling Disease Transmission and Its Prevention by Disinfection*. C.J. Hurst, editor. Cambridge: Cambridge University Press.

Stewart, M.H., and N.I. Lieu. 1997. "Nitrification in Chloraminated Drinking Water and Its Association with Biofilms." In: *Proc. of the AWWA Water Quality Technology Conference*. Denver: AWWA.

Su, J.F., K. Zhang, T.L. Huang, G. Wen, L. Guo, and S.F. Yang. 2015 "Heterotrophic nitrification and aerobic denitrification at low nutrient conditions by a newly isolated bacterium, *Acinetobacter* sp. SYF26." *Microbiology*, 161:Pt 4:829–837.

Sundermeyer-Klinger, H., W. Meyer, B. Warninghoff, and E. Bock. 1984. *Arch. Microbiol.*, 140:153.

Suwa, Y., Y. Imamura, T. Suzuki, T. Tashiro, and Y. Urushigawa. 1994. "Ammonia-Oxidizing Bacteria with Different Sensitivities to $(NH_4)_2SO_4$ in Activated Sludges." *Water Research*, 28:7:1523–1532.

Suzuki, I., U. Dular, and S.C. Kwok. 1974. "Ammonia or Ammonium Ion as Substrate for Oxidation by *Nitrosomonas europaea* Cells and Extracts." *Journal of Bacteriology*, 120:1:556–558.

Tan, B., C. Ng, J.P. Nshimyimana, L.L. Loh, K.Y. Gin, and J.R. Thompson. 2015. "Next-generation Sequencing (NGS) for Assessment of Microbial Water Quality: Current Progress, Challenges, and Future Opportunities." *Frontiers in Microbiology*, 6:1027.

Tanaka, Y., Y. Fukumori, and T. Yamanaka. 1983. "Purification of cytochrome a 1 c 1 from *Nitrobacter agilis* and characterization of nitrite oxidation system of the bacterium." *Archives of Microbiology*, 135: 265–271.

Tarre, S., and M. Green. 2004. "High-Rate Nitrification at Low pH in Suspended- and Attached-Biomass Reactors." *Applied and Environmental Microbiology*, 70:11: 6481–6487.

Tourna, M., M. Stieglmeier, A. Spang, M. Könneke, A. Schintlmeister, T. Urich, M. Engel, M. Schloter, M. Wagner, A. Richter, and C. Schleper. 2011. "*Nitrososphaera viennensis*, an Ammonia Oxidizing Archaeon From Soil." *PNAS*, 108:20:8420–8425.

USEPA (United States Environmental Protection Agency). 1985. *Ambient Water Quality Criteria for Ammonia*. EPA/440/585/001. Washington: USEPA.

USEPA (United States Environmental Protection Agency). 1993. *Manual: Nitrogen Control*. EPA/625/R-93/010 Washington: USEPA Office of Water.

Valentine, R.L., K. Ozekin, and P.J. Vikesland. 1998. *Chloramine Decomposition in Distribution System and Model Waters*. Denver, Colo.: AWWA Research Foundation and American Water Works Association.

van der Aa, L.T.J., L.C. Rietveld, and J.A.M.H. Hofman. 2000. "Modeling Nitrification in Rapid Sand Filters: Phosphate Limitation at Low Temperatures." Poster presentation at AWWA Water Quality Technology Conference, Salt Lake City, Utah, November 5–9.

van der Wielen, P.W.J.J., S. Voost, and D. van der Kooij. 2009. "Ammonia-oxidizing Bacteria and Archaea in Groundwater Treatment and Drinking Water Distribution Systems." *Applied and Environmental Microbiology*, 75:14:4687-4695.

van Kessel, M.A.H.J., D.R. Speth, M. Albertsen, P.H. Nielsen, H.J.M. Op den Camp, B. Kartal, M.S.M. Jetten, and S. Lücker. 2015. "Complete Nitrification by a Single Microorganism." *Nature*, 528:7583:555–559.

Verstraete, W., and D.D. Focht. 1977. "Biochemical Ecology of Nitrification and Denitrification." In: *Advances in Microbial Ecology*, Vol. 1. M. Alexander, editor. New York: Plenum Press: pp. 135–198.

Verstraete, W., and M. Alexander. 1973. "Heterotrophic Nitrification in Samples of Natural Ecosystems." *Environmental Science & Technology*, 7:1:39–42.

Verstraete, W., and M. Alexander. 1986. "Heterotrophic Nitrification." In: *Nitrification*. J.I. Prosser, editor. Washington: IRL Press.

Völsch, A., W.F. Nader, H.K. Geiss, G. Nebe, and C. Birr. 1990. "Detection and Analysis of Two Serotypes of Ammonia-Oxidizing Bacteria in Sewage Plants by Flow Cytometry." *Applied and Environmental Microbiology*, 56:8:2430–2435.

Voytek, M.A., and B.B. Ward. 1995. "Detection of Ammonium-Oxidizing Bacteria of the Beta-Subclass of the Class Proteobacteria in Aquatic Samples with the PCR." *Applied and Environmental Microbiology*, 61:4:1444–1450.

Waak, M.B., R.M. Hozalski, C. Hallé, and T. M. LaPara. 2019. "Comparison of the Microbiomes of Two Drinking Water Distribution Systems—with and Without Residual Chloramine Disinfection." *Microbiome*, 7:1:87.

Wagner, M., G. Rath, R. Amann, H.P. Koops, and K.H. Schleifer. 1995. "In Situ Identification of Ammonia-Oxidizing Bacteria." *Systematic and Applied Microbiology*, 18:2:251–264.

Wagner, M., G. Rath, H.P. Koops, J. Flood, and R. Amann. 1996. "In Situ Analysis of Nitrifying Bacteria in Sewage Treatment Plants." *Water Science & Technology*, 34:1–2:237–244.

Wahman, D.G., A.E. Henry, L.E. Katz, and G.E. Speitel Jr. 2006. "Cometabolism of Trihalomethanes by Mixed Culture Nitrifiers." *Water Research*, 40:18:3349–3358.

Wahman, D.G., K.A. Schrantz, and J.G. Pressman. 2010. "Determination of the Effects of Medium Composition on the Monochloramine Disinfection Kinetics of *Nitrosomonas europaea* by Propidium Monoazide Quantitative PCR and Live/Dead *Bac*Light Methods." *Applied and Environmental Microbiology*, 76:24:8277–8280.

Wahman, D.G., G.E. Speitel Jr., and L.E. Katz. 2005. "Cometabolism of Trihalomethanes by *Nitrosomonas europaea*." *Applied and Environmental Microbiology*, 71:12:7980–7986.

Wahman, D.G., and J.G Pressman. 2014. "Nitrification in Chloraminated Drinking Water Distribution Systems: Factors Affecting Occurrence." In: *Comprehensive Water Quality and Purification*. Ahuja S, editor. Waltham: Elsevier:283–294.

Wahman, D.G., K.A. Wulfeck-Kleier, and J.G. Pressman. 2009. "Monochloramine Disinfection Kinetics of *Nitrosomonas europaea* by Propidium Monoazide Quantitative PCR and Live/Dead *Bac*Light Methods." *Applied and Environmental Microbiology*, 75:17:5555–5562.

Wang, H., C.R. Proctor, M.A. Edwards, M. Pryor, J.W. Santo Domingo, H. Ryu, A.K. Camper, A. Olson, and A. Pruden. 2014. "Microbial Community Response to Chlorine Conversion in a Chloraminated Drinking Water Distribution System." *Environmental Science & Technology*, 48:18:10624–10633.

Wang, Y., L. Ma, Y. Mao, X. Jiang, Y. Xia, K. Yu, B. Li, and T. Zhang. 2017. "Comammox in Drinking Water Systems." *Water Research*, 116:332–341.

Ward, B.B. 1987. "Kinetic Studies on Ammonia and Methane Oxidation by *Nitrosococcus oceanus*." *Archives of Microbiology*, 147:126–133 .

Watson, S.W. 1965. "Characteristics of a Marine Nitrifying Bacterium, *Nitrosocystis oceanus* sp. n.1." *Limnology and Oceanography*, 10:R274–R289.

Watson, S.W. 1971. "Re-isolation of *Nitrosospira briensis*, S. Winogradsky, and H. Winogradsky, 1933." *Archiv für Mikrobiologie*, 75:179–188.

Watson, S.W., E.E. Bock, H. Harms, H.P. Koops, and A.B. Hooper. 1989. "Nitrifying Bacteria." In: *Bergey's Manual of Systematic Bacteriology*, Vol. 3. J.T. Saley, M.P. Bryant, N. Pfennig, and J.G. Holt, eds. Baltimore, Md.: Williams and Wilkins.

Watson, S.W., L.B. Graham, C.C. Remsen, and F.W. Valois. 1971. "A Lobular, Ammonia-Oxidizing Bacterium, *Nitrosolobus multiformis* nov. gen. nov. sp." *Archiv für Mikrobiologie*, 76:3:183–203.

Watson, S.W., and M. Mandel. 1971. "Comparisons of the Morphology and Deoxyribonucleic Acid Components of 27 Strains of Nitrifying Bacteria." *Journal of Bacteriology*, 107:2:563–569.

Watson, S.W., F.W. Valos, and J.B. Waterbury. 1981. "The Family *Nitrobacteraceae*." In: *The Prokaryotes*. M.P. Starr, H. Stolp, H.G. Trüper, A. Balows, and H.G. Schlegel, eds. New York: Springer-Verlag.

Watson, S.W. and J.B. Waterbury. 1971. "Characteristics of Two Marine Nitrite-Oxidizing Bacteria, *Nitrospina gracilis* nov. gen. nov. sp. and *Nitrococcus mobilis* nov. gen. nov. sp." *Archiv für Mikrobiologie*, 77:203–230.

White, G.C. 1999. *Handbook of Chlorination and Alternative Disinfectants*, 4th ed. New York: John Wiley & Sons.

White, B., and M. Adams. 2010. "Battling Nitrification With Blacklights." *Opflow*, 99:2:10–13.

Wilczak, A., J.G. Jacangelo, J.P. Marcinko, L.H. Odell, G.J. Kirmeyer, and R.L. Wolfe. 1996. "Occurrence of Nitrification in Chloraminated Water Distribution Systems." *Jour. AWWA*, 88:7:74–85.

Wolfe, R.L., and N.I. Lieu. 2001. "Nitrifying Bacteria in Drinking Water." In: *Encyclopedia of Environmental Microbiology*, Vol. 4. G. Bitton, ed. New York: John Wiley & Sons.

Wolfe, R.L., N.I. Lieu, G. Izaguirre, and E.G. Means III. 1990. "Ammonia-Oxidizing Bacteria in a Chloraminated Distribution System: Seasonal Occurrence, Distribution, and Disinfection Resistance." *Applied and Environmental Microbiology*, 56:2:451–462.

Wolfe, R.L., E.G. Means III, M.K. Davis, and S.E. Barrett. 1988. "Biological Nitrification in Covered Reservoirs Containing Chloraminated Water." *Journal AWWA*, 80:9:109–114.

Wolfe, R.L., and B. H. Olson. 1985. "Inability of Laboratory Models to Accurately Predict Field Performance Disinfectants." In: *Water Chlorination: Environmental Impact and Health Effects*. R.L. Jolley et al., eds. Chelsea, Mich.: Lewis Publishers.

Wong-Chong, G.M., and R.C. Loehr. 1975. "The Kinetics of Microbial Nitrification." *Water Research*, 9:12:1099–1106.

Wu, Y., X. Ke, M. Hernández, B. Wang, M.G. Dumont, Z. Jia and R. Conrad. 2013. "Autotrophic Growth of Bacterial and Archaeal Ammonia Oxidizers in Freshwater Sediment Microcosms Incubated at Different Temperatures." *Applied and Environmental Microbiology*, 79:9: 3076-3084.

Yoshioka, T., and Y. Saijo. 1984. "Photoinhibition and Recovery of NH_4^+-Oxidizing Bacteria and NO_2^--Oxidizing Bacteria." *Journal of General and Applied Microbiology*, 30:3:151–166.

You, J., A. Das, E.M. Dolan, and Z. Hu. 2009. "Ammonia-Oxidizing *Archaea* Involved in Nitrogen Removal." *Water Research*, 43:7:1801–1809.

Yu, C., L. Hou, Y. Zheng, M. Liu, G. Yin, J. Gao, C. Liu, Y. Chang, and P. Han. 2018. "Evidence for Complete Nitrification in Enrichment Culture of Tidal Sediments and Diversity Analysis of Clade A Comammox *Nitrospira* in Natural Environments." *Applied Microbiology and Biotechnology*, 102:21:9363–9377.

Zhang, Y., and M. Edwards. 2009. "Accelerated Chloramine Decay and Microbial Growth by Nitrification in Premise Plumbing." *Journal AWWA*, 101:11:51–62.

Zhang, Y., and M. Edwards. 2010. "Nutrients and Metals Effects on Nitrification in Drinking Water Systems." *Journal AWWA*, 102:7:56–66.

Zhang, Y., M. Edwards, A. Pinto, N. Love, A. Camper, M. Rahman, and H. Baribeau. 2010. *Effect of Nitrification on Corrosion in the Distribution System*. Denver, Colo.: Water Research Foundation and U.S. Environmental Protection Agency.

Zhang, Y., A. Griffin, and M. Edwards. 2008. "Nitrification in Premise Plumbing: Role of Phosphate, pH and Pipe Corrosion." *Environmental Science & Technology*, 42:12:4280–4284.

Zhang, Y., A. Griffin, M. Rahman, A. Camper, H. Baribeau, and M. Edwards. 2009. "Lead Contamination of Potable Water Due to Nitrification." *Environmental Science & Technology*, 43:6:1890–1895.

Zhang, Y., and W.-T. Liu. 2019. "The Application of Molecular Tools to Study the Drinking Water Microbiome—Current Understanding and Future Needs." *Critical Reviews in Environmental Science and Technology*, 49:13:1188–1235.

Zumft, W.G. 1992. "The Denitrifying Prokaryotes." In: *The Prokaryotes*, 2nd ed. A. Balows, H.G. Trüper, M. Dworkin, W. Harder, and K.H. Schleifer, eds. New York: Springer-Verlag.

Chapter 6

Monitoring for Nitrification Prevention and Control

Hélène Baribeau, Ashley Pifer, Matt Umberg, Amlan Ghosh, Ken Thompson

INTRODUCTION

The primary goals of nitrification monitoring are to detect the onset of a nitrification incident as early as possible, contain the incident within the smallest possible area, and take response actions to restore distribution system water quality to normal levels as quickly as possible. These programs can also help to preserve general water quality in the distribution system, maintain compliance with the drinking water regulations, and inform treatment and distribution system operations to prevent or limit the number of future incidents from occurring.

This chapter on nitrification monitoring begins with a discussion on the regulatory implications of nitrification episodes and regulatory requirements for development and maintenance of nitrification monitoring and control plans in states where such requirements are in place. The chapter discusses the components of a nitrification monitoring program, including parameters to monitor and at which locations and frequencies. In addition to grab samples, recent advances in monitoring strategies, including online analyzers to provide data from the field in near real-time, are discussed, followed by analytical methods for the proposed chemical and microbial monitoring parameters. Case studies of effective plans that have been developed by water systems are presented at the end of the chapter.

The following concepts should be considered when developing or reviewing nitrification prevention and control plans. With the exception of nitrite concentrations and nitrifying bacteria counts, changes in nitrification indicators can also result from challenges other than nitrification, such as increasing water age, microbial growth, corrosion and metal release, and others. Thus, nitrification monitoring should rely on a combination of chemical and/or microbial indicators monitored at key locations throughout the distribution system. In addition, development of baseline conditions in the absence of nitrification is necessary to compare monitoring data with reference levels and assess whether nitrification is occurring. Lastly, each nitrification monitoring and control plan needs to be tailored to its distribution system's historical water quality, operational data, and individual characteristics.

Table 6-1 summarizes the key points from this chapter.

Table 6-1 Key points from Chapter 6

Regulatory requirements	• Although nitrification monitoring is not addressed at the federal level, several states require formal plans for monitoring and control of nitrification in chloraminated distribution systems. • Nitrite and nitrate present acute health effects and are regulated by federal and state regulations. However, compliance is based on monitoring at the points of entry (POE) in the distribution system and not in the distribution system or storage facilities where nitrification typically occurs. • The Surface Water Treatment Rule requires water systems to maintain at least a 0.2 mg/L-Cl_2 disinfectant residual at the POE and a detectable disinfectant residual in the distribution system or a heterotrophic plate count value of less than 500 colony forming units/mL in 95 percent of the distribution samples analyzed each month. Some states have chosen to set a definition for "detectable residual," with minimum disinfectant residual requirements greater than 0.2 mg/L-Cl_2. • There are no regulatory standards associated with either free or total ammonia,* although certain states require monitoring of free ammonia as part of a nitrification monitoring and control program.
Monitoring parameters–water quality	• Because of the difficulties in detecting and identifying nitrifying microorganisms, most water providers continue to rely on chemical indicators to detect nitrification. • The existence of nitrite (and nitrate in systems where the background nitrate concentration is low and stable) is the most direct indicator of nitrification in drinking water systems. Nitrite and nitrate are the products of nitrification, so nitrification is already in progress when increases in nitrite or nitrate can be measured. • Total chlorine or monochloramine, free ammonia, and nitrite are the most useful parameters for detection of conditions conducive to nitrification and for early detection of nitrification. High or increasing free ammonia concentrations mean that more substrate is available for nitrifying organisms. A decrease in free ammonia and total chlorine or monochloramine residual combined with an increase in nitrite and/or nitrate concentration can confirm the onset of nitrification. • In low-alkalinity waters, pH may decrease during nitrification events. • Increased water temperature, which creates ideal conditions for nitrifying microorganisms, may be used as a warning that nitrification is more likely to occur, particularly in water storage tanks where thermal stratification is occurring.
Monitoring parameters–operations	• Operational data—such as water age (which is linked to water demand and flow), storage tank cycling, pipe flushing, and valve position history—can provide information to help interpret nitrification monitoring results. • Operational or real-time distribution system hydraulic model output can supplement and inform interpretation of measured water quality data.
Data management	• Because nitrification can escalate rapidly, responses need to be rapid and based on proper interpretation of monitoring data, which requires prompt analysis of these data as soon as they become available. • Trending water quality parameters over time at each sampling site can provide an easy visual assessment of the onset of nitrification. • Data should be archived and accessible for analysis and interpretation if needed.

(continued)

Table 6-1 Key points from Chapter 6 (*Continued*)

Monitoring frequency	• Although monitoring frequency is largely system specific, weekly sampling for the most useful nitrification indicators is typically conducted, with more frequent sampling when nitrification is detected. • Monitoring frequency can vary by sampling location and season.
Monitoring locations	• At a minimum, nitrification monitoring should be conducted at POEs, in storage facilities, and locations in the distribution system where water age is high. Critical customers such as hospitals and active connections with neighboring water systems should also be considered. • Distribution system sampling sites are largely system specific but should capture the various pipe diameters, pipe materials, water ages, and pressure zones if the system is considered to have a high risk or history of nitrification. • Mixed zones—areas influenced by more than one water source or areas with a blend of disinfectants (free chlorine and chloramine) should be included. • An assessment of water quality at individual locations in the distribution system is essential for proper interpretation of nitrification data. • Designated sample taps are critical to the integrity of the monitoring program, data quality, and ability to view, interpret, and compare with historical data.
Water quality monitoring	• Monitoring can be accomplished with grab sampling or online analyzers based on a water system's budget and availability of staff to conduct sampling. Continuous online analyzers are available for pH, temperature, free and total chlorine, and other nitrification related parameters. • Grab sampling allows water systems flexibility in selecting monitoring locations and frequency and may be the most cost-effective option for some systems. However, online water quality analyzers can provide substantial benefits to water systems by facilitating rapid detection of and response to nitrification. • Online analyzers are available for several of the same parameters addressed by grab sampling, including nitrite, nitrate, chlorine residual, free ammonia, water temperature, pH, and many others. • The cost of online analyzers often limits their installation to sites that give the most information, offer multiple benefits, and present the highest risks. Typically, these sites include POEs, storage facilities, downstream control features, and distribution system sites with high water age. • Because of the high frequency of data collection provided by online analyzers, data management can become significant. Appropriate databases, algorithms to process the data, dashboards to visualize the information, and alerts when anomalies are detected should be part of the data management system of online analyzers. • Maintenance of online analyzers is critical; they should not be installed if routine maintenance cannot be ensured.
Analytical methods	• Analytical methods used to monitor nitrification indicators are easy to perform. A number of parameters (e.g., chlorine residual, water temperature, pH) should be analyzed in the field to ensure data integrity.

* Total ammonia includes free ammonia (NH_3 and NH_4^+) and ammonia that is combined with chlorine in all species of chloramine.

REGULATORY REQUIREMENTS

Nitrification in itself is not regulated at the federal or state level, but it can impact compliance with several regulatory requirements. Compliance with existing US Environmental Protection Agency (USEPA) rules, including the Surface Water Treatment Rule (SWTR), Long-Term 2 Enhanced Surface Water Treatment Rule (LT2ESWTR), Revised Total Coliform Rule (RTCR), and Stage 2 Disinfectants and Disinfection Byproducts Rule (DBPR), can be compromised if nitrification is occurring within the distribution system. In recent years, several states have implemented more stringent disinfectant residual requirements and are now requiring monitoring and management of nitrification in chloraminated distribution systems. Some examples of state-specific nitrification related regulations include the following.

- California: permit amendment process that allows addition of ammonia for chloramination and requires some water systems to develop and submit nitrification monitoring and control programs
- Illinois: Joint Committee on Administrative Rules, Administrative Code Section 604.140 Nitrification Action Plan
- Louisiana: Louisiana Administrative Code, Section 367, disinfectant residual monitoring and record keeping
- Pennsylvania: Pennsylvania Safe Drinking Water Regulations, Section 109.716(a): water systems that use chloramine or purchase water that contains chloramine required to develop a nitrification control plan
- Texas: Drinking Water Standards Governing Drinking Water Quality and Reporting Requirements for Public Water Systems, Texas Administrative Code, Rule 290.46(z): water systems that distribute chloraminated water required to create a nitrification action plan, more detailed monitoring requirements provided in Rule 290.110

Other state and local primacy agencies may have additional monitoring and reporting requirements related to nitrification. Public water systems should refer to their local regulatory agencies for compliance requirements and additional guidance related to disinfectant residuals and nitrification management.

Nitrification Action/Control Plans (NAPs/NCPs) generally require the development of water quality goals, system-specific monitoring protocols, identification of critical conditions (i.e., alert or action levels), and corresponding response actions. Specifically related to monitoring, a NAP/NCP typically includes the following requisites.

- Sampling sites: Sites may include routine coliform, disinfectant residual, or disinfection byproduct monitoring sites; average water age and maximum water age locations; sites that are impacted by blending of multiple sources; or special sites such as water storage tanks or system dead ends.
- Water quality parameters: Parameters to be monitored at each site may include total chlorine, monochloramine, free chlorine, free ammonia, total ammonia, nitrite, nitrate, heterotrophic plate count (HPC), pH, temperature, etc. To monitor specifically for nitrification, the HPC-R2A method is recommended. The advantages and drawbacks of different HPC analytical methods are discussed in the Analytical Methods section of this chapter. While chloraminated systems will not typically have free chlorine in their distribution system water, some chloraminated systems temporarily convert to free chlorine either in response to a nitrification event or as a scheduled (typically annual) preventive maintenance procedure. During those times, free chlorine, in addition to total chlorine, should be monitored.
- Analytical methods: For regulatory compliance, analyses of water quality parameters need to be performed by state-approved methods for drinking water.
- Sampling frequency: This includes baseline sampling frequency when there is no evidence of nitrification occurrence in the distribution system and enhanced frequency when an alert or action level condition has been identified. Baseline sampling frequencies also vary based on monitoring location. Sampling frequency is discussed at length later in this chapter.

As an example of NAP/NCP requirements, Texas's guidance can be downloaded from Texas Commission of Environmental Quality website.

Some states require water systems to submit NAPs/NCPs and corresponding monitoring results for review and approval. Other states require that water systems retain NAPs/NCPs onsite and make them available for inspection by the regulatory agency, upon request. Some states, such as Louisiana, also require that public water systems retain the NAP/NCP results onsite for a minimum of 5 years.

If there are source water, treatment, physical or operational changes that impact water flow, quality or residence time (e.g., main replacements, distribution system expansion, or addition of a new storage tank), the NAPs/NCPs need to be revised such that the most up-to-date and accurate system-specific information is reflected within these plans.

Individual Water Quality Parameter Regulations

Chlorine residual. The SWTR requires water systems to maintain at least a 0.2-mg/L as Cl_2 disinfectant residual at the point of entry (POE) and a detectable disinfectant residual in 95 percent of the distribution samples analyzed each month. If a detectable residual is not found, an HPC value of less than 500 colony forming units (cfu) per mL (using standard plate count agar) serves as the equivalent to a detectable residual for regulatory purposes. The disinfectant residual is monitored at the same locations and at the same frequency as the RTCR samples and is based on system population. Some states have chosen to define "detectable residual": for example, 0.2 mg/L-Cl_2 for chlorinated water and 0.6 mg/L-Cl_2 for chloraminated water in Florida or 0.2 mg/L-Cl_2 for chlorinated water and 0.5 mg/L-Cl_2 for chloraminated water in Texas. Wahman and Pressman (2015) summarized the various states that have directly or indirectly defined "detectable" free chlorine and/or total chlorine residual as a numeric concentration and have developed associated rules and regulations.

It is possible that while the water system complies with minimum disinfectant residual regulatory requirements, locations with low residuals could still experience nitrification or could be vulnerable to growth of nitrifying microorganisms.

There are no regulatory requirements to monitor at dead ends or finished water storage facilities for total chlorine, considering that the water at those locations does not represent water consumed by most customers. However, some states, such as Louisiana, now require daily monitoring of total chlorine residuals at maximum residence time (MRT) locations in the distribution system. Although these requirements may have been established for reasons other than nitrification prevention and control, they can contribute to limiting the onset of nitrification.

Ammonia. There are no regulatory requirements associated with either free or total ammonia at the federal level. However, some states require monitoring of free ammonia at the POE and in the distribution system on a weekly basis, without a specific numerical standard, with the overall goal of minimizing free ammonia concentration at the POE and throughout the distribution system.

Nitrite and nitrate. Current state and federal regulations require monitoring of nitrite and nitrate at the POE, not within the distribution system. The maximum contaminant level (MCL) for nitrite-N is 1.0 mg/L, the MCL for nitrate-N is 10 mg/L, and the regulatory limit for nitrite-N plus nitrate-N is 10 mg/L. These MCLs are based on public health concerns, primarily methemoglobinemia or "blue baby syndrome." Nitrite and nitrate are considered acute contaminants, and violation of these MCLs requires immediate public notification. For systems that add ammonia to form chloramines for distribution system residual disinfectant, the likelihood of reaching these levels is low but not impossible. High chloramine doses and low chlorine-to-ammonia-N mass ratios may lead to nitrite levels that may exceed 1.0 mg/L-N. From a regulatory perspective, the concern is generally the decrease in the total chlorine residual due to nitrification and resulting increases in microorganisms, including the possible presence of coliform bacteria and pathogens.

Heterotrophic plate counts. For compliance with federal regulations, if a system sample does not have a detectable chlorine residual, an HPC of less than 500 cfu/mL serves as a detectable chlorine residual per SWTR 40 CFR 141.74. The pour plate method with incubation at 35°C for 48 h should be used for regulatory compliance (additional information about analytical methods for HPCs is presented below). The advantages and drawbacks of this analytical method are discussed in the Analytical Methods section of this chapter.

MONITORING PARAMETERS

Various chemical, microbial, and operational parameters can be included in a nitrification monitoring plan to identify conditions conducive to the onset of nitrification or determine the extent of nitrification. These parameters can be categorized as

- extreme usefulness—provides data that can be used to make decisions;
- usefulness—aids in the interpretation of data; or
- limited usefulness—can support the conclusions derived from monitoring data results.

The usefulness of any parameter is also related to the accuracy of measurement and timeliness of the results. Time is critical for the early detection, response, and mitigation of a nitrification episode, so frequent grab sampling or the use of online water quality analyzers may be warranted, as discussed later in this chapter. If analytical methods are not applicable to field measurements and the turnaround time in the laboratory is long, the usefulness of the parameter is diminished. Parameters categorized as "very useful" should be monitored more frequently than other parameters and/or at more sampling sites than parameters considered "useful."

This section describes the monitoring parameters frequently included in nitrification monitoring plans. Tables 6-2 and 6-3 list the parameters described in this chapter to be monitored either at the treatment plant effluent/POE(s) or in the distribution system, respectively. At the treatment plant effluent and/or POE, monitoring for nitrification indicators can establish a baseline or reference level that can be used to evaluate data collected downstream in the distribution system. The usefulness of monitoring parameters at water storage facilities is system specific. For example, a water treatment plant practicing clearwell chlor(am)ination would be monitored as shown in Table 6-2, whereas Table 6-3 would better characterize the usefulness of monitoring parameters for water storage facilities in high-water age areas.

Note that the designations related to the usefulness of each parameter only relate to nitrification prevention and control. Some parameters, such as pH, would be characterized much differently if the discussion were about coagulation, corrosion control, or another water quality issue.

Excluding the parameters that are directly related to nitrification (e.g., nitrite, total chlorine or monochloramine, free ammonia), the designation is also largely system specific. A parameter considered of "limited usefulness" for one water system may be considered "useful" in another system. For example, total alkalinity could be considered "useful" in water systems with poorly buffered water because alkalinity is linked to pH stability, which in turn influences monochloramine stability.

When comparing the analytical methods that can be used to measure HPC, the low-nutrient R2A agar provides higher counts and is more representative of the drinking water environment (additional information is provided in the Analytical Methods section). HPC-R2A is therefore recommended over other HPC methods as a nitrification indicator. Nevertheless, the ability of HPC to indicate the onset or occurrence of nitrification is limited.

Table 6-2 Usefulness of monitoring parameters at the treatment plant effluent/points of entry for nitrification prevention and control

	Extreme Usefulness	Usefulness	Limited Usefulness
Water quality parameters	• Total chlorine or monochloramine • Free ammonia • pH • Free chlorine*	• Nitrate • Nitrite • Water temperature • Total ammonia† • Organic carbon‡	• Total alkalinity • Bromide • Oxidation reduction Potential • Dissolved oxygen • heterotrophic plate count (HPC) or HPC-R2A • Adenosine triphosphate
Operational parameters	Point of entry usage and flow		

* Extreme usefulness during breakpoint chlorination but not for routine monitoring
† To calculate the chlorine-to-ammonia-N ratio
‡ Total or dissolved organic carbon to evaluate disinfectant residual stability and/or assimilable or biodegradable organic carbon to evaluate the availability of substrate for heterotrophs that live in biofilms alongside nitrifiers

Table 6-3 Usefulness of monitoring parameters for distribution system nitrification monitoring

	Extreme usefulness	Usefulness	Limited usefulness
Water quality parameters	• Total chlorine or monochloramine • Free ammonia • Nitrite • Nitrate* • Temperature • Free chlorine†	• pH • Total Ammonia • heterotrophic plate count (HPC) or HPC-R2A • adenosine triphosphate	• Dissolved oxygen • Organic carbon • Total alkalinity • Nitrifying microorganisms§
Operational parameters		• Water age‡ • Storage facility cycling history • Distribution system flushing history • Valve position	

* Extreme usefulness if background nitrate-N is consistent and low
† Extreme usefulness during breakpoint chlorination but not for routine monitoring
‡ Limited usefulness unless available from operational or real-time distribution system hydraulic model
§ Limited usefulness until rapid, simple, and inexpensive enumeration methods are available

Chlorine-to-Ammonia Ratio

When selecting appropriate monitoring parameters (i.e., free chlorine versus total chlorine, monochloramine, and free ammonia), it is important to consider the chlorine-to-ammonia ratio and how it relates to the speciation of chlorine and ammonia. The chlorine-to-ammonia mass ratio is the ratio of the total chlorine residual to the total ammonia concentration and is traditionally calculated in units of mg/L-Cl_2 to mg/L-N. Total chlorine includes all species of chloramine (mono-, di-, tri-, and organo-) and free chlorine. Total ammonia includes free ammonia (NH_3 and NH_4^+) and ammonia that is combined with chlorine in all species of chloramine. Monochloramine is the desired species for drinking water disinfection due to its stability relative to the other inorganic chloramines. Water treatment facilities typically use a chlorine-to-ammonia mass ratio of 4.5:1–5.0:1.

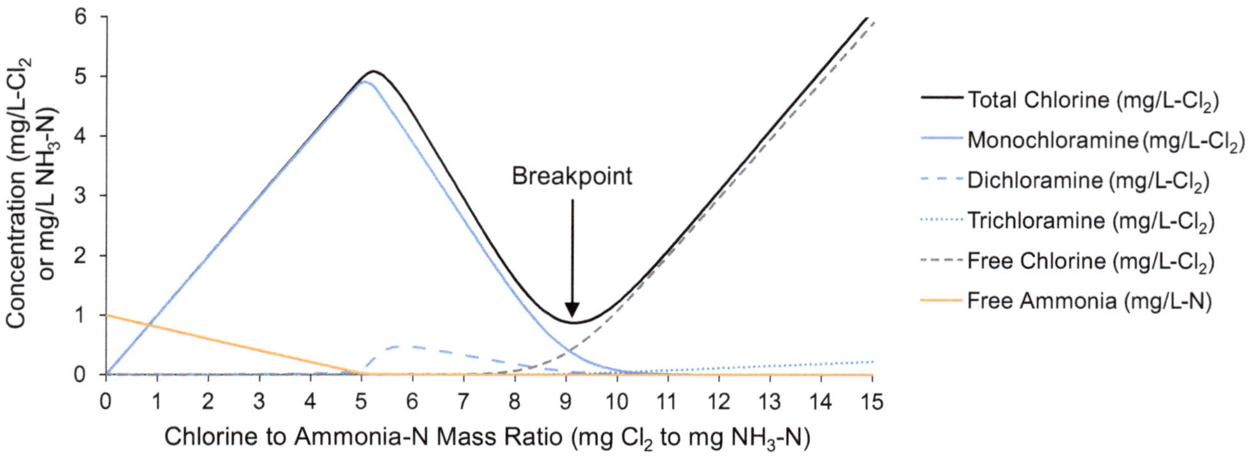

Figure 6-1 Sample breakpoint curve

NOTE: Total chlorine, monochloramine, dichloramine, trichloramine, and free chlorine are reported as mg/L-Cl_2, and free ammonia is reported as mg/L-NH_3-N. The curve was developed using the Chlorine Breakpoint Curve Simulator, Version 0.25 (Wahman 2017) with an initial free ammonia concentration of 1 mg/L-N, total alkalinity of 50 mg/L-$CaCO_3$, pH of 8.0, water temperature of 20°C, and a 30-min reaction time.

The breakpoint curve (see Figure 6-1) describes the equilibrium concentrations of the inorganic chloramine species in relation to the ratio of ammonia to chlorine present in organic free water. Although pH, water temperature, and inorganic constituents are not represented in this figure, they play an important role in determining chloramine speciation. Changes in total chlorine concentrations as part of the breakpoint chlorination curve are further discussed in Chapter 8. According to the breakpoint curve, as chlorine is added to water containing ammonia,

- At chlorine-to-ammonia-N (Cl_2:NH_3-N) mass ratios of 5:1 or less, and at the pH values found in drinking water systems, the predominant species will be monochloramine. This species has the lowest taste and odor threshold of the chloramine species and is the most stable. At mass ratios below 5:1, excess ammonia will be present.
- At mass ratios greater than 5:1 (up to the breakpoint), most of the total chlorine will still exist as monochloramine, but dichloramine will also be present.
- At mass ratios greater than 5:1, trichloramine (also called nitrogen trichloride) may form, depending on pH. At or below pH 4.4, trichloramine may form at lower mass ratios (i.e., closer to 5:1) (Palin 1950; Chapin 1929). Trichloramine may also form within the more typical range of drinking water pH (6–9) at mass ratios above the breakpoint (Wahman 2018; Kirmeyer et al. 2004).
- At mass ratios greater than the breakpoint, the chloramine species will be oxidized, and the additional chlorine added will be free chlorine, with trace amounts of trichloramine and nitrogen present.
- These theoretical ratios generally apply to the mixing of chlorine and ammonia in distilled water or other laboratory-grade waters but will vary in natural waters. Higher mass may be necessary to achieve breakpoint and obtain a free chlorine residual in field samples.

Furthermore, it should be noted that the breakpoint curve is time dependent. Until the reactants (chlorine and ammonia) and products (e.g., monochloramine) have reached equilibrium, various concentrations of reactants and products can occur simultaneously. This explains why free chlorine, ammonia, and monochloramine are observed when chlorine and ammonia are first blended. Dichloramine and/or trichloramine may also be present, depending on the chlorine and ammonia concentrations, pH, and reaction time.

In the presence of organics, free chlorine and inorganic chloramines also react with organic nitrogen to form low levels of organochloramines, which are not effective disinfectants or oxidants (AWWA 2006). Because of this, they are often referred to as "nuisance residuals." Organochloramines can form within the treatment plant and the distribution system, and some are more resistant to further oxidation, making them more stable than inorganic chloramines. As a result, the proportion of organochloramine to total chlorine residual increases as water age increases and the monochloramine residual approaches the detection limit. These compounds can be detected by total chlorine measurements and could indicate that a low but measurable disinfectant residual is present when it is not (APHA, AWWA, and WEF 2017; Wahman and Pressman 2015). Additional information is needed to better understand the prevalence of this issue in chloraminated distribution systems.

Ideally, chloramines would be formed such that only monochloramine would exist and there would be no free ammonia (i.e., exactly at a 5:1 mass ratio). This is rarely possible due to challenges with chemical feed, mixing limitations, and changes in background water quality that impact chlorine and/or ammonia demand. This being the case, most chloraminating systems target chlorine-to-ammonia ratios between 4.5:1 and 5:1 to minimize free ammonia and dichloramine concentrations. Chlorine-to-ammonia ratios are based on measured chlorine residual just prior to the point of ammonia addition and are typically used to calculate chlorine and ammonia doses in water treatment plants and booster chlor(am)ination facilities. The chlorine-to-ammonia ratio is not particularly useful for nitrification monitoring in the distribution system or storage facilities without booster chlor(am)ination stations; it is simpler to evaluate both parameters directly, as described in the following sections.

Total Chlorine and Monochloramine

In chloraminated distribution systems operating at or below a 5:1 chlorine-to-ammonia-N mass ratio, total chlorine and monochloramine residuals are approximately equal. Field methods are available and widely used for both total chlorine and monochloramine; however, only total chlorine methods are currently approved for regulatory compliance monitoring. Total chlorine residuals decrease over time due to decay and demand reactions. Free ammonia released from chloramine decomposition, coupled with lower disinfectant residuals, creates an environment more conducive to nitrification. Nitrification can begin in the presence of a total chlorine residual and accelerates the rate of total chlorine decay.

There is no universal threshold total chlorine residual required to prevent nitrification. Long-term monitoring is needed in chloraminated systems to identify system-specific target residuals, above which nitrification is unlikely, and lower residuals to be taken as alert and action levels. Preferably, the water system would determine summer and winter targets, alert, and action levels and account for typical water age at each sampling location. Target levels can be used to set chlorine and ammonia doses at treatment facilities. Alert levels should trigger increased monitoring, and action levels should trigger an operational response, such as increased reservoir/tank turnover or flushing. Chapter 7 presents detailed information on target total chlorine residuals to prevent nitrification.

Free chlorine

According to the breakpoint curve in Figure 6-1, free chlorine and monochloramine do not coexist at mass ratios below 5:1 or greater than the breakpoint. However, as mentioned earlier, the breakpoint curve represents a condition in which the reactants and products have reached equilibrium. Before equilibrium is reached (e.g., immediately after chlorine and ammonia are injected into the water), both free chlorine and monochloramine will be present at mass ratios greater than the breakpoint, as shown in Figure 6-2 for reaction times of 0, 3, and 30 min. The time to reach equilibrium will vary based on the time to achieve

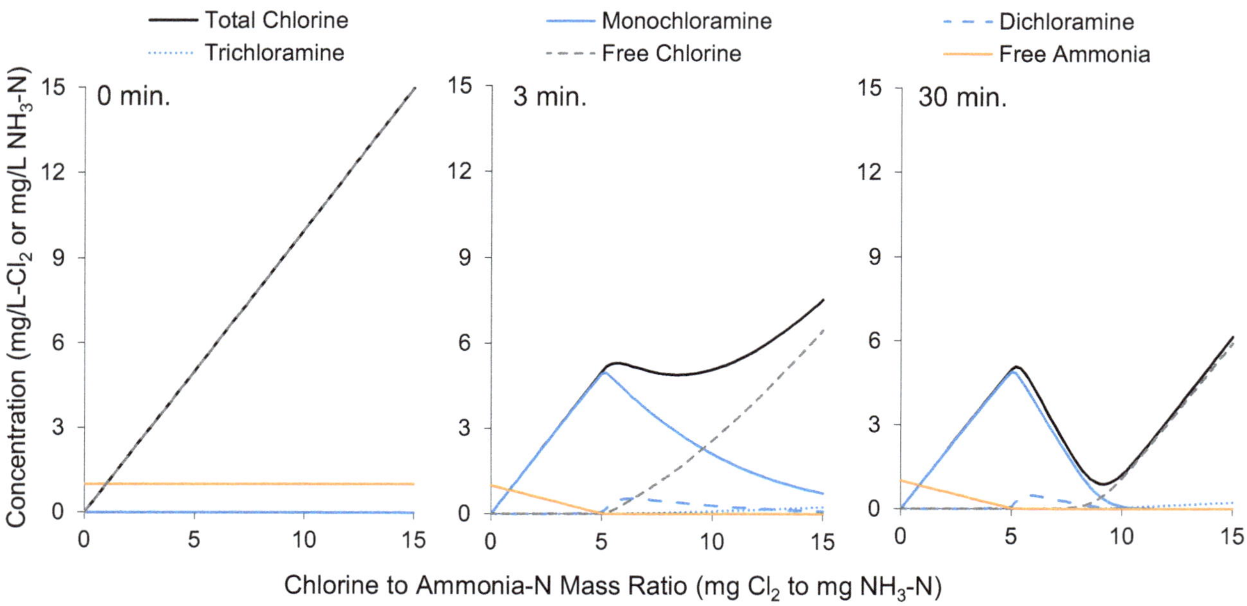

Figure 6-2 Sample breakpoint curve at 0-, 3-, and 30-min reaction times

NOTE: Total chlorine, monochloramine, dichloramine, trichloramine, and free chlorine are reported as mg/L-Cl_2, and free ammonia is reported as mg/L-NH_3-N. The curve was developed using the Chlorine Breakpoint Curve Simulator, Version 0.25 (Wahman 2017) with an initial free ammonia concentration of 1 mg/L-N, total alkalinity of 50 mg/L-$CaCO_3$, pH of 8.0, and water temperature of 20°C.

mixing and water quality, both of which are site specific. The Chlorine Breakpoint Curve Simulator (Wahman 2017) can provide insight into the expected time to reach equilibrium based on water quality.

As a routine practice, water systems practicing chloramination should not monitor for free chlorine in the distribution system. Commonly used free chlorine measurement methods (e.g., N,N-diethyl-p-phenylenediamine [DPD] colorimetric) are not accurate when high levels of monochloramine are present in the sample. Free chlorine should be monitored when breakpoint chlorinating the distribution system, a section of the system, or a storage facility. In this case, typical free chlorine residuals vary from 0.5–1.5 mg/L-Cl_2. At the treatment plant, the measurement of free chlorine immediately upstream of ammonia addition is critical to understanding how much chlorine is present to properly dose ammonia and form chloramines. Uncontrolled mixing of free chlorinated and chloraminated water in distribution systems (e.g., in blending zones) is not recommended and should not be performed because free chlorine can destroy monochloramine if the chlorine-to-ammonia-N ratio of the chloraminated water is near 5:1 and free ammonia residual is low (refer to the breakpoint curve on Figure 6-1). Water systems that use both disinfectants should analyze for free chlorine in the free chlorinated portions of the distribution system, and total chlorine and/or monochloramine in the chloraminated portions. Free ammonia monitoring, discussed in detail in the following section, would also be recommended to minimize the possibility for formation of dichloramine, trichloramine, or breakpoint chlorination in blending zones.

Ammonia, Free and Total

Ammonia is introduced into chloraminated drinking water systems through source water or chemical addition to form chloramines. Ammonia is typically measured in chloraminated systems as free ammonia (NH_3 + NH_4^+) or total ammonia (free ammonia + NH_2Cl + $NHCl_2$ + NCl_3 + organochloramines). Total ammonia is useful for chlorine-to-ammonia ratio and chemical dosing calculations. Free ammonia serves as the energy source for nitrifying microorganisms and is a useful parameter for chloramine formation and nitrification monitoring. Measuring free ammonia in finished water at the treatment plant allows verification of chemical dosing and adjustment of chlorine and/or ammonia doses to minimize the free ammonia concentration entering the distribution system. A consideration for free ammonia monitoring in finished water is that in water treatment plants or booster chloramination facilities targeting a 5:1 chlorine-to-ammonia-N mass ratio, the free ammonia concentration would be at or below 0.10 mg/L-N, and preferably 0.05 mg/L-N or less, which can be difficult to measure.

At the treatment plant or booster chloramination facility effluent where free ammonia concentrations are low, total ammonia can be used to identify chemical feed problems. Examples of calculations are presented in the text box below.

Consider adding free chlorine and ammonia to form monochloramine at a 4.8:1 chlorine-to-ammonia-N mass ratio at a target total chlorine residual of 4 mg/L-Cl_2. If chlorine and ammonia doses are applied correctly, a sample taken downstream would be expected to have a total chlorine concentration of 4 mg/L-Cl_2 and a monochloramine concentration of approximately 4 mg/L-Cl_2. The total ammonia concentration would be

$$Total\ ammonia = 4\frac{mg}{L}\ as\ Cl_2 \times \frac{1\frac{mg}{L}\ as\ N}{4.8\frac{mg}{L}\ as\ Cl_2} = 0.83\frac{mg}{L}\ as\ N$$

Knowing that monochloramine forms at approximately a 5:1 chlorine: ammonia-N mass ratio, the portion of total ammonia in monochloramine (i.e., combined ammonia) would be

$$Combined\ ammonia = 4\frac{mg}{L}\ as\ Cl_2 \times \frac{1\frac{mg}{L}\ as\ N}{5\frac{mg}{L}\ as\ Cl_2} = 0.8\frac{mg}{L}\ as\ N$$

The difference between total and combined ammonia would be the expected free ammonia concentration. In this case, the difference is 0.03 mg/L-N, which could be below the reporting limit for the free ammonia method used. Measuring total ammonia, even rounded to 0.8 mg/L-N, could confirm that sufficient ammonia addition occurred as intended more reliably than attempting to measure free ammonia at 0.03 mg/L-N.

A similar approach could be taken to compare total ammonia and total chlorine residual to calculate the free ammonia concentration entering the distribution system. For example, if a total chlorine residual of 2.5 mg/L-Cl2 and a total ammonia concentration of 0.6 mg/L-N are measured in an entry point sample, the chlorine-to-ammonia-N ratio is

$$Chlorine: ammonia\ as\ N\ ratio = \frac{2.5\frac{mg}{L}\ as\ Cl_2}{0.6\frac{mg}{L}\ as\ N} = 4.2$$

Knowing that monochloramine forms at approximately a 5:1 chlorine-to-ammonia-N ratio, the combined ammonia concentration would be approximately

$$Combined\ ammonia = 2.5\frac{mg}{L}\ as\ Cl_2 \times \frac{1\frac{mg}{L}\ as\ N}{5\frac{mg}{L}\ as\ Cl_2} = 0.5\frac{mg}{L}\ as\ N$$

and the free ammonia concentration would be approximately 0.1 mg/L-N.

Total ammonia can be also used for calculation of the breakpoint chlorine dose, but free ammonia should be analyzed to verify breakpoint results.

In the distribution system, monitoring free ammonia is very useful. An increase in free ammonia concentration above normal levels indicates overdosing at the plant or chloramine decay and suggests that nitrification could possibly occur. A decrease in total chlorine residual that is not due to decreased total chlorine entering the system and is not accompanied by an increase in free ammonia indicates that nitrification is occurring. In some systems, a slight increase followed by a decrease in free ammonia concentrations can be used to predict the onset of nitrification.

Compared to free ammonia, total ammonia data are more ambiguous and of limited usefulness for distribution system monitoring prior to the onset of nitrification. For example, as chloramines decompose, the total ammonia concentration decreases only slightly over time because ammonia is primarily being converted from chloramines to free ammonia, whereas the free ammonia concentration increases significantly. The release of free ammonia makes conditions more conducive to nitrification but would not be detected by monitoring total ammonia alone since the increase is so slight. In the event that the chlorine dose increased by supplemental dosing in the system, it would be associated with a decrease in free ammonia but would not be reflected by the total ammonia concentration until the 5:1 mass ratio was exceeded.

For most systems, free and total ammonia are both useful for detecting ongoing nitrification because both decrease when ammonia oxidation rates exceed chloramine decomposition rates. However, if the rate of ammonia oxidation is less than or equal to the rate of free ammonia release from chloramine decomposition (e.g., in the earliest stages of nitrification), total ammonia decreases but free ammonia increases or remains constant. Comparisons of free and total ammonia with total chlorine or monochloramine and/or nitrite can be helpful for interpreting the data.

Nitrite and Nitrate

As detailed in Chapter 5, ammonia-oxidizing microorganisms (AOMs) oxidize free ammonia. Ammonia-oxidizing bacteria (AOB) like *Nitrosomonas*, which oxidize free ammonia to nitrite (NO_2^-), may be the most studied to date. Nitrite-oxidizing bacteria (NOB) like *Nitrobacter* oxidize nitrite to nitrate (NO_3^-). In recent years, the presence of ammonia-oxidizing archaea (AOA), which oxidize free ammonia to nitrite, and complete ammonia oxidation (comammox) bacteria, which oxidize free ammonia to nitrate, have been identified in drinking water distribution systems. Nitrite can be oxidized to nitrate through a chemical pathway as well (Connell 1996), which may occur in water systems that boost free chlorine without adding ammonia. Most natural waters do not contain nitrite, so its presence indicates nitrification is occurring. Without the activity of AOMs, there would be no nitrite above the system's background (nondetectable) level.

When ammonia is oxidized by AOMs, the ammonia concentration decreases as ammonia is converted to nitrite. Generally, the nitrite concentrations initially increase and then either level off when ammonia is no longer available or decrease as NOB consume nitrite and convert it to nitrate. Nitrate concentrations are source water dependent and are stable in drinking water distribution systems (unless denitrification is occurring, which is rare in these systems). In some systems, nitrate has been used to identify or confirm nitrification events or to characterize the severity of a nitrification event. Smith and Smith (1996) showed increased levels of nitrate well above background in a system that uses surface water. However, NOB grow very slowly in drinking water, so many water systems observe nitrite but not nitrate during nitrification. Also, surface water supplies experience fluctuations in background nitrate, making it difficult to interpret whether an increase in nitrate concentration is associated with nitrification or increased runoff into the source water. When the background nitrate concentration is much higher than the amount of nitrate formed by nitrification, an increase in nitrate concentration due to nitrification can go unnoticed. In addition to regular quarterly measurements, more frequent monitoring at several key locations, such as POEs, outlets of storage facilities, and a few ends of the system, would be useful to maintain a database of baseline nitrate levels and account for seasonal variations.

Nitrate testing may be performed as a laboratory test or as a field test. Laboratory tests add expense and potentially a time delay before results are received, which may discourage some water systems (especially smaller systems) from incorporating this parameter into their routine nitrification monitoring program. Also, there are no easy field tests that can measure low nitrate concentrations, such as those that would result from nitrification. Generally, systems in warmer climates find nitrate a more useful parameter than systems in colder climates because complete nitrification is more likely to occur in warmer water than in colder water.

Water Temperature

Water temperature is a critical factor in the growth of nitrifying microorganisms. As explained in Chapter 5, the optimum growth temperature for AOB and NOB has been reported in the range of 20–30°C (68–86°F). Increases in water temperature are also linked to seasonal changes in organic material characteristics and accelerated chloramine decay, which may enhance nitrification. This explains why most water systems experience nitrification during summer months. For this reason, temperature can be used as an indicator that nitrification is more or less likely to occur.

Generally, there is little fluctuation in temperature in distribution system pipelines on a daily basis. However, as described in Chapters 8 and 9, storage facilities can experience enough temperature variation to create stratification within the water column (Grayman et al. 2000). In the absence of effective mixing, thermal stratification can occur within storage facilities, thereby encouraging the decay of chloramine and increasing the potential for nitrification. Temperature monitoring can detect the onset of thermal stratification and prompt corrective action.

pH

Adjustments to drinking water pH are made to optimize water treatment plant processes and for corrosion control. For example, coagulation is optimized at an acidic pH, whereas corrosion control is optimized under basic pH conditions. Disinfection with free chlorine is more effective at low pH because hypochlorous acid (predominant at low pH) is 300 times stronger as a disinfectant than the hypochlorite ion (predominant at high pH). However, following ammonia addition, higher pH (above ~8) is preferable to help maintain chloramine stability.

As discussed in Chapter 5, pH also affects nitrifying microorganisms. The optimal pH range for most AOB and NOB is 7.5–8.0, although growth has been observed within a pH range of 6.5 to nearly 9.0. Thus, pH can promote or inhibit nitrification. Additionally, nitrifiers can decrease the pH of poorly buffered water, making decreasing pH a possible indicator of ongoing nitrification. Studies have reported that a target pH value above 8.5 can help limit nitrification by minimizing the decay rate of chloramine residual, although this may not be possible in water with very low alkalinity (e.g., Wilczak et al. 2003).

Nitrifying Microorganisms

Detection of nitrifying microorganisms—particularly AOB and NOB—during nitrification events in drinking water distribution systems has been well documented. Nitrifying microorganisms must be present for nitrification to occur, and microbial density would be expected to increase during nitrification episodes. However, nitrifying microorganisms can be present in the absence of detectable indicators of nitrification (e.g., increase in nitrite concentration), and thus, the detection of these microorganisms does not predict the onset of nitrification. This can be confirmed by the results of Ike and colleagues (1988), who observed that nitrite was usually not detected until the AOB concentration was approximately 1 MPN/mL. Also, relationships between nitrifying microorganisms and other parameters are system dependent; they may be observed during nitrification episodes in some systems but not in other systems.

Attached nitrifying microorganisms have been detected on pipes of various materials, including polyvinyl chloride and cast-iron pipes and in bench- and full-scale distribution systems. In some studies (e.g., Carrico et al. 2008), attached nitrifying microorganisms were even detected in the absence of chemical indicators of nitrification.

As detailed in Chapter 5, AOB and NOB are not the only microorganisms responsible for nitrification in drinking water distribution systems. Chapter 5 documents the speciation of nitrifying bacteria, including AOB, NOB, AOA, and comammox (i.e., single microorganisms that can complete the nitrification process from ammonia to nitrite and then to nitrate), in the natural environment and in drinking water distribution systems.

Although direct analysis of nitrifying microorganisms remains the most accurate method to confirm their presence, most methods for nitrifying microorganism detection, identification, and enumeration are complex, time consuming, and/or require sophisticated equipment and/or training, as detailed in Chapter 5. Because of the difficulties in detecting and identifying nitrifying microorganisms, most water providers continue to rely on chemical indicators to detect nitrification.

Heterotrophic Plate Count

Several studies have shown increases in HPCs during nitrification events (e.g., Zhang et al. 2009; Powell 2004; Wolfe and Lieu 2003; Odell et al. 1996; Wilczak et al. 1996; Skadsen 1993; Wolfe et al. 1988;) in part because the nitrite produced during nitrification consumes chloramine residual (Valentine 1984), thereby allowing bacterial growth. This trend was also observed in building plumbing, for both suspended and attached microorganisms, as determined by HPCs with incubation on R2A media (Zhang and Edwards 2009) (information about analytical methods that can be used to measure HPCs is provided below). The effect of nitrification on disinfectant residuals and microorganism density, for both suspended and attached microorganisms, is so important that it can reverse the well-recognized advantage of chloramine over free chlorine to inactivate attached microorganisms (Zhang and Edwards 2009).

Although heterotrophic bacteria have often been used as surrogates for nitrifying bacteria in distribution systems, heterotrophic bacteria differ from autotrophic bacteria (like most AOB and NOB) in that they use organic carbon for their carbon source, whereas AOB and NOB use carbon dioxide. Thus, increases in HPCs in distribution systems also can be attributed to factors other than nitrification (e.g., Odell et al. 1996; Pintar et al. 2005; Skadsen 1993; Volk and LeChevallier 2000). Although Donlan and Pipes (1988) observed correlations between the density of attached microorganisms and chloramine concentrations, their study did not find correlations between the density of attached microorganisms and ammonia, nitrite, or nitrate concentrations. Donlan and Pipes (1988) used HPCs with incubation on R2A medium at 20°C for 7 days to assess microorganism density, with confirmation by direct enumeration using acridine orange dye. For this reason, using HPCs as indicators of nitrification needs to be done in conjunction with monitoring of parameters that more closely represent nitrification, such as nitrite and nitrate concentrations. Also, full-scale distribution system results indicated that HPCs could not be used as an early indicator of nitrification (Gupta and Akagi 2014).

Competition between heterotrophic and nitrifying bacteria and benefits that may result from these interactions are detailed in Chapter 5. Although heterotrophic and nitrifying bacteria compete for dissolved oxygen, nitrogen, and space, they also benefit from each other in several ways.

- Each group of microorganisms produces organic compounds that can favor the growth of the other group. However, Zhang and Edwards (2009) observed that in the absence of disinfectant and in natural waters with measurable concentrations of organic material, nitrification did not increase suspended and attached HPCs because the concentrations of organic compounds produced by nitrifiers were insignificant compared to the organic concentrations in natural waters.
- Heterotrophs can biodegrade organic compounds that can inhibit the growth of nitrifying bacteria.
- Heterotrophs produce extracellular polymers that improve aggregation of both heterotrophic and nitrifying bacteria into flocs and biofilms.
- The faster-growing heterotrophs can protect the slower-growing nitrifiers from detachment, which can result in stratification of microorganisms in multispecies biofilms.

Adenosine Triphosphate

Adenosine triphosphate (ATP) is a compound found in all living cells in which energy is stored in high-energy phosphate bonds; i.e., ATP is an energy carrier. Intracellular ATP (ATP contained inside living cells) can be considered potential energy. Thus, higher concentrations of cellular ATP indicate greater biomass and cell volume. Extracellular ATP represents ATP released into the environment and indicates weak and/or lysed cells (Whalen et al. 2018).

The ATP firefly bioluminescence assay is a rapid detection method that uses ATP to estimate the total density of viable bacteria in a sample (the method is discussed later in this chapter). Thus, ATP concentration is a good indicator of biomass and has been successfully applied in several studies to assess biological stability. Because ATP is contained in all living cells, ATP has been considered a superior indicator of the total biomass compared to culturing techniques such as HPC. HPCs typically recover only a small portion of metabolically active bacteria, and results vary based on method used, as discussed above. Also, results from culturing techniques tend to be influenced by aggregated cells (i.e., clusters of bacteria). Additional advantages of ATP over other indicators of microbial activity are that its measurement takes less than 1 h and it is generally field- and user-friendly.

Interest in using ATP analysis as an indicator of suspended microorganism activity has grown in recent years, and ATP has been incorporated in nitrification monitoring and prevention plans (e.g., Gupta and Akagi 2014; Sung 2011, 2017; Monks et al. 2010; Zhu et al. 2010). As explained by Arp et al. (2002), Chain et al. (2003), and Wood (1986), ATP can be a good indicator of AOB activity, particularly for *Nitrosomonas europea*. Ammonia oxidation can be broken down into two steps: ammonia oxidation to hydroxylamine (NH_2OH) by the ammonia monooxygenase (AMO) enzyme and NH_2OH oxidation to nitrite by the hydroxylamine oxidoreductase (HAO) enzyme. Some of the electrons released during the oxidation of NH_2OH into nitrite are used for cellular processes such as biosynthesis and ATP generation. This process would lead to correlations between ATP and nitrification. Certain studies, however, have not observed such correlations.

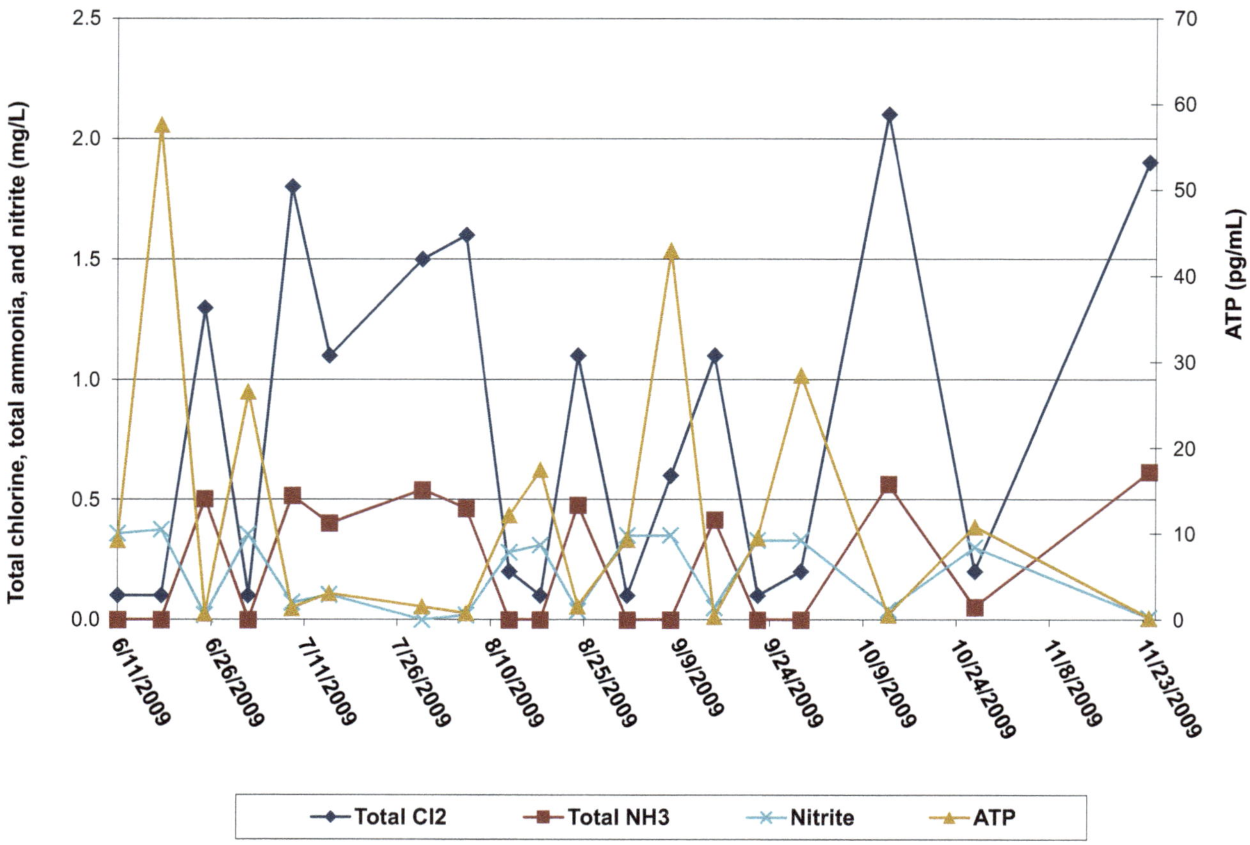

Figure 6-3 Relation between adenosine triphosphate (ATP) and chemical indicators of nitrification

When low chlorite concentrations (i.e., 0.32-0.42 mg/L) were used to inhibit nitrification in a full-scale distribution system, Zhu et al. (2010) observed ATP concentrations varying from 0.33–1.02 pg/mL, whereas an ATP concentration of 13.7 pg/mL was measured in the absence of chlorite and in presence of nitrification (nitrite concentration of 0.20 mg/L N). These authors also noted that ATP concentrations correlate well with other nitrification indicators, including total chlorine, total ammonia, and nitrite concentrations (Figure 6-3). Gupta and Akagi (2014), however, observed no or poor correlations between ATP concentrations and nitrite, nitrate, free ammonia, or chlorine concentrations. Good correlations between ATP concentrations and HPC-R2A were only observed with one type of ATP test, the LuminUltra's Quench-Gone Aqueous Test (Fredericton, N.B.).

Sung (2011) showed better correlations between nitrite and ATP concentrations (R^2 of 0.76) than between HPCs and ATP concentrations (R^2 of 0.29) in a full-scale distribution system. Based on these results, Sung (2017) recommended ATP measurement over HPC if monitoring of a biological component is desired. For water systems, however, this recommendation implies that an additional parameter that most systems do not analyze for would be added to the monitoring program (see Figure 6-3).

Additional Parameters

This section presents the relationship between nitrification, chloramine residual, and additional parameters indirectly related to nitrification.

Alkalinity. Fluctuations in pH in the distribution system are governed by the buffering capacity of the water. For example, in poorly buffered waters, pH often increases as water moves through the distribution system. A drop in pH during nitrification will be more apparent in water with very low alkalinity. The alkalinity data are helpful in interpreting the pH data.

Specific conductivity and total dissolved solids. This parameter does not influence nitrification and is not an indicator of nitrification but can be a useful parameter where multiple water sources blend in a distribution system. Where different waters exhibit different susceptibility to nitrification due to chloramine stability and/or bacterial growth potential, then conductivity can be used to identify water sources and blends of water sources throughout the distribution system.

Dissolved oxygen. This parameter is generally not routinely monitored, but a decline in dissolved oxygen (DO) concentration can support other data that indicate nitrification. DO should be considered as a parameter for nitrification studies but not for routine sampling.

Oxidation-reduction potential (ORP). While ORP by itself is not a direct measure of nitrification, it can be a useful indicator in combination with other parameters. For example, changes in total chlorine or dissolved oxygen concentrations will affect ORP of the water. Lower chlorine or oxygen can result in reduction of metals (iron, manganese, lead) or dissolution of metals from pipe scales, which further exacerbate lowering of chlorine residuals and may eventually lead to nitrification.

Organic carbon (total organic carbon [TOC] and dissolved organic carbon [DOC]). Organic carbon is linked to biostability in distribution systems. TOC and DOC exert disinfectant demand and can impact disinfectant residuals in the distal parts of the distribution system. The assimilable and biodegradable fractions of DOC (i.e., assimilable organic carbon [AOC] and biodegradable dissolved organic carbon [BDOC], respectively) serve as substrate for heterotrophic microorganisms. TOC and DOC are very useful parameters for monitoring the effectiveness of water treatment processes (e.g., enhanced coagulation) within the water treatment plant and at the POE. However, TOC and DOC concentrations typically change very slightly through a distribution system, and thus, they are not necessary for routine nitrification monitoring in the distribution system. TOC and/or DOC should be monitored during studies characterizing treated water quality or evaluating water treatment processes. Similarly, AOC and BDOC measurements are best suited for special studies, such as evaluating biological treatment processes or the impact of oxidation processes (e.g., ozonation) on biostability.

Bromide. The presence of bromide in the source water can lead to the formation of hypobromous acid, bromamines, and bromochloramines during disinfection with chlorine and chloramines. These species are unstable and can also result in rapid losses in monochloramine and total chlorine residual levels, facilitating nitrification in the distribution system. In source waters with high bromide, monitoring bromide is important, not only to determine the potential formation of DBPs but also for the development of appropriate chlorine residual management and nitrification control strategies.

Coliform bacteria. Coliform bacteria act as an indicator for potential occurrence of pathogens in the water. It is important to remember that nitrifying bacteria are not observed in the coliform tests. However, in severe nitrification occurrences when chloramine residual had dissipated, excessive bacterial counts, including the presence of coliforms and sometimes even the specific coliform *Escherichia coli* (*E. coli*), have been observed.

Microbial decay factor. The microbial decay factor is the ratio of the microbial decay rate to the chemical decay rate. It is an analytical tool that has been employed in studies aimed at assessing the microbial ecology of distribution systems (Sathasivan et al. 2010a, 2005; Sathasivan et al. 2010b). The tool may be useful for modeling and prediction efforts, although additional applications are required to fully assess its usefulness in a nitrification control program.

Nitrification potential curves. Nitrification potential curves were introduced to delineate the presence or absence of nitrification in a distribution system and to determine whether water quality may be conducive to nitrification (Fleming et al. 2008, 2005). They can be used as an operational strategy to detect and prevent nitrification. The concept is "based on the premise that two competing effects determine the presence of AOB in chloraminated systems: inactivation resulting from the presence of chloramine disinfectant (biocide) and growth caused by the presence of ammonia (food)" (Fleming et al. 2008, 2005). Nitrification potential curves require calculation of the half-saturation constant, which defines the ammonia uptake characteristics of AOB species and the ratio of growth to inactivation rates of AOB using total chlorine residuals and free ammonia concentrations. They can be used to assess the consequences of increasing total chlorine residuals or decreasing free ammonia concentrations on nitrification potential. In the example illustrated in Figure 6-4, nitrification would not likely occur at location 2 but would likely occur at location 1. Nitrification potential curves, however, may be difficult to use when evaluating the effect of operational changes. Also, they are specific to a given water quality and AOB species (see Figure 6-4).

Operational Parameters

As previously discussed, distribution system characteristics and operation influence nitrification potential. This section presents operational parameters associated with distribution system maintenance and water age that should be considered as part of nitrification monitoring and control strategies. Detailed discussion about these parameters is presented in Chapter 7.

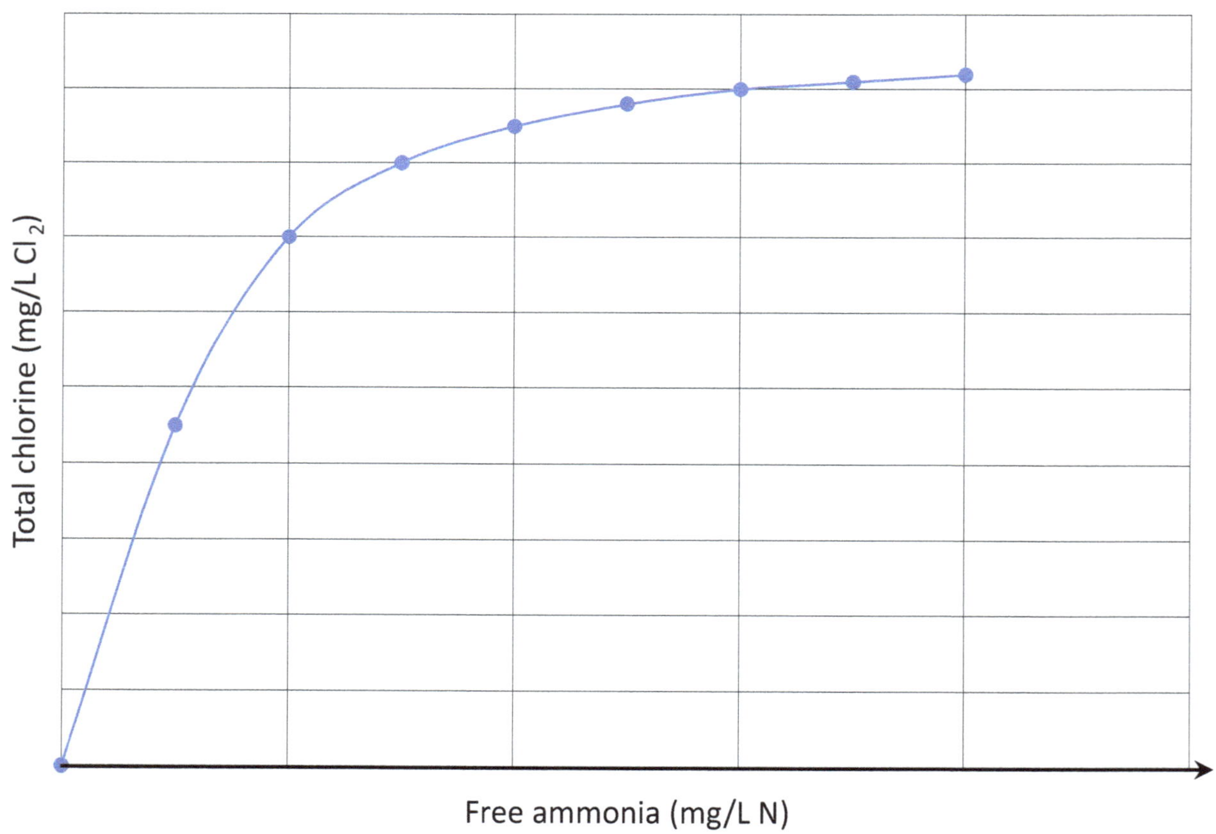

Figure 6-4 An example of nitrification potential curve

Entry point(s) usage. POEs can impact water quality throughout a distribution system by introducing water from different sources with different water quality parameters or by impacting flow patterns and thus water age or locations of blending zones. Monitoring should include tracking entry points in use and flow contributed by each POE.

Storage facility cycling data and mixing. Cycling and mixing are two strategies to limit water age in storage facilities. Cycling can be monitored based on tank level, which is often available in real-time via supervisory control and data acquisition (SCADA). The frequency and degree of cycling (i.e., change in tank level) depend on the distribution system configuration, the type and configuration of the storage facility, the presence of pump stations and their operation, and several other parameters (water demand, season, etc.). However, cycling should be maximized to the extent allowable while still maintaining adequate storage, system pressure, and emergency/fire flow requirements. The history of cycling at a storage facility can be used to identify conditions conducive to nitrification or that have been contributing factors to nitrification events in that facility and its zone of influence.

Thermal stratification in storage facilities increases water age by inhibiting mixing, thereby encouraging chloramine decay and increasing the potential for nitrification. Stagnant zones can also occur in storage tanks due to suboptimal flow patterns (e.g., due to common tank inlet/outlet). Mixing problems can be identified and characterized by collecting grab samples at different depths in different areas of a storage tank for temperature and/or disinfectant residual measurements. An example of a storage tank monitoring plan is included in Appendix C. Sensors that monitor water temperature at high frequency can be deployed at various depths and locations within a storage facility to assess mixing over a longer period. Computational fluid dynamic modeling can be used to evaluate the effectiveness of water mixing and design of the storage facility, as well as to evaluate mixing improvements. Additional information on cycling and mixing of storage facilities is presented in Chapter 7.

Storage facility inspection and cleaning records. Records associated with storage facility inspections and cleaning can be useful for identifying locations where accelerated disinfectant residual loss is likely to occur. Regular storage facility inspections (e.g., with a submersible robot) can be used to identify storage facilities with excess sedimentation and prioritize them for cleaning, recognizing that sediments tend to harbor microorganisms, including nitrifying microorganisms. Storage facility inspections would ideally be conducted annually, or at a minimum, every 5 years, as recommended by ANSI/AWWA G200, Distribution Systems Operation and Management. Storage facility inspections can also be triggered by changes in water quality in a tank or in its zone of influence.

Flushing data. Flushing programs are used to control water age in distribution systems and to remove lower-quality water, accumulated sediments, and excess biofilm. Tracking of flushing location, frequency, flow rate, and volume, as well as chlorine residual before and after flushing can be included in nitrification monitoring strategies. Flushing data can be useful to trigger proactive flushing based on time since the last flush or in response to the results of water quality monitoring (e.g., low disinfectant residual).

Valve positions. Distribution systems contain numerous valves, and over time, some valves that are intended to be open can be unintentionally closed and vice versa. In some cases, unintentionally closed or opened valves can negatively impact water age in an area of the distribution system and lead to a nitrification event. A valve exercising program or other checks of valve open/closed status can be a useful part of a nitrification monitoring strategy. Valve checks can be conducted on a time schedule or in areas where construction or repairs have recently been completed to prevent unintentional, avoidable increases in water age.

Distribution system hydraulic model output. Distribution system hydraulic models have long been used to predict water age for a limited number of specific operating conditions and demand scenarios. As computing tools (e.g., cloud-based computing, advances in software) have improved, distribution system models can more readily be used to predict water age under specific operational scenarios or even in real time using SCADA data as inputs. Operational and real-time models can be used to calculate water age throughout the distribution system, identify zones of influence for POEs and storage facilities, or predict concentrations of water quality parameters. Alert and action levels can be assigned to model output such as water age, and water quality monitoring locations can be shifted based on predicted locations of blending zones or high water age areas. Comparisons of model output and measured water quality data can be used to identify operational conditions that are conducive to nitrification. Additional information on distribution system modeling capabilities is available in AWWA Manual M32, *Computer Modeling of Water Distribution Systems*.

Data Management

Because nitrification can escalate rapidly, data analysis is critical to ensure that the onset of nitrification is captured as quickly as possible and responses are initiated without delay. Some systems use web-based collaborative platforms to enter field data in real time; others download data from field instruments on a regular (preferably daily) basis to facilitate data analysis.

Proper interpretation of nitrification monitoring parameters relies on trend graphs (i.e., graphs showing data for one or more parameters as a function of time) and comparison of data obtained with system-specific trigger levels for key indicators of nitrification that indicate the need to remain alert or initiate action. Typically, trend graphs are produced for individual sample sites and include parameters such as total chlorine and/or monochloramine, free and/or total ammonia, nitrite, and nitrate. These graphs should be made using individual datapoints, not with average values. Trend graphs provide context necessary to determine if a change in a parameter is likely due to nitrification or another cause. For example, a decrease in chlorine residual can be due to many factors, including high water age, corrosion, microbial growth, cross-connection, or others. An increase in nitrate concentration can be due to seasonal change in source water. However, a concurrent decrease of monochloramine and free ammonia with an increase in nitrite would indicate that nitrification is occurring. Trend graphs are only informative when monitoring is frequent and consistent.

Data compilation and production of trend graphs can be accomplished using spreadsheets or proprietary information management software. The data management warehouse or data lake can be integrated into an advanced platform that integrates data analytics, visualization, and real-time alerting. Data from different sources, like online water quality monitoring (OWQM) stations and Laboratory Information System (LIMS), can be used as part of the analysis.

Data records should be archived electronically and accessible for subsequent analysis and interpretation, if needed.

MONITORING FREQUENCY

This section presents the frequency of monitoring based on grab samples; online analyzers are discussed in a subsequent section.

Although the frequency of analysis of many parameters described in this chapter is governed by regulatory compliance, this chapter focuses on nitrification monitoring. Thus, the monitoring frequencies presented in these sections are based on the importance or priority ranking of the parameter for making operational decisions to prevent or control nitrification and the variability in the parameter over time and throughout the distribution system.

Monitoring should be more frequent for the parameters characterized as having "extreme usefulness" (and potentially also for the parameters characterized as having "usefulness") in Tables 6-2 and 6-3. These parameters are system specific but may include total chlorine, monochloramine, free ammonia, nitrite, water temperature, and pH. For these parameters, weekly sampling has been generally practiced, with more frequent sampling when nitrification is detected, using traditional grab samples. In circumstances where nitrification escalates rapidly, however, this frequency may not provide the granularity needed for achieving early detection, which can be achieved with online monitoring. Other parameters may be monitored at a lower frequency, with additional sampling when nitrification is detected. For example, nitrate may be monitored monthly at potentially fewer sampling sites, with more frequent and widespread monitoring when nitrite concentrations increase above baseline levels. Monitoring for HPCs weekly at all nitrification sampling sites may not be necessary. Instead, HPCs may be monitored only when nitrification is detected to ensure that the bacteriological quality of the water has not been compromised. Water systems are encouraged to review and revise their nitrification monitoring and control plans on a regular basis and assess the usefulness and frequency of sampling of their nitrification indicators. Parameters that show little variation over time may be monitored less frequently or in response to changes in other high-priority water quality parameters.

Monitoring frequency may vary depending on the sampling location. For example, a certain monitoring site may be used as an indicator of nitrification. When nitrification is detected at that site, monitoring is extended to other sites within the area or pressure zone. Monitoring at the outlets of storage facilities should not be relaxed, and monitoring at these sites should be conducted at the highest frequency. Additional information about monitoring locations is presented later in this chapter.

Monitoring frequency may also vary by season. Nitrification indicators may be monitored more often in the warmer season than in winter if nitrification data have shown that nitrification is absent in winter. This implies that southern US water systems may have to monitor all year around. Monitoring frequency should also consider water demand and the presence of seasonal residents. In certain systems, nitrification tends to occur in winter when water demand for irrigation and landscaping is lower or when summer residents are absent.

To facilitate data interpretation, samples for nitrification indicators should all be collected at the same time, considering that certain parameters are used to confirm results of the key indicators of nitrification.

Without compromising water quality, a water system's resources and system configuration should also be considered when establishing monitoring frequency. Whereas it may be possible to sample all nitrification monitoring sites of a small distribution system in a single day, this may not be possible in large or widespread systems. The type of analytical instrument used and therefore the rapidity of analysis also determines how many samples staff can analyze during a sampling round, which can influence monitoring frequency.

MONITORING LOCATIONS

Designated monitoring locations for grab sample collection or online monitoring stations are an important component of a monitoring program. Recognizing that certain locations are more informative of regulatory compliance while others are more informative of the occurrence or magnitude of nitrification, this section focuses on nitrification monitoring sites.

An adequate number of monitoring locations must be provided to fully represent the highest risk portions of the distribution system. Although the exact monitoring locations are largely system specific, water treatment plant effluents or POEs, storage facilities, high-water age areas, and stagnant areas should be captured by the nitrification monitoring program. Monitoring sites should also represent variations in pipe diameter, pipe materials, water age, and pressure zones if considered to have a high risk of nitrification. Sampling may also occur in blending zones—i.e., areas influenced by more than one source of water or areas where different disinfectants like free chlorine and chloramine come into contact—if there is a history of water quality problems. A well-calibrated distribution system hydraulic model can be used to identify the most adequate nitrification monitoring locations. These models can calculate water age throughout the system and identify high- and low-flow areas and areas where blends of water sources may occur under different water-demand conditions. Monitoring at high–water age sites is important because water age is an integral component of chloramine degradation.

Storage facilities are particularly important sites because nitrification tends to develop in these facilities, and many nitrification-control strategies are applied to tanks or reservoirs (e.g., mixing and cycling, booster chloramination, breakpoint chlorination). Preferably, the outlets of storage facilities should be sampled rather than the inlet, although both sites (and potentially different depths inside an affected storage facility) may need to be sampled if nitrification is suspected to be localized to a tank or reservoir only. Water quality data obtained from storage facilities should be correlated to operational information at the time of sampling: e.g., whether the facility was filling or draining, the rate at which water level was changing, the supply flows and blends, and whether active mixers were online or not. The data can be used to determine the necessary cycling or turnover to minimize the likelihood of nitrification. If a booster chloramination or chlorine (leveraging free ammonia) station is present at a storage facility, water quality data can also inform the operational targets for this station.

While system-wide averaging is required for regulatory compliance, an assessment of water quality at individual locations in the distribution system is essential for proper interpretation of data developed under a nitrification prevention and control program.

Sampling Taps

Designated sample stations or taps are critical to the integrity of the monitoring program and database. If sample taps are maintained to establish a historical database, data can be used to assess changes in treatment or operations with a greater degree of confidence. A compromise must be made between establishing new locations for adequate coverage of the system and maintenance of the historical database.

Sample taps should be protected from outside sources of contamination. Water system staff should periodically examine the sample taps for leaks or other potential sources of contamination. Dedicated sampling stations used for drinking water monitoring only are preferable. Sample taps should allow for adjustment of flow rate to flush the sample line before the sample is drawn. Narasimhan and colleagues (2004) found that flushing was as effective as flaming or disinfecting the sample tap (with chlorine or alcohol) before sampling for coliform bacteria. Flaming is discouraged if the sample valve components can be degraded by excessive heat. Because piping materials can impact certain parameters (e.g., certain metals can inhibit nitrifying microorganisms), materials used for taps and sample lines should be consistent with parameters to be analyzed.

APPROACH FOR ONLINE WATER QUALITY MONITORING

Drinking water systems have been increasingly using online water quality monitoring (OWQM) technologies over the last 20 years to rapidly detect in near real-time a wide variety of water quality anomalies in the distribution system, including nitrification episodes. This has been driven by advancements in monitoring technologies to make sensors smaller, more reliable, and easier to operate and maintain. USEPA's *Online Water Quality Monitoring in Distribution Systems* (2018) provides guidance on the design of OWQM systems for distribution monitoring.

While there are no regulatory requirements that require the use of OWQM in distribution systems, having access to real-time water quality information that is produced by properly calibrated and maintained monitoring technologies can help systems maintain compliance with provisions of rules, such as the residual disinfectant monitoring and reporting requirements of the SWTR. Additionally, it provides real-time knowledge about water quality in the distribution system. Incidents, such as nitrification, loss of chlorine residual, and other unusual water quality anomalies, can be detected and information management systems can be configured to send alerts to staff to respond and mitigate the problem, reducing the impact to customers.

This section builds on the discussions of monitoring parameters and monitoring locations from earlier in this chapter to address these topics from the perspective of OWQM in the distribution system.

Monitoring Locations

Due to the cost, maintenance requirements, and factors that impact installation of OWQM equipment, the number of locations where monitoring stations can be installed is limited. Therefore, it is important to focus on the locations that provide the most useful information to guide operational decisions to maintain water quality and prevent nitrification. Water systems should consider locations, whenever possible, that provide multiple benefits, such as regulatory compliance and detection of nitrification or intentional contamination that could impact large population centers. Examples of such locations include the following:

- **Entry point(s) to the distribution system** for process control at the water treatment facility, to provide a baseline for the Cl_2:NH_3-N ratio, as well as the chloramine and free ammonia concentrations in water entering the system
- **Storage facilities** to provide information at locations that can be susceptible to stratification, such as storage tanks and reservoirs. This information can be used to inform facility operation to prevent nitrification from occurring or respond to incidents that are in progress.
- **Downstream of control features (e.g., booster stations)** for process control to prevent or limit nitrification incidents in downstream areas
- **Low-flow or stagnant areas (e.g., areas with high water age)** provide information from dead ends and the furthest reaches of the distribution system where nitrification is most likely to occur (Note: Installing automatic flushing devices near online stations in these locations can enhance the value of monitoring, as water quality can be directly impacted in the area. If locations already have such devices, they should be prioritized.)

Monitoring Parameters

The advantages and limitations of several water quality parameters to assess nitrification were detailed above. High-performance sensor technology used to monitor key parameters at OWQM stations is readily available to provide real-time information that can be used to detect nitrification incidents, as shown below:

- **Nitrate/nitrite.** An increase in nitrate and measurable concentration of nitrite can signal that nitrification is occurring.
- **Chlorine residual.** A decrease in total chlorine or monochloramine residual levels can indicate the degradation of chloramines.
- **Free ammonia.** An increase in free ammonia concentration followed by a decrease can signal the beginning of a nitrification incident.
- **Water temperature and pH.** This combination of parameters, which is commonly monitored by a single sensor, provides information on the suitability of the water environment to enhance or inhibit a nitrification episode from occurring.
- **Conductivity.** For distribution systems supplied by multiple water sources with different conductivity levels, conductivity data can be used to determine which source is providing water to a given monitoring location. This can help in understanding more about water age, mixing zones, and other hydraulic information in the area and can provide insight into how system operations can impact nitrification incidents.
- **TOC/DOC.** TOC and/or DOC data are useful to understand the cause of unusual chlorine decay in the distribution system.

Analytical methods used to measure the above parameters are specific to the type of analyzer used and depend on the equipment provider. Water systems interested in installing online analyzers are encouraged to contact these providers and ensure that the analytical methods meet their needs and that measures are in place to ensure proper calibration and maintenance of analyzers. A list of available monitoring resources is available at www.epa.gov/waterqualitysurveillance/online-water quality-monitoring-resources.

Data Management

The frequency and quantity of data generated from OWQM stations can overwhelm the end user if advanced data management and analytics are not implemented. Typical data collection frequency for real-time monitoring for nitrification episodes is 2–5-min intervals to provide the granularity necessary to observe the early stages and progression of a nitrification incident. When a system deploys many stations with multiple sensors per station, the volume of data becomes quite significant and makes the system too complex to be managed manually. Key aspects of an effective data management system typically include

- a database suitable for the type of data collected and stored (typically NoSQL),
- algorithms to process the data into actionable information,
- a dashboard to provide spatial visualization of the information, and
- an alerting function to automatically send out alerts when water quality anomalies occur.

FIELD AND LABORATORY ANALYTICAL METHODS

This section presents the analytical methods that are available to analyze the water quality parameters discussed above. Several parameters (e.g., chlorine residual, water temperature, pH) should be analyzed in the field to ensure data integrity. Instruments are now available to facilitate field analyses (e.g., the HACH [Loveland, Colo.] SL1000 portable parallel analyzer), and these tools rely on the analytical methods presented in this section.

Although not strictly required for field samples, standard laboratory quality assurance/quality control (QA/QC) practices (e.g., analysis of blanks, spikes, and duplicates; regular instrument calibration; training; and periodic proficiency testing) should be employed.

Nitrite and nitrate. Nitrite and nitrate can be monitored in the field with colorimetric methods. A common method for determining nitrite-N is the diazotization method, which has a range of 0.003–30 mg/L. This method is also described in *Standard Methods,* 24th edition (APHA et al. 2022), which uses an absorbance of 543 nm. Field testing of nitrite along with the total chlorine is the most useful way to track nitrification in drinking water distribution systems. Because NOB grow very slowly in drinking water systems, nitrite will often persist. If the nitrite is oxidized either chemically or biologically, nitrate can be tested by a variety of methods, including a nitrate electrode or the colorimetric methods (*Standard Methods,* 23rd edition [APHA et al. 2017]). The cadmium reduction method has been adapted for field use; however, its use is limited by the fact that cadmium is a hazardous waste, so samples must be properly disposed. The colorimetric nitrite method is reliable and can reasonably measure differences from 0.005–0.010 mg/L-N.

Exposure to sun may cause erroneously high nitrite readings, so samples should be developed in the dark. A field method that uses two sample cells may yield a value slightly higher than the actual value if the cells are not perfectly matched. This can be tested by putting deionized water into both cells, zeroing one, and reading the value of the other as if it were a drinking water sample. The value imparted by the reagent can also be determined by adding reagent to deionized water in a matched sample cell and reading the value as if it were a drinking water sample. It may not be practical to perform these tests on a routine basis, but it is useful for the analyst to observe interferences, which may influence the interpretation of the nitrite results, and for the analyst to understand simple QA/QC procedures.

Total chlorine, free chlorine, and monochloramine residual. Various tests are available to measure chlorinated species, including amperometric titration (AT), spectrophotometry, high-performance liquid chromatography (HPLC), and membrane introduction mass spectrometry (MIMS).

Standard Methods 23rd edition (APHA et al. 2017) refers to AT as the "standard of comparison." Theoretically, AT should yield the most accurate and precise results. ATs are not affected by oxidizing agents; however, this method requires higher skill levels, is more time consuming than spectrophotometric methods, is not suitable for field analysis, and is prone to organochloramine interference. The cleanliness and condition of the electrodes or loss of residual during stirring may impact this method's accuracy and precision (APHA et al. 2017). The water tested by AT should be 0.2–2.0 mg/L-Cl_2. Alternatively, sample dilution using laboratory-grade water can be made to measure higher concentrations.

By contrast, methods using DPD are easier to perform, require simple and inexpensive equipment, and thus, have been widely used in the water industry. DPD can be used in a titration process to determine the chlorine concentration or used in the spectrophotometric method. The more common spectrophotometric method relies on the appearance of a red color (which absorbs light at 515-nm wavelength) to indicate the concentration of chlorine in the sample. *Standard Methods* lists the range for this method under ideal conditions as 0.010–4.0 mg/L-Cl_2 (APHA et al. 2017). Samples greater than 4.0 mg/L should be diluted with chlorine-demand-free water, or a sample cell with a reduced path length can be used.

Wahman and Pressman (2015) provide a brief review of advantages, disadvantages, and potential interferences of the different total chlorine, free chlorine, and combined chlorine analytical methods available.

Two types of analysis blanks should always be determined: a reagent blank and a matrix blank. The reagent blank is prepared in the laboratory by adding the reagents to distilled water and then measured. This test shows whether the reagents cause a positive interference to the test. The matrix blank is a sample measured without the reagents added. The result from the matrix blank identifies whether there is a compound present in the sample that absorbs light at 515 nm.

The use of the reagent blank allows for the subtraction of the color imparted by the reagent and improves accuracy of the test. This is especially important in dechlorination applications where sufficient removal of chlorine from the water must be ensured. Although equipment exists with preset calibration curves so that a calibration curve does not need to be generated by the analyst, verifying equipment with standards (as specified in *Standard Methods*) is still very important.

Monochloramine can be measured using the same amperometric titration (AT) method (4500-Cl D), DPD ferrous titrimetric method (4500-Cl F), and colorimetric method (4500-Cl G) as for total chlorine. As described earlier, the AT method is very accurate, but it requires great care and technical skills and is not appropriate for field analyses. The DPD method is the most common method in the field, but the presence of oxidized manganese, copper, or chromate may interfere with monochloramine analysis.

Additionally, a HACH indophenol method (MonochlorF, HACH Method 10200) is available for the analyses of monochloramine and free ammonia. While it is not a USEPA-approved method for monochloramine analyses, presence of free ammonia or metals (such as manganese) does not interfere with this method, which is an advantage over the traditional DPD method. DPD methods are not well suited to development of breakpoint curves because nitrogen trichloride (trichloramine) reacts as free chlorine and interferes with results. An evaluation of the HACH Method 10200 Monochloramine/Free Ammonia-N (Smith 2005) indicated that in some systems, monochloramine values measured with this method were consistently slightly higher than total chlorine values measured by DPD. The reason for this discrepancy was not identified. In one system, the free ammonia-N values matched the ammonia selective electrode method for ammonia-N (4500-NH_3; APHA et al. 2017), but this was not the case for all systems. Therefore, it is recommended that water systems perform their own comparison between field methods and reference methods (i.e., *Standard Methods*) before adopting new methods for their monitoring program.

El-Farra and Andrews (2000) and Andrews (2001) described a simple method to determine organochloramines by subtraction. The method relied on subtracting values of monochloramine and dichloramine (and/or free chlorine in water without inorganic chloramines) from the total chlorine measurement determined by amperometric titration. The results suggested that while AT indicated predominance of monochloramine in the water sample collected from a chloraminated distribution system, the subtraction procedure conversely could reveal that a significant fraction of total chlorine in the same sample may not be monochloramine or dichloramine but rather some unknown organospecies, especially for samples obtained from areas of high water age. They also developed an HPLC method employing amperometric detection (Amiri and Andrews 2007, 2008) and compared its measurements of chloraminated samples to those from the simple subtraction method to confirm the above results (Gagnon et al. 2008). Lee et al. (2007) reported that the HACH MonorchlorF test was a true measure of inorganic monochloramine without organochloramine interference when compared to MIMS.

Methods for ammonia-N. Total ammonia-N can be determined spectrophotometrically or with an ammonia-selective electrode (APHA et al. 2017), generally referred to as an ion-selective electrode (ISE). The free ammonia-N concentration can also be determined with an ammonia-selective electrode or spectrophotometer. The ammonia-selective electrode method is not practical for field measurements due to the lack of sensitivity of the electrode and the fact that the calibration standards should be the same temperature as the sample. Some analysts add a known concentration of ammonia to the sample to improve the accuracy of the ammonia-selective electrode method if the ammonia concentration is below 0.1 mg/L-N. The range of the ISE method is 0.03–1,400 mg/L-N. Less common methods for ammonia are the manual or automated phenate method (APHA et al. 2017).

As discussed above, HACH Method 10200 (indophenol method) allows monitoring of both free ammonia and monochloramine.

For measurement of total ammonia-N, a colorimetric method with a range of 0.01–0.50 mg/L is also available from HACH and other vendors. In the HACH method, ammonia reacts with chlorine to form monochloramine, and the monochloramine reacts with salicylate to form 5-aminosalicylate. The 5-aminosalicylate reacts with a sodium nitroprusside catalyst to form a blue color that is masked by excess reagent to form a green color. The concentration is determined by absorbance of the sample at 655 nm. Field measurements using this method have produced mixed results. This method distinguishes between free and combined ammonia.

Ammonia results should be documented as "ammonia as nitrogen" (ammonia-N or NH_3-N) because all chlorine-to-ammonia ratios are based on ammonia-N. Ammonia can be expressed as nitrogen (ammonia-N) by multiplying the ammonia concentration by 0.824 (or 14/17—the atomic mass of nitrogen divided by the molecular mass of ammonia). When performing an analysis for ammonia, it is important to note whether the test equipment provides the result as ammonia or ammonia as nitrogen. The same is true for nitrite and nitrate (results should be reported as nitrite-N or nitrate-N).

Nitrifying microorganisms. Although it is now recognized that microorganisms other than AOB and NOB can be present in drinking water distribution systems and be responsible for nitrification, most analytical methods used to date have focused on AOB and NOB. Available methods can be separated by type: culturing techniques that rely on growth of microorganisms followed by enumeration, nonculturing techniques that rely on microscopy, and molecular methods. The advantages and disadvantages of these methods are summarized in Table 6-4 and further detailed in Chapter 5.

Although it may be tempting to use the simpler culturing technique such as multiple tube fermentation with most probable number (MPN) to detect nitrifying bacteria in drinking water, results should be interpreted carefully because they only selectively detect 0.1–5% of the total community of nitrifying bacteria (e.g., Zhang et al. 2009; Pintar et al. 2005; Wolfe et al. 1988). The low sensitivity of the MPN technique has also been

Table 6-4 Advantages and disadvantages of analytical methods for nitrifying bacteria

Method	Advantages	Disadvantages
Culturing techniques	• Relatively easy to perform • Quantification of microbial density possible • Indicator of viable microorganisms	• High sample volume required • Low recovery (0.1–1 percent in environmental samples) • Underestimation of the number of microorganisms present because of the inability of any media to support the growth of all nitrifying bacteria, the presence of metabolically active but not culturable bacteria, and possible cell aggregation • Counts may vary depending on the culturing media used • Long turnaround time (the incubation period is 3–4 weeks)
Microscopy-based and serological techniques	• Faster than culturing techniques • Speciation possible	• May require initial isolation of pure cultures • Inability of any antibodies to capture all nitrifying bacteria • More complex than culturing techniques
Molecular methods	• Fastest technique • Speciation possible	• Different probes/primers are required to identify all nitrifying bacteria • Relatively complex; not readily usable by water systems • Unless a more complex molecular method is used, typically do not measure viability

documented. For example, Carrico and colleagues (2008) did not detect nitrifying bacteria in bulk water of full-scale distribution systems despite chemical evidence of nitrification. They suspected that nitrifying bacteria were mainly attached to pipe walls and may not have been "present in high enough concentrations in the bulk water to be detected." Pintar and colleagues (2005) came to similar conclusions when studying bench- and full-scale distribution systems.

Biological activity reaction tests (BARTs) have been used to detect various microorganisms, including blue-green algae, heterotrophic bacteria, iron- and sulfate-related bacteria, *Pseudomonas*, and slime-forming bacteria, as well as nitrifying and denitrifying bacteria. The BART for nitrifying bacteria (N-BART) is manufactured by Droycon Bioconcepts, Inc. (Regina, Saskatchewan) and distributed by HACH, Inc. N-BART is a simple test that uses nitrite as an indicator of the presence of nitrifying bacteria. Samples are incubated for 5 days, after which the bacterial density can be estimated by the intensity of the color that has developed. Results are semiquantitative. The rapidity and easiness of the N-BART are clear advantages.

Although N-BART has been successfully used in several other studies, one of its drawbacks is its low sensitivity, i.e., approximately 10^3 cfu/mL. Thus, positive signals are obtained only in presence of a substantial concentration of nitrifying bacteria, i.e., when nitrification is well underway. Another drawback is that the N-BART represents AOM only because it measures nitrite; it does not represent NOB because it does not measure nitrate. Finally, the N-BART estimates the density of AOM present in a sample, but it does not reflect microbial speciation.

Heterotrophic plate count. Several methods are available to monitor for heterotrophic bacteria in drinking water. Each method presents advantages and drawbacks (APHA et al. 2017):

- In the pour plate method, a sample is pipetted into a sterile plate, then warm (45–46°C) sterile growth medium is poured onto the plate and mixed with the sample before incubation. Colonies can be embedded into the solidified growth medium or present on the surface of the medium. Drawbacks include the limited sample size that can be processed to prevent dilution of the growth medium (typically 2.0 mL or less) and the addition of a warm growth medium that can induce heat shock to bacteria.
- In the spread plate method, a small volume of sample is pipetted onto the surface of a solidified growth medium in a sterile plate. The sample is spread evenly over the surface, and the plate is incubated. Colonies develop only on the surface of the growth medium. Compared to the pour plate method, colonies can be easily distinguished and can be transferred to other media as needed. One of the advantages of the spread plate method is the prevention of bacterial heat shock during sample processing. However, sample size is limited to 0.5 mL or less.
- As its name implies, the membrane filtration method involves sample filtration through a membrane, deposition of the membrane on top of a solidified growth medium, and incubation. Although higher sample volumes can be processed with the membrane filtration method, only samples with low turbidity can be processed, and the filtration process can damage bacterial cells if filtration pressure is too high.

Variations also exist for the media (APHA et al. 2017). Plate count agar, which is made of tryptone glucose yeast agar, is used for the pour plate and spread plate methods, whereas the m-HPC agar is only used for the membrane filtration method. R2A and NWRI agars can be used with all three methods. The plate count and m-HPC agars are high-nutrient growth media, whereas R2A and NWRI agars are low-nutrient media. Because heterotrophic bacteria present in drinking water are adapted to low-nutrient environments, several studies have shown lower counts with high-nutrient media compared to low-nutrient media.

The high-nutrient plate count and m-HPC agars are typically incubated at 35°C for 48 h, which is consistent with the SWTR requirements (40 CFR 141.74). The low-nutrient media are typically incubated at lower temperature and for a longer period: with R2A agar, best results are obtained at 28°C and a 7-day incubation, whereas a 20°C incubation for 7 days tends to yield higher bacteria counts with the NWRI agar (APHA et al. 2017). Because the lower incubation temperature is more representative of distribution systems, it tends to provide higher bacterial counts (APHA et al. 2017).

After sample plating and incubation, colonies of bacteria are counted. Each colony represents thousands of individual bacterial cells that are counted as a single cfu/volume of the sample.

Because the low-nutrient R2A agar provides higher counts and is more representative of the drinking water environment, it is recommended as a nitrification indicator. However, the long incubation period of the HPC-R2A method means that results are not available until 7 days after sample incubation. Therefore, HPC-R2A results should be used to monitor long-term trends in the distribution system, in conjunction with other nitrification monitoring parameters. Because factors other than nitrification (such as increased water age and decrease in disinfectant residual) can lead to HPC increases within a distribution system, HPC can be used as an indicator of nitrification but should not be the sole identifying factor.

Adenosine triphosphate (ATP) bioluminescence assay. This assay measures total density of viable bacteria in a sample. Tests can be completed in less than 1 h. In the ATP firefly bioluminescence assay, the enzyme luciferase, luciferin, magnesium ions, and ATP react together to produce light. Because the level of light that is emitted from the sample is proportional to the ATP concentration, the light intensity is quantitatively measured and directly correlated with a quantity of ATP extracted from a known concentration of bacteria (APHA et al. 2017; AWWA, 2010). The pulse of light is typically detected using a luminometer as relative light units (RLU), and results are reported in picograms of cellular ATP per milliliter (pg cATP/mL).

ATP analysis typically relies on commercially available kits, such as the Clean-Trace Water–Total ATP Test (3M Food Safety, St. Paul, Minn.) or the Quench-Gone Aqueous Test Kit (LuminUltra Technologies, Ltd., Fredericton, New Brunswick). The method sensitivity is approximately 0.1 pg cATP/mL. In drinking water, LuminUltra proposes ATP concentrations of less than 0.5 pg cATP/mL for waters with minimal biological activity. ATP concentrations ranging from 0.5–10 pg cATP/mL indicate that water quality may be compromised, and preventive actions should be considered, whereas concentrations greater than 10 pg cATP/mL indicate that biological control has been compromised and corrective actions should be undertaken. ATP analysis can also be performed on biofilm samples, in which case results are reported as pg cATP/cm^2. ATP concentration is affected by the bacterial density, microbial species present, and physiological states of those species (ASTM 2015). Interferences include high concentrations of dissolved organic carbon and heavy metals (ASTM 2015). The ATP tests do not differentiate between ATP from different microbial sources, such as bacteria, fungi, algae, and protozoa (ASTM 2015).

Like many other parameters, ATP needs to be monitored for a sufficient period in absence of nitrification to determine baseline (or reference) ATP concentrations to which subsequent results can be compared.

Water temperature. Some pH instruments with temperature compensation display the temperature of the water when pH is measured. If such instruments are used, temperature can easily be included if pH is measured in the field. In addition to the use of thermistors (probes), a mercury-filled thermometer with 0.1-degree intervals is typically used for temperature measurement (APHA et al. 2017). Temperature measurements should be calibrated periodically according to *Standard Methods* (APHA et al. 2017).

pH. pH is determined by an electrometric method (APHA et al. 2017). The pH meter should be calibrated at the beginning of each series of tests. Either a two-point or three-point calibration may be used. The fact that calibration standards match the sample temperature when pH analyses are performed in the laboratory rather than in the field accounts in part for increased precision and accuracy in laboratory analyses. However, pH needs to be measured within 15 min of sample collection, which dictates the need to measure pH in the field. In any case, the level of accuracy of field pH measurements is generally sufficient for both process control and system monitoring. In low-alkalinity waters, additional quality assurance measures, such as use of low ionic strength pH probes, can improve the accuracy and precision of the measurements.

CASE STUDIES

Two case studies are described below, highlighting utilities that had historically experienced nitrification in their distribution systems and were proactive in adopting measures to successfully mitigate nitrification in their systems.

Philadelphia Water Department

Philadelphia Water Department (PWD) is a combined water and wastewater urban utility that serves treated drinking water to 1.6 million customers in Philadelphia, Pa. Chloramine is used as a secondary disinfectant in the distribution system.

PWD's approach to nitrification detection and prevention is based on utility experience with nitrification prevention and distribution system optimization guidance available through AWWA's Partnership for Safe Water. During warm weather months, the following actions are implemented to prevent nitrification and maintain total chlorine residual.

- Total chlorine residual levels leaving the treatment plant are incrementally boosted as source water temperatures increase. This boost is implemented to prevent or counteract the onset of nitrification incidents and to maintain total chlorine residual goals within the drinking water distribution system.
- Finished water storage facilities are closely monitored for total chlorine residual and turnover rates to prevent degradation of water quality.
- Online monitoring for total chlorine residual, pH, and water temperature, along with other general water quality parameter (including nitrite, ammonia, and HPC) grab sampling, is routinely conducted throughout the distribution system, with more frequent monitoring in areas where nitrification is detected.
- Flushing is initiated or increased in areas where nitrification has historically been an issue.

PWD maintains an extensive network of water quality sampling locations that include POEs, inlets/outlets of storage facilities, and various public and private facilities within the drinking water distribution system. Among all water quality parameters that are monitored, total chlorine residual, nitrite levels, and water temperature are used to detect nitrification incidents. Total chlorine residual, HPC, and water temperature are monitored at all locations; nitrite is monitored at locations within areas that typically experience higher water age. Increases in nitrite concentration, along with corresponding decline in total chlorine residual, are typically the first indication that a nitrification incident is occurring.

In addition to collecting grab samples, permanent OWQM stations are located at key water quality monitoring locations within the distribution system to continuously monitor total chlorine residual, pH, ammonia, conductivity, and temperature that provide general water chemistry trends using data analytics in PWD's data dashboard. Examples of OWQM and grab sample data that have been collected are illustrated in Figure 6-5.

During warm weather months, several portable OWQM stations are deployed in problem areas that do not have existing permanent OWQM stations. PWD primarily monitors the total chlorine residual and temperature trends at these sites to identify nitrification. Conductivity data produced at these stations can also be useful for determining which water source(s) are supplying the site with water at a given time. Source identification and hydraulic modeling help determine water age at individual locations to better aid in establishing follow up actions.

Flushing volumes can be increased based on total chlorine residual trends (both grab and online), nitrite, and HPC data. Nitrification specific thresholds have been established to generate automatic alerts when total chlorine residual falls below 1 mg/L Cl_2, HPCs are detected, or nitrite levels rise above the detection limit.

Investigations of water quality involve hydraulic investigations of the affected area, field measurements of total chlorine residual and nitrite with handheld instruments, collection of grab samples (total chlorine residual, nitrite, ammonia, and HPC), and implementation of response actions (e.g., flushing).

Investigations can be triggered in the following cases for individual sites.

- Daily water quality reports from the online systems and laboratory alerts generated from the analysis of routine sampling indicate that total chlorine residual has fallen below 1 mg/L Cl_2.

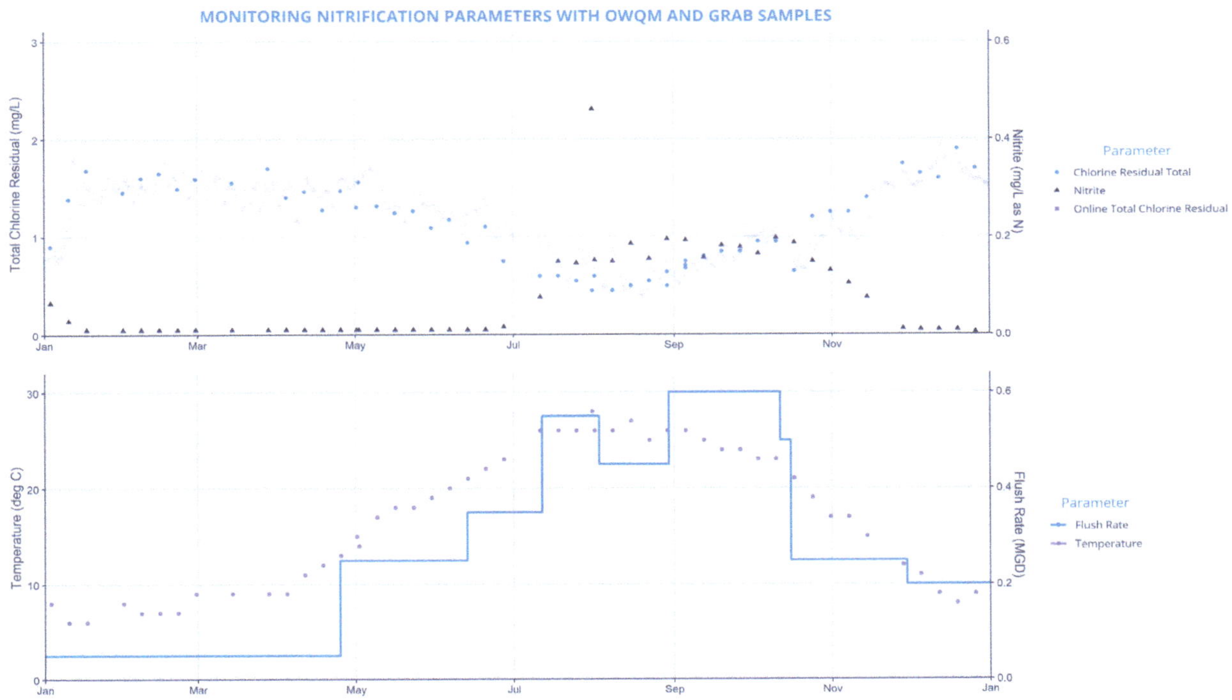

Source: Philadelphia Water Department

Figure 6-5 Typical water quality trend at a site experiencing nitrification in the summer months. The top figure shows both online water quality monitoring and grab total chlorine residual results as well as nitrite. The bottom figure shows the trend in temperature and the continuous flushing rate employed at the site.

- OWQM trends indicate a decline in total chlorine residual over time.
- Nitrite, ammonia, and HPC trends indicate escalating nitrification activity.
- Team reviews of water quality (which occur biweekly during warm weather months) identify potential issues with water age or nitrification.

The availability of OWQM data provides PWD with a better understanding of the changing conditions in the distribution system to better inform system operations and improve flushing activities.

East Bay Municipal Utility District

The East Bay Municipal Utility District (EBMUD) serves 1.4 million customers in Oakland, Calif., and surrounding areas in Alameda and Contra Costa counties. Chloramine is used as a secondary disinfectant in the finished water. Due to the large number of storage facilities (more than 150) and pressure zones (more than 120) over the 332-square mile service area, residence times in some parts of the system can be very long, so localized nitrification can occur.

In the late 1990s, when EBMUD first converted from free chlorine to chloramine, a large project was conducted to install online chlorine analyzers in distribution reservoirs. A total of 26 units were purchased, installed, and connected to EBMUD's SCADA system. Unfortunately, these devices had many problems.

- The units required frequent maintenance and calibration, and reagents needed to be replenished frequently, taxing limited staff resources.
- Some were located on the inlet/outlet lines of reservoirs or in nearby pumping plants, resulting in nonrepresentative samples.
- Disposal of waste streams containing spent reagents was problematic since many reservoirs are located in areas with limited or no utility power and no local sewer access.
- Operators never developed confidence in the accuracy of the numbers displayed on SCADA, and maintenance was deferred on most units.

By about 2005, the online chlorine analyzers had been taken offline. Periodic grab samples were used to monitor water quality in the reservoirs and respond to nitrification events.

Starting in 2018, a new monitoring program was started to install newer-generation chlorine analyzers in distribution reservoirs. These analyzers were better suited for remote distribution locations, since they do not use reagents and require less frequent maintenance and calibration compared to the older units. EBMUD used several years of grab sample data that had been collected to select 15 OWQM locations at reservoirs that periodically experienced nitrification problems. EBMUD applied important lessons learned from the initial analyzer installations to locate sample taps in areas that were representative of water quality in particular zones. Some of the monitoring stations are operated via solar power.

For all in-service reservoirs, grab samples are analyzed twice monthly for total chlorine, nitrite, pH, water temperature, and conductivity. For the reservoirs with online analyzers, these grab samples help to verify the accuracy of the online devices and trigger maintenance orders when the readings are outside of tolerance.

Because these new units have been properly sited and regularly maintained, operators have developed confidence in the results and use them for making operational decisions. Before the installation of the online analyzers, EBMUD was deciding whether to breakpoint-chlorinate each reservoir based on the analysis of two grab samples per month. Now, the OWQM data are used along with the grab sample data to determine if other less-drastic options, such as deep cycling, might be more appropriate.

KEY TAKEAWAYS

A system-specific nitrification monitoring program can track trends in water quality and operational parameters related to nitrification and trigger rapid and appropriate response when needed. The program should be evaluated periodically for its suitability and ability to rapidly reveal any deviations from baseline or reference levels. If responses cannot be undertaken rapidly, nitrification may escalate and become difficult to control, requiring disruptive actions such as breakpoint chlorination of a storage reservoir, for example. Nitrification monitoring and control plans should be revised periodically to capture changes in treatment processes, operations, water quality, or others. A review of a nitrification plan provides an opportunity to reassess monitoring parameters, sampling locations and frequency, analytical methods, and baseline levels. Data should be analyzed and plotted to develop long-term trends and then safely archived.

Distribution system water quality is often assessed on a system-wide basis and is useful for process control. This approach, however, is counterproductive to nitrification control because nitrification is rarely a system-wide phenomenon, occurring instead at low-flow or high water–age areas like remote sites and storage facilities. Nitrification monitoring and control should consider the trends of individual sampling locations, taking into account that some sites may have different water sources at different times.

REFERENCES AND BIBLIOGRAPHY

American Society for Testing and Materials (ASTM). 2015. *ASTM D4012-15 Standard Test Method for Adenosine Triphosphate (ATP) Content of Microorganisms in Water*. West Conshohocken, PA: ASTM.

Amiri, F., and S.A. Andrews. 2007. Quantitative Analysis of Organic and Inorganic Fractions of Chloramines in Drinking Water. *Proceedings, AWWA Distribution Research Protection Symposium*, Reno, NV, March 2–3, 2007.

Amiri, F., and S. Andrews. 2008. "Development of a Size Exclusion Chromatography—Electrochemical Detection Method for the Analysis of Total Organic and Inorganic Chloramine." *Journal of Chromatographic Science*, 46:7:591–595.

Andrews, S.A. 2001. Influence of Organic Chloramines in Distribution System Chlorine Residuals. In the Special Session "Chloramination Practices: Current Status" at the American Water Works Association Annual Conference and Exhibition, Washington, D.C., June.

APHA, AWWA, and WEF (American Public Health Association, American Water Works Association, and Water Environment Federation). 2017 (23rd ed). *Standard Methods for the Examination of Water and Wastewater*. Washington, D.C.: APHA.

ANSI/AWWA (American National Standards Institute/American Water Works Association). 2021. Distribution Systems Operation and Management. ANSI/AWWA G200. Denver: AWWA.

Arp, D.J., L.A. Sayavedra-Soto, and N.G. Hommes. 2002. "Molecular Biology and Biochemistry of Ammonia Oxidation by *Nitrosomonas europea*." *Archives of Microbiology*, 178:4:250–255.

AWWA. 2006 (2nd ed.). *Water Chlorination/Chloramination Practices and Principles*. Denver: AWWA.

AWWA. 2010. *The Water Dictionary: A Comprehensive Reference of Water Terminology*, 2nd Edition. N.E. McTigue, J.M. Symons, editors. Denver: AWWA.

Carrico, B.A., F.A. Digiano, N.G. Love, P. Vikesland, K. Chandran, M. Fiss, and A. Zaklikowski. 2008. "Effectiveness of Switching Disinfectants for Nitrification Control." *Journal AWWA*, 100:10:104–115.

Chain, P., J. Lamerdin, F. Larimer, W. Regala, V. Lao, M. Land, L. Hauser, A. Hooper, M. Klotz, J. Norton, L. Sayavedra-Soto, D. Arciero, N. Hommes, M. Whittaker, and D. Arp. 2003. "Complete Genome Sequence of the Ammonia-Oxidizing Bacterium and Obligate Chemolithoautotroph *Nitrosomonas europea*." *Journal of Bacteriology*, 185:9:2759–2773.

Chapin, R.M. 1929. "Dichloro-Amine." *Journal of the American Chemical Society*, 51:7:2112–2117.

Connell, G.F. 1996. *The Chlorination/Chloramination Handbook*. Denver: AWWA.

El-Farra, S., and S.A. Andrews. 2000. Extent of Organochloramine Formation in a Drinking Water Distribution System. Proceedings, American Water Works Association Water Quality Technology Conference, Salt Lake City, Utah, November.

Donlan, R.M. and W.O. Pipes. 1988. "Selected Drinking Water Characteristics and Attached Microbial Population Density." *Journal AWWA*, 80:11:70–76.

Fleming, K.K., G.W. Harrington, and D.R. Noguera. 2005. "Nitrification Potential Curves: A New Strategy for Nitrification Prevention." *Journal AWWA*, 97:8:90–99.

Fleming, K.K., G.W. Harrington, and D.R. Noguera, 2008. "Using Nitrification Potential Curves to Evaluate Full-scale Drinking Water Distribution Systems." *Journal AWWA*, 100:10:92–103.

Gagnon, G.A., H. Baribeau, S.O. Rutledge, R. Dumancic, A. Oehmen, C. Chauret, and S. Andrews. 2008. "Disinfectant Efficacy in Distribution Systems: A Pilot-scale Assessment." *Journal of Water Supply: Research and Technology*, 57:7: 507–518.

Grayman, W.M., L.A. Rossman, C. Arnold, R.A. Deininger, C.D. Smith, J.F. Smith, and R. Schnipke. 2000. *Water Quality Modeling of Distribution System Storage Facilities*. Project #260. Denver: AWWA Research Foundation.

Gupta, K., and Y. Akagi, 2014. "Nitrification Monitoring and Control: A Utility Case Study." *Proceedings* 2014 AWWA WQTC, New Orleans, Louisiana.

Ike, N.R., R.L. Wolfe, and E.G. Means. 1988. "Nitrifying Bacteria in a Chloraminated Drinking Water System." *Water Science and Technology*, 20:11–12:441–444.

Kirmeyer, G., K. Martel, G. Thompson, L. Radder, W. Klement, M. LeChevallier, H. Baribeau, and A. Flores. 2004. *Optimizing Chloramine Treatment*, 2nd edition. Project #2760. Denver: AWWA Research Foundation.

Lee, W., P. Westerhoff, X. Yang, and C. Shang. 2007. "Comparison of Colorimetric and Membrane Introduction Mass Spectrometry Techniques for Chloramine Analysis." *Water Research*, 41:14:3097–3102.

Monks, V., R. Song, and E. Zhu, 2010. Managing Chloramines as a Secondary Disinfectant in the Distribution System. *Proceedings 2010 AWWA WQTC*, Savannah, Georgia.

Narasimhan, R., J. Brereton, M. Abbaszadegan, A. Alum, and P. Ghatpande. 2004. *Sample Collection Procedures and Locations for Bacterial Compliance Monitoring*. #2676. Denver: Awwa Research Foundation.

Odell, L.H., G.J. Kirmeyer, A. Wilczak, J.G. Jacangelo, J.P. Marcinko, and R.L. Wolfe. 1996. "Controlling Nitrification in Chloraminated Systems." *Journal AWWA*, 88:7:86–98.

Palin, A.T. 1950. "A Study on the Chloro-Derivatives of Ammonia and Related Compounds with Special Reference to their Formation in the Chlorination of Natural and Polluted Waters." *Water and Water Engineering*, 151(Oct. 1950), 189 (Nov. 1950), 248 (Dec. 1950).

Pintar, K.D.M., W.B. Anderson, R.M. Slawson, E.F. Smith, and P.M. Huck. 2005. "Assessment of a Distribution System Nitrification Critical Threshold Concept." *Journal AWWA*, 97:7:116–129.

Powell, R., 2004. Implementation of Chloramination by a Florida Utility: The Good, The Bad and The Ugly. *Proceedings 2004 AWWA WQTC*, San Antonio, Texas.

Sathasivan, A., I. Fisher, and G. Kastl. 2005. "Simple Method for Quantifying Microbiologically Assisted Chloramine Decay in Drinking Water." *Environmental Science & Technology*, 39:14:5407–5413.

Sathasivan, A., I. Fisher, and G. Kastl. 2010a. "Application of the Microbial Decay Factor to Maintain Chloramine in Large Tanks." *Journal AWWA*, 102:4:94–103.

Sathasivan, A., K.C.B. Krishna, and I. Fisher. 2010b. "Development and Application of a Method for Quantifying Factors Affecting Chloramine Decay in Service Reservoirs." *Water Research*, 44:15:4463–4472.

Skadsen, J. 1993. "Nitrification in a Distribution System." *Journal AWWA*, 85:7:95-103.

Smith, C.D. 2005. *Evaluation of Hach Method 10200 for Monochloramine and Free Ammonia-N*. Prepared for the Hach Company by Charlotte Smith & Associates, Inc. Jan. 27, 2005.

Smith, C.D., and J.F. Smith. 1996. Excessive Loss of Chloramine Residual: A Case Study. *In Proc. of the AWWA Annual Conference*. Denver, Colo.: AWWA

Sung, W., 2011. Nitrification Monitoring at MWRA. *Proceedings 2011 WQTC*, Phoenix, Arizona.

Sung, W. 2017. "Recommendations for Nitrification Monitoring: MWRA's Experience." *Journal of New England Water Works Association*, CXXXI:3:151–155, September.

USEPA (United States Environmental Protection Agency). 2018. *Online Water Quality Monitoring in Distribution Systems*. Washington, D.C.: USEPA.

Valentine, R.L. 1984. "Disappearance of Monochloramine in the Presence of Nitrite." *Water Chlorination, Volume V: Chemistry, Environmental Impact and Health Effects*. Lewis Publisher, Inc., Chelsea, Mich.

Volk, C.J., and M.W. LeChevallier. 2000. "Assessing Biodegradable Organic Matter." *Journal AWWA*, 92:5:64–76.

Wahman, D.G. 2018. "Web-Based Applications to Simulate Drinking Water Organic Chloramine Chemistry." *Journal AWWA*, 110:11: E43–E61.

Wahman, D.G., and J.G. Pressman. 2015. "Distribution System Residuals—Is "Detectable" Still Acceptable for Chloramines?" *Journal AWWA*, 107:8:53–63.

Wahman, D.G., 2017. Chlorine Breakpoint Curve Simulator. Version 0.25. https://usepaord.shinyapps.io/Breakpoint-Curve/ (accessed February 27, 2025).

Whalen, P.A., D.R. Tracey, and J. Duguay. 2018. "Adenosine Triphosphate (ATP) Measurement Technology." In: *Microbiological Sensors for the Drinking Water Industry*. T.L. Skovhus and B. Hojris, Editors. London: IWA Publishing:137–154

Wilczak, A., L.L. Hoover, and H.H. Lai. 2003. "Effects of Treatment Changes on Chloramine Demand and Decay." *Journal AWWA*, 95:7:94–106.

Wilczak, A., J.G. Jacangelo, J.P. Marcinko, L.H. Odell, and G.J. Kirmeyer. 1996. "Occurrence of Nitrification in Chloraminated Distribution Systems." *Journal AWWA*, 88:7:74–85.

Wolfe, R.L. and N.I. Lieu. 2003. "Nitrifying Bacteria in Drinking Water." *Encyclopedia of Environmental Microbiology*. G. Britton, editor. New York: Wiley & Sons: doi.org/10.1002/0471263397.env228.

Wolfe, R.L., E.G. Means III, M.K. Davis, and S.E. Barrett. 1988. "Biological Nitrification in Covered Reservoirs Containing Chloraminated Water." *Journal AWWA*, 80:9:109–114.

Wood, P. 1986. "Nitrification as a Bacterial Energy Source." In: *Nitrification*. Prosser, J.I., editor. Oxford: Oxford University Press:39–62.

Zhang, Y., and M. Edwards. 2009. "Accelerated Chloramine Decay and Microbial Growth by Nitrification in Premise Plumbing." *Journal AWWA*, 101:11:62–51.

Zhang, Y., N. Love, and M. Edwards. 2009. "Nitrification in Drinking Water Systems." *Critical Reviews in Environmental Science and Technology*, 39:3:153–208.

Zhu, E., V. Monks, and R. Song, 2010. Effectiveness of Early Application of Sodium Chlorite in Nitrification Prevention–A Demonstration-Scale Study. *Proceedings 2010 AWWA WQTC*, Savannah, Georgia.

Chapter 7

Operational and Treatment Practices to Prevent Nitrification

Susan Teefy, Eric Zhu

INTRODUCTION

This chapter focuses on operational and treatment measures to prevent nitrification in chloraminated systems. Prevention strategies can be characterized as those that limit excess ammonia in water delivered to the systems (limiting food), those that reduce water age (less time for ammonia-oxidizing microorganisms [AOM] to grow, less chance of low disinfectant residuals), and those that limit the growth of AOM (pH adjustment, chlorite, ultraviolet [UV] light). The choice of strategy or combination of strategies is site specific.

The recommendations in this chapter are based on real situations encountered by water utilities throughout the world. Some of the recommended approaches may or may not work at each utility due to differences in water quality, system materials, system demand, system configuration, and other utility specific factors. Many of the prevention approaches discussed in this chapter coincide with general water quality management practices that most utilities are familiar with and use, such as distribution system flushing, storage facility turnover and cleaning, and good chemical feed control. It is important to keep in mind that AOM are ubiquitous in the environment and quite resilient. Each of the nitrification prevention approaches described in this chapter is intended to manage nitrification, not "cure" it. Drinking water in a distribution system is disinfected; it is not sterilized, and some AOM will inevitably survive. The goal of these measures is to keep nitrification in check and prevent adverse water quality impacts, not to completely eliminate AOM from the system.

This chapter also discusses several specific approaches to nitrification prevention, including monochloramine residual boosting, blending of different water qualities, elevation of pH, application of chlorite/chlorine dioxide, and use of UV. Table 7-1 summarizes the recommendations.

Table 7-1 Key points from Chapter 7

Chemical dosing and feed control	• Limit excess free ammonia leaving the treatment plant, at least <0.10 mg/L as N, preferably <0.05 mg/L as N. However, even tight control of excess free ammonia in the finished water has not always been effective and must be combined with other practices to minimize chloramine decay or ammonia release in the distribution system. • Maintain chlorine-to-ammonia-N weight ratio leaving the treatment plant between 4.5:1 and 5:1 to limit free ammonia. • Design ammonia injection ports for redundancy, promotion of rapid, thorough mixing, and prevention or removal of scaling. • Provide good mixing of both chlorine and ammonia to properly control monochloramine formation. • When carrier water is used, it should be softened to avoid precipitation and clogging of the ammonia and sodium hypochlorite feed systems. • Monitor the source water for ammonia concentrations; if ammonia is present in the source water, adjust the chlorine and ammonia feeds accordingly.
Monochloramine residual	• Maintain a minimum concentration of monochloramine leaving the treatment plant: recommended >2.0 mg/L. (The optimal value is site specific.) • Maintain a goal of the monochloramine concentration at all monitoring points in the distribution system at 1.0 mg/L or higher, depending on system and site requirements. Based on experience, several utilities have established internal goals of 1.0–2.0 mg/L monochloramine or higher for all distribution system sites.
Storage facility operation	• Minimize water age in storage facilities. Storage turnover is accomplished when a percentage of the tank volume is being drawn and filled every day or every few days. The amount of turnover must be sufficient to remove old water and provide thorough mixing of the contents of the storage facility. It is generally recommended that at least 30 percent turnover be achieved daily. Utilities may adjust this goal based on experience and system requirements. • Maximize mixing and cycling in storage reservoirs and eliminate short-circuiting and dead spaces. There are several effective methods to improve mixing in storage facilities, including adding mixers, modifying inlet and outlet structures, and increasing the fill time while maintaining sufficient inlet velocity to improve mixing. Modeling can be beneficial in understanding storage performance and water age issues. These actions are discussed in more detail in Chapter 9. • Routinely inspect and clean storage tanks and reservoirs, ideally at least once every 5 years. • Analyze necessary versus available storage in facilities that are not cycling sufficiently. Decommissioning excess storage can be a long-term solution to nitrification problems. Seasonal decommissioning may also be a practical approach to minimizing water age during periods of the year when demands are lower. • Install, maintain, and monitor online chlorine residual analyzers where possible. Measuring chlorine residual can provide an early detection of water quality issues at a storage facility.

(continued)

Table 7-1 Key points from Chapter 7 (*Continued*)

Distribution system operation	• Decrease water age in the distribution system. Based on hydraulic modeling and water quality testing, areas of high water age can be identified, and approaches to reducing age can be developed. Examples of such actions include looping of water mains to eliminate dead ends and proper sizing of water mains. • Practice systematic flushing. System-wide unidirectional flushing (UDF) is better than conventional (limited areas) flushing as it improves removal of biofilms, removes a greater amount of sediment, and is less expensive per mile of flushed pipeline. • Routinely flush dead-end and other low-flow areas. Routine flushing is an established practice for maintaining distribution system water quality in areas of high water age. Automatic flushing may be an appropriate approach to minimize staff time. In a utility survey, hydrant flushing was rated as one of the most important perceived methods for prevention of nitrification, although this is site specific. However, once nitrification is established, flushing alone is unlikely to control nitrification. • Evaluate periodic switch to free chlorine. Free chlorination is an effective strategy for controlling and even stopping nitrification episodes once they have begun, but it has been met with mixed success for nitrification prevention and can result in increases in disinfection byproduct concentrations, increased turbidity, and biofilm sloughing in some cases. It is required by some states. It is likely to be more beneficial if combined with UDF. • Provide good corrosion control. Corroded pipes can harbor ammonia-oxidizing microorganisms, shielding them from disinfectants and making them more difficult to flush out. • Evaluate increasing the pH. The use of elevated pH to prevent nitrification has met with mixed success, and its impact on disinfection byproduct formation, stability of monochloramine residual, chemical costs, corrosion, and scale formation should be assessed. • Consider blending waters. Different water sources may have different water quality characteristics that impact nitrification: e.g., stability of monochloramine, pH, organic carbon concentrations, etc. Avoid blending waters with different disinfectants. • Evaluate booster chloramination or booster chlorination if sufficient ammonia is present. Recombining free ammonia released from chloramine decay and demand can increase the monochloramine residual and reduce the ammonia available for nitrifying bacteria.
Alternative nontraditional approaches	• Evaluate use of chlorite, either through direct application of chlorite or indirect chlorite feed via chlorine dioxide treatment. Chlorite may be effective for nitrification prevention but has been met with mixed success. The issue of adding a regulated substance to drinking water for prevention of nitrification needs to first be resolved by the appropriate regulatory agencies, the utility, and its customers. • Evaluate use of ultraviolet (UV) light in storage facilities. Limited work suggests that application of UV may be a promising emerging approach for waterborne AOM, particularly in storage facilities. UV will not penetrate biofilm, so it will only reduce waterborne organisms.

UTILITY PRACTICES IMPACTING NITRIFICATION

In a survey of more than 50 US utilities, Kirmeyer et al. (2004) examined the most common methods for prevention of nitrification and utilities' rating of effectiveness of the methods. Distribution system flushing, increasing the monochloramine residual, and modifying the chlorine-to-ammonia-N ratio were found to be the most common preventive methods and were perceived to be the most essential (Figure 7-1). However, the actual effectiveness of these approaches has been demonstrated to vary with the utility. Not all surveyed utilities had tried all these approaches.

In a 2017 disinfection practices survey conducted by the AWWA Disinfection Systems Committee, more than 70 utilities responded to questions about chloramine applications and related nitrification control measures (AWWA 2018a). About half (48 percent) of respondents indicated that they never experience problems with meeting target total chlorine residuals in their distribution systems. The rest reported difficulty meeting target residuals occasionally, one to three times per year (41 percent) or frequently/always (10 percent). Thirty-one of 70 (44 percent) utilities reported temporary free chlorine conversions. Of those reporting nitrification control plans and remediation actions, increased monitoring and periodic flushing were the most commonly reported actions (see Figure 7-2).

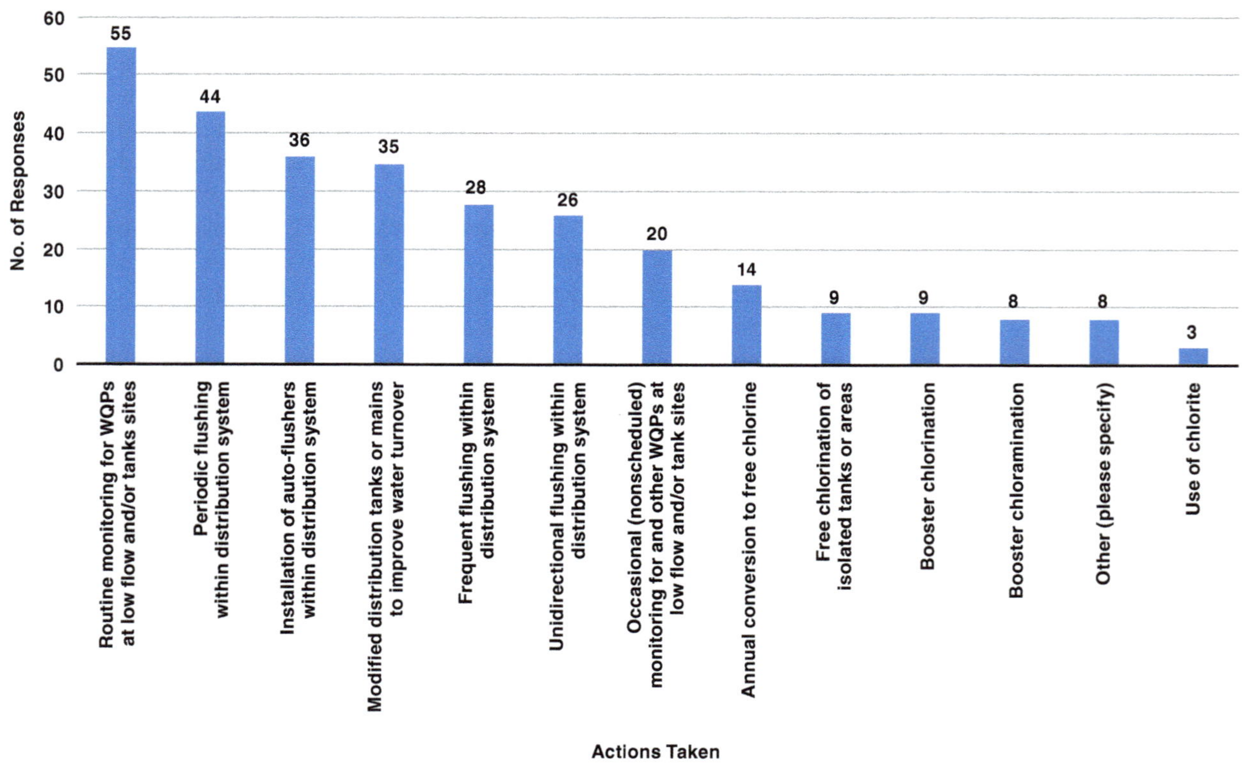

WQPs = water quality parameters

Figure 7-1 Actions taken by respondents to combat nitrification in distribution system

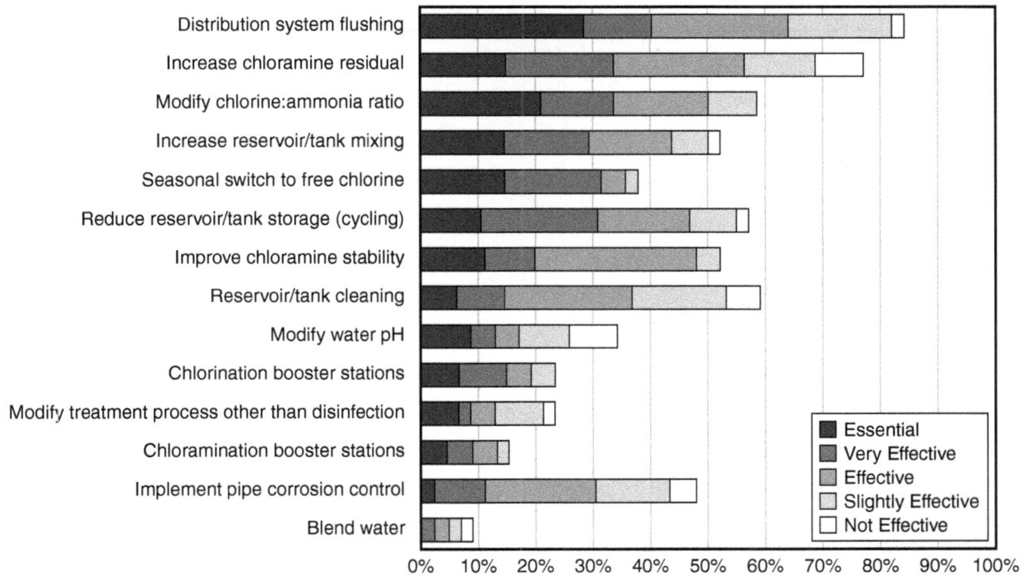

Source: Kirmeyer et al. 2004

Number of responding utilities = 50.

Figure 7-2 Utility practices and perceptions regarding prevention of nitrification

TRADITIONAL APPROACHES TO NITRIFICATION PREVENTION

Ammonia Control

Proper control of ammonia is necessary to establish good monochloramine formation and to avoid excess ammonia that will provide a nutrient source for growth of AOM.

Concentration and Ratio

Proper control of ammonia and chlorine dosages is essential for the prevention of nitrification. Because AOM thrive on ammonia, limiting the residual free ammonia available for these bacteria can be one of the major control strategies for preventing nitrification. It is recommended that excess free ammonia should be kept at less than 0.10 mg/L as N, preferably less than 0.05 mg/L as N. There have been numerous examples of ammonia overfeed contributing to nitrification (Kirmeyer et al. 1995; Skadsen 1993; Davis 1990). In one study, reducing the free ammonia level from 0.11 mg/L as N to 0.05 mg/L as N correlated with a reduction in nitrite concentration, which was used as an indicator of nitrification (Holt et al. 1995).

Holt et al. (1995) further examined the relationship between free ammonia and nitrification (as indicated by nitrite concentration, total chlorine concentration, and temperature) in a series of experiments in a test rig. Holt et al. observed simultaneously a decrease in the total chlorine residual, an increase in temperature, a decrease in ammonia concentration to below the method detection limit, and a spike in nitrite concentration. Results from one of the test rigs depicting these water quality changes are shown in Figure 7-3.

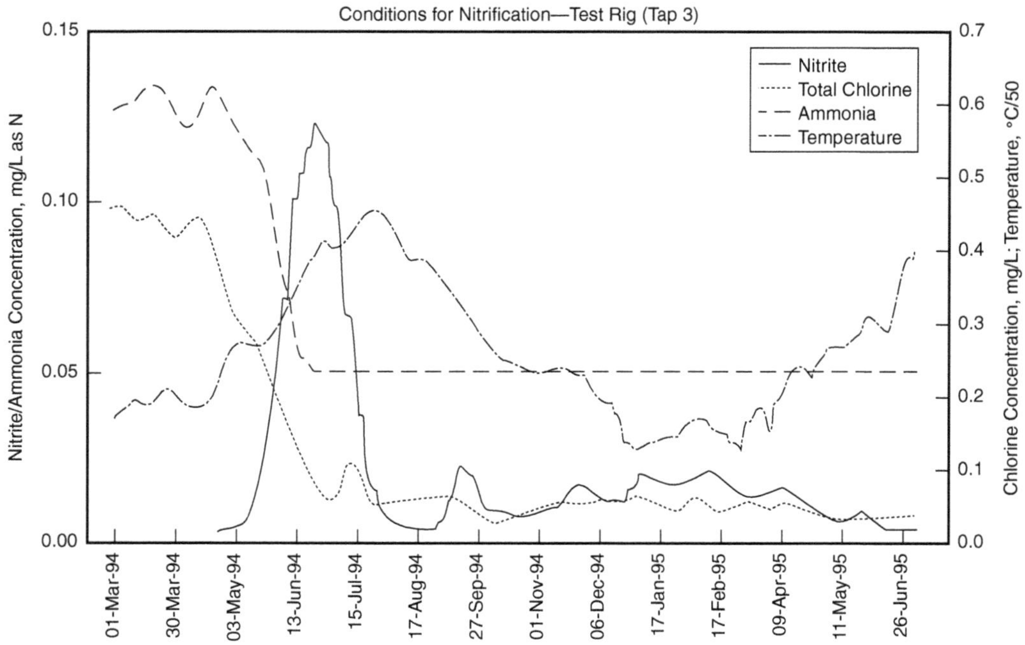

Source: Holt et al. 1995

Figure 7-3 Correlations between free ammonia, temperature, total chlorine residual, and nitrite concentration

To limit excess free ammonia, the proper weight ratio of chlorine to ammonia-N must be fed at the treatment plant when forming monochloramine. A weight ratio of 4.5:1–5:1 as chlorine to ammonia-N is recommended. In the 2008 AWWA survey of disinfection practices, the average and median chlorine-to-ammonia-N feed ratios were 4.6:1 and 4.5:1, respectively, with a mean/median free ammonia target level entering distribution systems of 0.3 and 0.1 mg/L, respectively, in over 70 chloraminated systems. A lower ratio around 3:1 has been implicated in some occurrences of nitrification (Davis 1990; Negrin et al. 1990; Wolfe et al. 1988). At the Metropolitan Water District of Southern California (MWDSC), maintaining a ratio of nearly 5:1 has reduced the incidence of nitrification (Kirmeyer et al. 2004). However, care must be taken to avoid overdosing; data from the East Bay Municipal Utility District (EBMUD) indicated a significant rise in distribution system disinfection byproducts (DBPs) and more rapid decay of monochloramine residual when the ratio slightly exceeded 5:1 (Work et al. 2019).

Note that even tight control of excess free ammonia and maintenance of proper chlorine-to-ammonia-N ratio have not always been effective at preventing nitrification (Kirmeyer et al. 2004; Skadsen 1993; Negrin et al. 1990). This is because monochloramine will decay or react with substances in the distribution system and release free ammonia, which will then support nitrification. Therefore, while maintaining a proper ratio and limiting excess ammonia at the treatment plant is a necessary operational practice, it must be combined with other practices that minimize the extent and rate of monochloramine decay and demand to consistently help prevent nitrification.

Types of Ammonia

Various types of ammonia chemicals have different benefits and drawbacks that may impact the formation of monochloramine and thereby increase the potential for nitrification. In a 2003 survey of more than 50 chloraminated utilities, only 4 percent used dry ammonium sulfate, 2 percent added no ammonia as sufficient concentrations were present in the source water, 44 percent used aqua ammonia (AA), and 50 percent used anhydrous ammonia (Kirmeyer et al. 2004). In the 2008 disinfection survey, just under half of utilities (8 of 18) using ammonia used anhydrous, with just over half (10 of 18) using AA and none using dry ammonium sulfate (AWWA Disinfection Systems Committee 2008). In the 2017 survey, over half (40 of 74) used AA, less than a third (22 of 74) used anhydrous, and 16 percent (12 of 74) used liquid ammonium sulfate (LAS). None reported using dry ammonium sulfate. Due to problems with clumping, dry feed is not recommended.

LAS feed systems are still relatively new but are being used more often in the water industry. LAS is an odorless liquid with a vapor pressure similar to water (Kirmeyer et al. 2004). It is relatively stable and does not undergo chemical degradation over time. LAS systems have multiple advantages over AA, including elimination of the need for a pressurized tank and off-gas adsorption and a higher ammonia content of 38 percent versus 20 percent. Furthermore, LAS systems do not need softened carrier water as they do not precipitate hardness (Kirmeyer et al. 2004). The handling and conveyance of LAS are similar to that of AA (Kirmeyer et al. 2004). The cost of liquid ammonium sulfate is variable; in some cases, it may be advantageous and in others more expensive. A cost analysis should be performed to assess this variable, taking into account potential impact on pH adjustment chemicals since LAS is a weak acid while AA is a base. The main advantage of LAS is its lower hazard and associated safety and code requirements.

Both anhydrous ammonia and AA are relatively easy to use, but they do have potentially significant safety considerations. Off-gassing may occur and must be managed with ventilation, scrubbers, and dedicated storage rooms. Industry surveys support that anhydrous and AA were used about equally through the 2000s; however, AA has overtaken anhydrous in more recent installations (AWWA Disinfection Systems Committee 2008). In general, AA is safer than anhydrous because it consists of a lower 15 to 30 percent ammonia concentration by weight. A 20 percent concentration is typical in most systems (Harms and Owen 2004). An example of an AA feed system is shown in Figure 7-4. The system in this figure includes a double-walled steel tank for AA storage, stainless steel fill pipes with quick connects for the delivery truck, and a chemical injection pump.

Anhydrous ammonia is a hazardous substance that requires special handling and extra equipment such as explosion-proof electrical switches and sensors to monitor ambient air ammonia level (Kirmeyer et al. 2004). Depending on the quantity stored, concentration, and local regulatory requirements, process safety management and/or a hazardous materials business plan may be required. An example of an anhydrous ammonia feed system is shown in Figure 7-5. In the depicted system, a 1,000-gal nurse tank (bulk storage) with an ammonia vapor transfer pump (not shown) feeds two day tanks. The 500-gal/day tanks are located on a platform scale for monitoring ammonia usage. Ammonia vapor is withdrawn through a vacuum regulator and applied to the water through ammonia injectors. Rotameters are used for flow control. Icing can occur on piping downstream of the regulator, and it may be necessary to locate the storage tanks in a heated room or provide heat tracing on outdoor piping. With increasing emphasis on security concerns, most utilities now favor the use of aqueous or LAS rather than anhydrous ammonia. The use of anhydrous ammonia may be addressed at some point in the future under the US Environmental Protection Agency (USEPA) Anti-Terrorism Standards, Chemical Facilities Anti-Terrorism Standards (CFATS) Program.

Both high and low temperatures can create problems for control of ammonia feed. High ambient temperature may cause ammonia to off-gas, potentially causing vapor locks and altering the ammonia concentration. Therefore, both temperature control and proper exhaust ventilation are important for ammonia storage. The ammonia supply lines need to be insulated when the ammonia tank is located outside to prevent the ammonia in the lines from overheating. In one utility using 19.4 percent aqua ammonia, vapor lock was observed in pumps, and air gaps occurred in feed lines when the temperature was 85°F (29°C) due to

Source: Manchester (N.H.) Water Works

Figure 7-4 Example of an aqua ammonia (ammonium hydroxide) storage tank and metering pump

vaporization at high temperature (Harms and Owen 2004). Both ambient temperature and sunlight exposure are important factors to consider. Low temperatures can cause pressure drops and restrict feed rates (Harms and Owen 2004). Therefore, where low temperatures are experienced, indoor storage and temperature control are important to maintain the proper ammonia feed rate. Alternatively, the tank and effluent line must be heated if stored outdoors in areas where the temperature may drop below the ammonia freezing point. It is recommended that the utility check with the local fire department on safety requirements before starting to handle ammonia in any form.

Ammonia Feed Systems: Operations and Maintenance

There are several approaches used for forming monochloramine, including introducing either chlorine or ammonia first or adding both simultaneously. In a survey of chloraminated utilities with 54 responses, it was found that 4 percent add ammonia first, 11 percent add chlorine and ammonia simultaneously, and 81 percent add chlorine first (Kirmeyer et al. 2004). In the same survey, it was also found that 4 percent of the utilities use multiple scenarios depending on water temperature. In the 2008 AWWA disinfection survey of chloraminated utilities with over 70 responses, it was found that 20 percent add ammonia

Source: Ann Arbor (Mich.) Water Plant

Figure 7-5 Example of an anhydrous ammonia feed system

first, 18 percent add chlorine and ammonia simultaneously, and 62 percent add chlorine first (AWWA Disinfection Systems Committee 2008). In the 2017 version of this survey AWWA 2018a), results were similar (13 percent add ammonia first, 21 percent add both simultaneously, and 61 percent add chlorine first).

Application of free chlorine before ammonia addition is the most frequently used approach. An appropriate free chlorine contact time is often provided to achieve the necessary disinfectant residual multiplied by the effective disinfectant contact time (CT) to meet disinfection requirements of the Surface Water Treatment Rule (SWTR) and/or Groundwater Rule (GWR). This is a common practice for water utilities, provided that DBPs can still be controlled, particularly for compliance with the Stage 2 Disinfectants and Disinfection Byproducts (D/DBP) Rule.

Depending on the treatment process, some utilities have found that a short free chlorine contact time (from a few minutes to a few hours) before ammonia application results in a more stable final monochloramine residual (Wilczak 2003b). In the 2008 AWWA disinfection survey, the median free chlorine contact time prior to ammonia addition was found to be 1 min (AWWA Disinfection Systems Committee 2008). This operational procedure has been recommended for plants using ozone followed by granular activated carbon (GAC) filtration before chloramination. Also, a free chlorine contact period ahead of ammonia addition has been shown to reduce subsequent N-formation (Krasner et al. 2013). Further discussion of regulatory implications is provided in Chapters 1, 4, and 8.

In the case of simultaneous addition of chlorine and ammonia, disinfection efficacy will be solely dependent on monochloramine contact time and concentration. Monochloramine is a less effective disinfectant than free chlorine; therefore, many utilities that operate surface water treatment plants find this approach more challenging due to the need for additional contact time to meet disinfection requirements. In some cases, water utilities use strong disinfectants such as ozone or chlorine dioxide to meet the CT requirements and then add monochloramine prior to entry to the distribution system. Kirmeyer et al. (2004) recommended that even when adding chlorine and ammonia together, at least 5 ft of distance should separate the feed points to prevent the formation of nitrogen trichloride.

Although uncommon, ammonia is sometimes fed ahead of chlorine. For example, in waters with very high total organic carbon (TOC), any free chlorine contact time may result in unacceptable levels of DBPs. In some membrane applications, monochloramine is used to limit biofouling, and a free chlorine residual must be avoided to prevent damage to the membranes. Also, when boosting the monochloramine residual, ammonia is sometimes added ahead of the chlorine to avoid breakpointing the water.

After 9 years of successful chloramination, nitrification in the distribution system was observed for the first time in Ann Arbor, Mich., following a filtration media change from sand to GAC. It was theorized that the AOM, which colonized the GAC filters, entered the distribution system and led to distribution system nitrification as well (Skadsen 1993). In the Ann Arbor case, ammonia had been added upstream of the filters. Once nitrification began in the filters, sometimes all the ammonia was consumed, but in other instances, some ammonia passed through the filters. Under these circumstances, it is possible to have inconsistent and variable concentrations of ammonia prior to chlorine application. This situation will greatly complicate process control and thereby make a utility more vulnerable to water quality problems and nitrification. In such cases, ammonia addition after filtration is preferred.

Scale buildup can be a continuous occurrence when using ammonia feeders, particularly when using AA. When the ammonia is injected into the water stream, the resulting pH increase can initiate a calcium carbonate precipitation reaction (Kirmeyer et al. 2004). This results in solid calcium carbonate forming on the injection equipment and in chemical feed lines. This typically occurs when the water contains 30 mg/L or more of hardness (as calcium carbonate) (Kirmeyer et al. 2004). The rate of scale buildup is proportional to the hardness of the water and must be considered when developing plant-specific maintenance guidelines (Kirmeyer et al. 2004). Examples of calcium carbonate buildup and removal are shown in Figure 7-6. This figure shows mineral buildup on the injection line fittings. The continuous buildup on the injectors will decrease the ammonia feed rate, potentially leading to unintentional breakpoint chlorination and a decrease in total residual chlorine.

The 2007 Great Lakes Upper Mississippi River Board of State Public Health and Environmental Managers (GLUMRB) "Interim Standard-Use of Chloramine Disinfectant for Public Water Supplies" explains that addition of ammonia gas or ammonia solution will increase the pH of the water and addition of ammonium sulphate depresses the pH. The actual pH shift may be small in well-buffered water, but the effects on disinfectant power

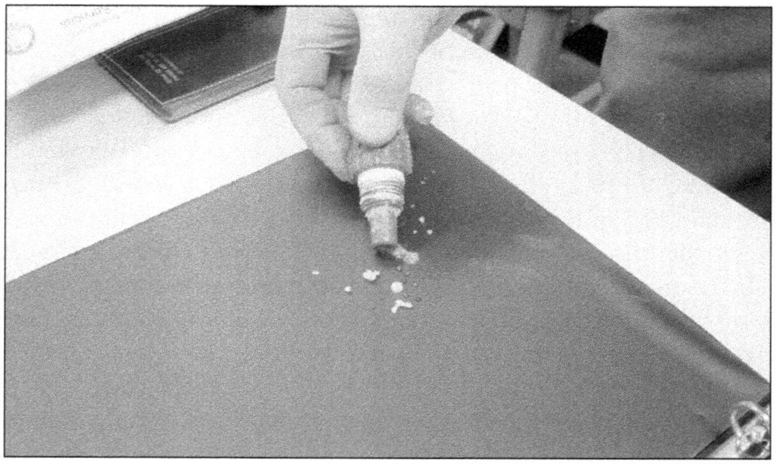

Source: City of San Bruno, Calif.

Figure 7-6 Calcium carbonate precipitation removed from an ammonia injector

and corrosiveness of the water may require consideration. Ammonia gas forms alkaline solutions which may cause local plugging with lime deposition. Where hard water is to be treated, a side stream of presoftened water may be needed for ammonia dilution to reduce plugging problems (GLUMRB 2007). Therefore, ammonia injectors should be routinely removed and rinsed with acid, such as hydrochloric acid, until the deposits are dissolved. In some cases, weekly cleaning may be necessary for proper operation. If possible, it is desirable to provide multiple feed locations such that they can be rotated in and out of service and one can be taken offline, if necessary, for cleaning without compromising operations.

The use of flexible feed lines can be beneficial for breaking up scale formation, and retractable injection quills have been found to be beneficial. Likewise, multiple injection points at the same location may be useful. The redundancy of multiple injection points allows for flexibility if feed problems are encountered. Spare injectors should be available so that when one is removed for cleaning, it is immediately replaced without interrupting the disinfection process. Alternatively, it may be more beneficial to soften the water supply that is used as ammonia chase water. The use of softened carrier/dilution water will minimize the scale buildup.

Softened water can usually be supplied by a point-of-use softening device or small household reverse osmosis or ion exchange unit. In this case, scaling may be prevented. When the operator observes decreased ammonia feed, decreased formation of monochloramine, or the presence of free chlorine, one of the first actions should be to check the ammonia injectors and mixers. Mixers and feed lines exhibit the same problems and symptoms due to hard water. Mechanical propeller mixers suffer from similar issues. They can be quickly covered with a precipitate, leading to equipment malfunction. Therefore, they are not typically used for feeding ammonia.

Utilities that use direct ammonia gas feed systems can avoid the calcium carbonate precipitation problems by using special diffusers that are designed to prevent deposit buildup. If a utility has severe calcium buildup problems, converting to a direct gas feed system may be a viable option. There are few direct feed systems in use in the United States, so experience and availability are limited (Harms and Owen 2004).

EBMUD used to feed 19 percent AA "neat" (i.e., no chase/carrier water is used) up to the injection point, at which point it was chased through a T-connection with unsoftened water. Chase water provides for better ammonia mixing, which is important if a limited run of pipe is available prior to the entry to the distribution system or sampling station. A vertical straight injection line should be opened periodically (e.g., every 6 months at EBMUD) to break up brittle precipitates with a steel rod. A spare injection port should be provided to allow for redundancy. This type of ammonia injection was used at EBMUD to provide for ammonia addition prior to entry to the distribution system and promote rapid monochloramine formation.

Ammonia Control Systems

To control chlorine and ammonia feed rates, control loops may be used to adjust the chemical pumps. The most common approaches are control based on the total chlorine residual level, pacing of chemical pumps based on water flow, or a combination of the two. These approaches have different advantages and disadvantages and must be evaluated before implementation.

Controlling the chemical injection pumps based on total chlorine residual level is applicable when the utility's goal is to maintain a target residual concentration. This allows for adjustments based on total chlorine demand but will not respond to changes in the level of free ammonia in both the raw water and finished water. If a varying level of ammonia in the water is an issue, a compound control loop system that utilizes both total chlorine residual and free ammonia levels may be an option.

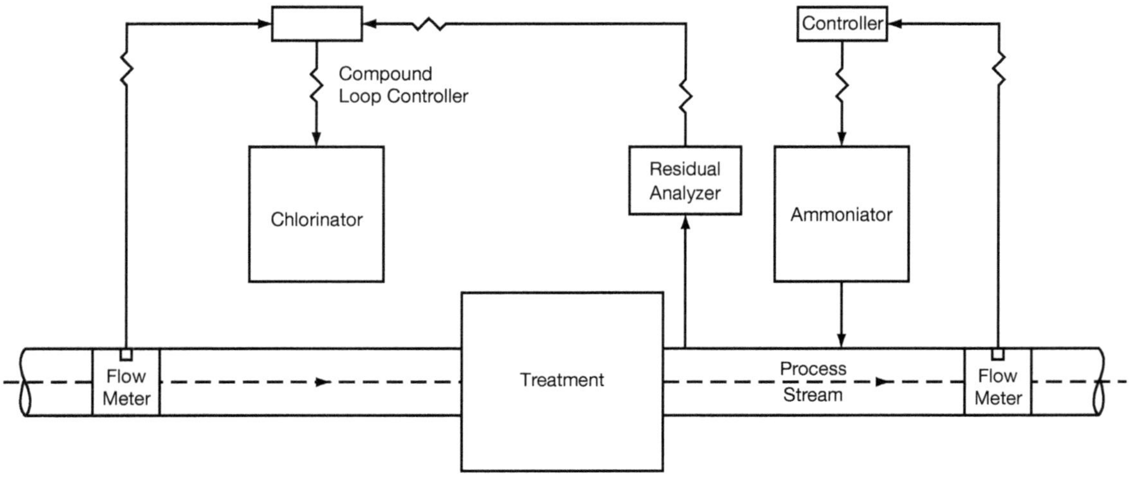

Source: Kirmeyer et al. 2004

Figure 7-7 Example of control schematic for monochloramine formation

Flow pacing will provide a means to adjust chemical pump feed rates as water treatment flow rates change. This type of control is based on a simple relationship whereby the chemical feed rate increases or decreases directly proportionally to the flow rate. If the chlorine demand is constant, this approach can provide a stable formation of monochloramine residual and adequately limit excess ammonia. Flow pacing alone will not be a good option when the influent ammonia concentration varies or the chlorine demand is not constant.

The most common approach is to base ammonia feed rates on both chlorine residual and water flow. An example of such a control system is shown in Figure 7-7. To properly control both ammonia feed and monochloramine formation, good mixing of both chlorine and ammonia into the water stream and subsequent reaction are essential. Similarly, a well-mixed, representative sample point downstream of chemical injection is essential. Most water treatment facilities do not operate at a constant flow rate throughout the course of a day. For this reason and because controlling the chlorine-to-ammonia-N ratio during the formation of monochloramine is such a critical factor in the prevention of nitrification, it is recommended that, at a minimum, both total chlorine residual and flow be used to pace ammonia addition.

If the utility also experiences free ammonia in its source water, a compound control system that includes total chlorine, flow, and free ammonia may be needed. A compound control loop will consider all three parameters and make the necessary adjustment(s) to the chemical feed rates. This type of control system may be difficult to operate due to lag time between sampling, analysis, and actual chemical pump adjustments; therefore, it must be designed and operated properly.

All this information should be available in real time, necessitating the use of online analyzers. Both reagent-based and reagentless online analyzers for total chlorine and ammonia are available. Additional information about control systems for chloramination can be found in Kirmeyer et al. (2004).

Source Water Impacts

Surface water sources can contain varying concentrations of ammonia based on activities within the watershed. Surface water ammonia levels may vary based on storm events, sewage discharges/upsets, industrial discharges, fertilizer applications (residential and/or agricultural), and domestic activities. During these events and activities, the ammonia concentration in the source water can vary suddenly and dramatically. Therefore, it is important to know the land use activities within the watershed and monitor for events that will affect influent ammonia levels. If primary disinfection is accomplished with free chlorine and breakpoint chlorination, the source water ammonia will not impact secondary disinfection with monochloramine. Likewise, if biological ammonia removal is practiced, depending on the concentration, the source water ammonia will not usually be a concern for secondary monochloramine disinfection. However, if other treatment practices are used, the source water ammonia may carry through the treatment process. In the latter case, as the ammonia level fluctuates in the source water, the ammonia feed rate must be adjusted to prevent either overfeeding or underfeeding. If the ammonia feed rate is not adjusted to compensate for increasing source water ammonia, the plant may overfeed free ammonia and make the distribution system more vulnerable to nitrification. Conversely, unexpected low source water ammonia can lead to underfeed and inadvertent breakpoint chlorination.

Some groundwater sources may also contain naturally occurring ammonia or ammonia that has entered the groundwater due to industrial or agricultural activities. In this case, ammonia addition may or may not be needed. In Minnesota, a survey of 39 public water supply wells found ammonia (as N) concentrations of 1.7–7 mg/L (Rezania 2010). Such levels of ammonia may create other challenges in preventing nitrification. In southern California, some wells have naturally occurring ammonia levels of 0.2–0.5 ppm.

In Germany, a river was found to have varying concentrations of ammonia of 0.05–4.2 mg/L (Uhl and Gimbel 2000). These high and variable ammonia levels will impact monochloramine formation. If the source water ammonia concentration is sufficient, the utility may just add the appropriate ratio of chlorine and obtain the desired monochloramine residual. However, if CT requirements under the USEPA Ground Water Rule must be met, the overall disinfection strategy will need to be considered, as monochloramine alone may not be sufficient to provide 4-log virus inactivation. Because groundwater quality is typically more stable than surface water quality, it is usually easier to control chlorine and ammonia dosages.

It is also possible that groundwater systems that are not disinfected but contain ammonia will nitrify. There are many groundwater systems that are not disinfected, and some of these are in areas that have significant ammonia sources (Henrie et al. 2012). If sufficient ammonia is present, nitrification can occur in these distribution systems given proper conditions, like temperatures above the threshold needed for AOM and NOB growth. The symptoms of nitrification will be similar to chloraminated distribution systems, with the exception of disinfectant changes. This type of nitrification has not been well documented in drinking water. However, more complete nitrification from ammonia to nitrate will more likely occur under this condition, unlike in chloraminated systems where nitrite typically accumulates because NOB are more susceptible than AOM to chloramine disinfection (Wolfe and Lieu 2003).

If ammonia concentrations in source water are above the level needed to form monochloramine, it may be necessary to breakpoint-chlorinate the water. Very high concentrations of ammonia would require high doses of chlorine for the breakpoint operation. In this scenario, ammonia removal becomes an issue before final monochloramine disinfection can be properly achieved. In Lansing, Mich., this situation occurred when the groundwater became contaminated with high and variable concentrations of ammonia. The company responsible for the contamination was originally required to reduce the ammonia concentration only to 34 mg/L as ammonia. Therefore, the water utility would have been receiving 34 mg/L ammonia in their source water. The cost to breakpoint this groundwater was prohibitive. Negotiations between the utility, regulators, and the responsible party resulted in an agreement to lower the ammonia concentration to 1.2 mg/L-N—the maximum amount desired by the utility to properly form monochloramine.

In addition to the cleanup of the groundwater and removal of contaminated wells from service, the utility continued to successfully chloraminate and meet water demands (Maier 2003). This approach has continued to prove successful (Maier 2011).

In Dubuque, Iowa, alluvial wells contain up to 1.9 mg/L of ammonia nitrogen (Rodriguez 2011). By blending these wells with other wells that contain lower ammonia levels, the utility was able to reduce the total concentration. However, breakpoint chlorination still resulted in an average chlorine demand of 6.8 mg/L, and booster chlorination was needed in the distribution system to maintain an adequate residual. The utility converted to biological filtration for ammonia removal that reduced the chlorine demand and slightly improved the breakpoint chlorination process (Rodriguez 2011).

Monitoring

It is important to monitor water quality parameters to ensure that proper levels of ammonia and chlorine are injected and the correct chlorine-to-ammonia-N ratio is achieved. By obtaining the concentration of total chlorine, free ammonia, and total ammonia in the finished water, the ratio can be calculated. Achieving the proper ratio is important because if the chlorine-to-ammonia-N ratio increases beyond 5:1, formation of dichloramine can occur, which can lead to customer complaints due to tastes and odors, increases in DBP formation, and accelerated decay of monochloramine in the distribution system. If the ratio is below 4:1, excess free ammonia will enter the system and may lead to nitrification.

Monitoring can also directly detect overfeed of free ammonia in the finished water. Grab sample or online free ammonia residual monitoring can be used for determination of both source water and finished water free ammonia. Based on the free ammonia level, the operator can adjust the feed rate accordingly. Grab-sample monitoring should be done daily at the entry point to the distribution system and at least weekly at the source water if there is ammonia in the water. There are several methods available for bench analysis of lab samples.

Monitoring chemical storage tank level or weight, precise chemical flow measurement, and routine calibration of chemical metering pumps are additional methods used to determine chemical dosages and ensure that overdosing of free ammonia is not occurring. Daily checks of ammonia inventory will allow an independent verification of feed rates. By comparing the dosage and the changes in quantity of stock, the operator can verify that the two measurements are comparable. Discrepancies between feed rate and inventory can indicate a potential ammonia feed problem that should be investigated and corrected. An example of feed rate versus stock usage is depicted in Figure 7-8. A hypothetical divergence in the data is shown beginning at day 25, which might indicate a feed problem. Monitoring of chemical inventory should be used in addition to methods such as residual analysis, calculation of chlorine-to-ammonia ratios, and feed control loops.

MONOCHLORAMINE RESIDUAL

Residual at Point of Application

The concentration of monochloramine entering the distribution system is important for the prevention of nitrification. Schrantz et al. (2013) performed a comprehensive lab-scale evaluation of simultaneously increasing monochloramine concentrations and chlorine-to-nitrogen mass ratios on nitrification at USEPA. By monitoring viable AOM under different chloramine concentrations, the study found that nitrification was significantly reduced with a 1.5 mg/L monochloramine residual and stopped when the monochloramine residual reached 3 mg/L. In a Water Research Foundation (WRF) survey, a sufficient concentration

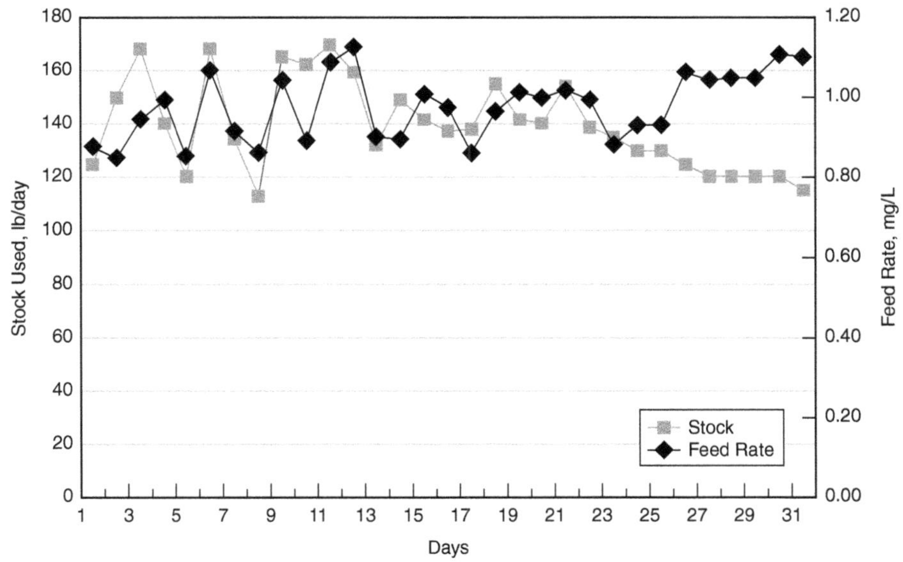

Data modified from City of Ann Arbor (Mich.) records

Figure 7-8 Comparison of ammonia feed rates (as pumped) versus stock used (as weight): A problem is indicated by the diverging hypothetical data late in the month.

of monochloramine was found to be a significant factor in preventing nitrification (Kirmeyer et al. 2004). The concentration of monochloramine needed in the finished water will vary with the extent of the distribution system and water quality characteristics. Various sources recommend different concentrations: for example, "Ten States Standards" (Great Lakes Upper Mississippi River Board [GLUMRB] 2012) states that "Chloramine concentrations should be maintained higher than for chlorine to avoid nitrifying bacterial activity. A range of 1–2 mg/L, measured as combined chlorine, on entry to the distribution system and greater than 1 mg/L at the system extremities is recommended." Kirmeyer et al. (1995) and Davis (1990) recommend maintaining 2.0–3.0 mg/L of monochloramine residual at the entrance to the distribution system; some water agencies operate at even higher concentrations of 3–4 mg/L.

However, once the nitrifier population is established, these doses may not be inhibitory. Nitrifying bacteria have been found at monochloramine residuals >5.0 mg/L (Cunliffe 1991) and increases in nitrite and nitrate concentrations observed at monochloramine residuals of 3–6 mg/L entering the system (Wilczak et al. 1996). Generally, whenever the AOM growth rate exceeds the AOM inactivation rate, the utility will be vulnerable to nitrification (Harrington et al. 2003). Increasing the monochloramine dose will also increase the ammonia from monochloramine decay that will be available to AOM (Harrington et al. 2002). It can also increase available free ammonia as a result of the chlorine-to-ammonia-N feed ratio. For example, maintaining a chlorine-to-ammonia-N feed ratio of 4.5:1 but increasing the monochloramine dose from 2–4 mg/L will result in a 100 percent increase in the available free ammonia entering the distribution system.

Because temperature is an important factor, seasonal adjustments of monochloramine residual may be appropriate. Some utilities increase the monochloramine residual entering the distribution system when the temperature is higher. Some researchers have recommended that a residual of 1.5 mg/L in the winter and 2.0 mg/L in the summer at the entry point to the distribution system may be appropriate (Palacios and Smith 2003). In Newport News, Va., the utility's monochloramine residual goal is 3.5 mg/L in the winter and 4.0 mg/L in the summer at the entry point to the distribution system. The distribution goal is to have at least 90 percent of the sample sites used for Total Coliform Rule compliance with a monochloramine residual of at least 2.0 mg/L (Hotaling 2005). The elevation of the residual in the summer should help maintain distribution system residuals when decay rates are likely to be higher and thereby help prevent nitrification.

Distribution System Residual

The monochloramine residual leaving the treatment plant must also take into consideration the monochloramine decay rate and expected losses in the distribution system. The rate of decrease will depend on multiple factors, including type of water main, corrosion control, water quality characteristics, distribution system demands, and residence times. Systems with extensive service areas and/or long residence times may experience greater decreases in monochloramine residuals and therefore may be more vulnerable to nitrification. Such systems may want to consider a relatively higher initial residual to maintain an adequate residual throughout the distribution system. However, the Stage 1 D/DBP Rule limits the use of chloramine to ≤4 mg/L average residual in the distribution system calculated on a rolling annual average basis of all sites.

The SWTR requires a minimum entry point total chlorine residual of 0.2 mg/L and "detectable" total chlorine residual in 95 percent of the distribution system sample sites (sampled monthly). Some states have different, more stringent minimum disinfectant residual requirements that need to be taken into consideration when setting residual targets, action triggers, and overall system optimization approaches (Friedman 2010).

Utilities typically set system-specific minimum distribution system residuals and nitrification trigger or target values, taking into consideration regulatory requirements, system water quality, and configuration. A monochloramine residual in distribution system storage tanks of >2.0 mg/L has been found to prevent nitrification (Harrington et al. 2002; Odell et al. 1996; Kirmeyer et al. 1995). Palacios and Smith (2003) examined over 120 samples from a chloraminated distribution system and found that nitrification was more likely to occur when the total chlorine residual was less than 1.5 mg/L. In Ann Arbor, the target monochloramine residual leaving the treatment plant is 3.0 mg/L, with a distribution goal of all sample sites over 1.5 mg/L (Skadsen 2002) that is now current practice. There appears to be a trend toward higher minimum residuals among many utilities, with a minimum of 1.0–2.0 mg/L at all distribution locations. The Ten States Standards (GLUMRB 2012) now recommends at least 1.0 mg/L at the distribution system extremities.

Influence of Treatment Processes

Some treatment processes affect the formation and stability of monochloramine. At EBMUD, when ozonation followed by biologically active GAC filtration was implemented, the monochloramine demand substantially increased despite lower concentrations of TOC (Wilczak et al. 2003a), presumably due to the demands of a less biologically stable water. At a minimum, one h of free chlorine contact time was necessary to see good monochloramine stability following the ozone/biological activated carbon treatment process (Wilczak et al. 2003a). In general, total chlorine demand that was satisfied at the treatment plant was not exerted in the distribution system. At this utility, increases in the concentration of hydrogen peroxide resulted in a decrease in monochloramine stability followed by nitrification in the distribution system (Wilczak et al. 2003a).

In Arlington, Tex., similar results were observed with decreased monochloramine residuals in the distribution system following installation of ozone and GAC filtration (Vokes 2007). Improved monochloramine stability was achieved with increased free chlorine contact time at pH 8.3 and improved biofiltration by nurturing the biomass in the filters. Therefore, it is important to assess the impact of treatment changes on monochloramine demand and stability as well as nitrification potential.

Other treatment processes may also impact the potential for nitrification. In one pilot evaluation, it was found that enhanced coagulation delayed the onset of nitrification (Harrington et al. 2002). Theoretically, this could have been the result of more effective (potential monochloramine demand) removal of NOM before the water entered the distribution system.

Temperature

Understanding the effects of water temperature on nitrification is important in developing preventive measures. In climates where seasonal temperature fluctuations occur, the decrease in water temperature in winter months may act to reduce nitrifier activity. In one system, AOM were isolated during the summer but not the winter (Wolfe et al. 1990). In some cases, evidence of nitrification subsided as water temperatures declined (Skadsen 1993). However, the literature supports that, while their activity may not be evident, nitrifiers can survive in extreme cold (Pintar et al. 2000). In warm climates, such seasonal variations (decreased nitrifier activity) will not be realized because temperatures remain high enough for year-round nitrification to be evident. Thermal stratification in storage can lead to disinfectant residual loss and nitrification in the warmest, noncirculated areas of storage. Mixing of storage facilities can help to reduce thermal stratification. If colder waters are available, such as from underground well sources, preferential use of those waters might reduce nitrification as well. A water temperature of <15°C is likely necessary to provide any inhibition of AOM, but AOM have been observed at temperatures of 5°C (Kirmeyer et al. 1995).

Sediments and Biofilm

Microorganisms are routinely associated with sediments and biofilms in storage facilities and in the distribution system (Lipponen et al. 2002; Kirmeyer et al. 1995). In addition, it was found that both sediment and biofilm contribute to nitrifying bacteria survival by shielding them from disinfectant residuals (Wolfe et al. 1990). Higher numbers of AOM have been observed in sediment than in biofilm (Wolfe et al. 1990; Ike et al. 1988). EBMUD found that deposits on the bottom of a reservoir were contributing to nitrification (Wilczak 2004).

In one utility, sand leaking through malfunctioning underdrains was transported into a finished water reservoir that was nitrifying. When the sand was removed, the utility was able to maintain a monochloramine residual in that reservoir and nitrification stopped (Vestal and Wilczak 2005). A study of covered reservoirs in southern California detected AOM in both liquid and biofilm samples (Baribeau et al. 2001), and AOM have been isolated from sludge and wall surfaces in a coated steel tank (Stewart and Lieu 1997).

Routine internal inspection of reservoirs will detect both floor and wall deposits. Routine sediment removal and cleaning is a good operational practice, which may decrease the frequency of nitrification. The frequency of this inspection and cleaning is recommended every 3–5 years (AWWA 2018b; ANSI/AWWA 2015; USEPA 2002). At one utility, a combination of raising the monochloramine residual to 3.0 mg/L at the entry to the distribution system and cleaning the storage tanks to remove sediment accumulation was effective at controlling nitrification (Harms and Owen 2004). At another utility, cleaning of storage reservoirs improved the general hygiene but did not resolve nitrification (Song et al. 1999). When cleaning is performed, it is important to provide adequate disinfection prior to returning the unit to service. Alternatively, the use of properly disinfected divers or remotely operated equipment can avoid dewatering of the facility and subsequent chlorination.

Corrosion and Distribution System Piping Materials

Distribution system cleaning, pipe replacement, and corrosion control will all help reduce bacterial populations, including nitrifying bacteria in the sediment and biofilms. Nitrifying bacteria have been observed in all types of distribution system pipe materials. Friedman et al. (2010) have shown that corrosion related metals (zinc and copper) and nutrients (phosphate and inorganic and organic carbon) interactively affect (promote and/or limit) nitrification. In one distribution system, AOM have been observed as high as 100,000/ cm^2, although the levels of AOM found in cast-iron pipes were inconsistent (Camper et al. 1996). In another study, the numbers of AOM were found to be inconsistent in unlined iron pipe but were the lowest in concrete pipe (Stewart and Lieu 1997). In a survey of Finnish drinking water systems, high concentrations of AOM were observed in pipe sediments (Lipponen et al. 2002).

STORAGE FACILITY OPERATION

There are numerous factors that may contribute to nitrification in storage facilities. The major factor is high water age due to poor mixing, inadequate turnover, and stratification combined with oversized capacity. The accumulation of sediments and development of biofilm are also contributors. Application of good storage operations and design to minimize water age and prevent accumulation of sediment in storage are essential nitrification-prevention measures for chloraminated systems. Monitoring is also critical; in all cases discussed below, monitoring is necessary to inform operational decisions.

Water Age

High water age has been implicated as a major factor in the occurrence of nitrification (Friedman 2010, USEPA Nitrification White paper 2002, Kirmeyer 1995; Skadsen 1993; Davis 1990; Negrin et al. 1990; Barrios and Stone 1989; Wolfe 1988). For example, at the MWDSC, nitrification was observed after the residence time in Orange County reservoirs increased from 3.3–4.5 days (Wolfe et al. 1988). Therefore, it is important to implement operational controls to reduce water age. This can be done by increasing turnover, improving mixing, reducing storage capacity, or even taking tank(s) offline if possible. Elimination of dead zones in any storage facility is critical. Such dead zones can contribute to high water age.

Storage Facility Turnover/Cycling

Turnover is the exchange of an existing volume of water with a new, fresh one so that water age is minimized. Storage tanks with inadequate turnover can cause depletion of monochloramine residual as well as increases in microbial activity and nitrification. Changes in facility operation, such as increased turnover, can help avoid or minimize future nitrification events in storage facilities. For example, the Philadelphia Water Department lowered their tank levels for a number of years to improve turnover and prevent nitrification (Odell et al. 1996).

Turnover is accomplished when a percentage of the tank volume is being drawn and filled every day or few days. There are no established standards for water turnover in tanks; it is based on several factors such as tank volume, inlet/outlet (I/O) configuration, mixing, demand, and water quality. Some states have made recommendations for optimal turnover that range from 20 percent of the volume per day (Ohio) to 50-percent daily turnover (Georgia). A target of 30 percent turnover per day, adjusted based on individual utility needs and water age, is often used as a goal. Increased turnover of up to 50 percent per day may be appropriate (deep cycling) when early indicators of nitrification are observed. Management of water age and storage facility turnover is utility specific and depends on water quality and distribution system configuration. Consideration must also be given to emergency storage volumes available for firefighting and pressure management.

Storage Facility Stratification

Even with frequent turnover, stagnation can still occur if stored water is not mixed. Turnover does not assure good tank mixing, only that a substantial portion of water is exchanged daily. In poorly mixed storage tanks, the volume of water that is not exchanged can experience significant decreases in water quality (including nitrification), particularly in tanks with a common I/O line. Water storage that floats on the system can thermally stratify and allow water in the unused portion of the tank to reach significant age and higher than normal temperatures. A storage tank with a common I/O pipe at the bottom and 20 percent turnover can still stratify and develop nitrification.

Nitrification in the unused portion of the tank can potentially go undetected over time until that water is used or tested. An example of this practice would be the use of the lower portion of an elevated tank, where the water in the upper portion of the tank is not turned over or used. This is common practice where an elevated tank or a standpipe tank is used primarily for pressure control (Grayman et al. 2000). Stratification is usually detected by temperature monitoring (Figure 7-9) or by modeling and can be confirmed by water quality monitoring.

A utility may find it beneficial to reduce stratification by deliberately drawing down storage facilities to increase circulation and decrease water age. Drawdown of tanks to remove "old" water can help improve water quality but may only be a short-term fix if the tank is poorly mixed. Exercising tanks to minimize detention time can control and eliminate nitrification (Harms and Owen 2004). The ability to draw down storage is more feasible in pumped systems than in ones that float on the system and will be constrained by the need to keep sufficient water storage for firefighting requirements.

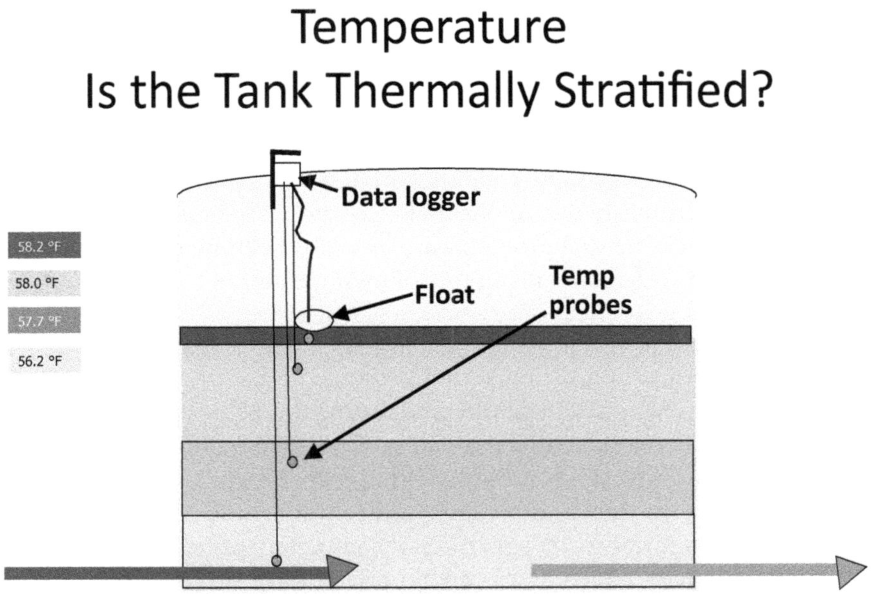

Source: Charlotte D. Smith & Associates, Inc.

Figure 7-9 Example of stratification and temperature monitoring in a storage facility

Recirculating water in standpipes and elevated storage can create a uniform water age, reduce/eliminate stratification, and thereby reduce the occurrence of nitrification (Kirmeyer et al. 1995). In Fort Worth, Tex., the use of recirculation pumps in tanks resulted in an improved monochloramine residual and provided promising results for preventing nitrification (Kirmeyer et al. 1995). Cascaded pumping operations and/or scheduled concurrent pumping operations can be beneficial for reducing water age in remote storage facilities. In addition, changes in operational control set points to increase turnover and decrease water age may be feasible. In all cases, operators should strive to use fill-and-draw operating strategies to maximize tank turnover.

Hydraulic Mixing/Cycling and Filling

Ensuring sufficient and regular mixing of stored finished water is as important to maintaining water quality in storage facilities as providing adequate turnover. Improving mixing in storage tanks will generally improve water quality by equalizing water age and reducing stratification. It is recommended that analysis of tank mixing characteristics and water age be done to determine if such stratification is occurring. It is recommended to optimize both design of tanks and operational strategies to ensure stratification is avoided.

There are two extremes by which water flows through storage tanks: completely mixed flow or plug flow (Grayman et al. 2000). Most facilities operate somewhere between these two extremes. Plug flow is preferred for disinfection, but completely mixed flow is ideal for maintenance of water quality in distribution system storage facilities. A completely mixed-flow tank is better equipped to handle variations in input quality and flow rate than a plug-flow tank. Therefore, the water storage facility should be designed to encourage good mixed-flow behavior rather than plug-flow behavior (Grayman et al. 2000). Completely mixed flow results in a lower average water age. A tracer study can be used to illustrate how close a tank is to either extreme. A tracer study can also illustrate how much dead volume is in a tank.

Fortunately, from the perspective of preventing nitrification, it is easier to achieve good mixed flow than plug flow. The main deterrents to achieving well-mixed conditions in tanks appear to be improper placement, orientation, and size of I/O pipes; inadequate inlet water momentum and mixing time; and the potential temperature differences between the inflow and tank contents. Each of these factors can be dealt with through minor structural or operational changes (Mahmood 2005; Grayman et al. 2000).

Mixing time primarily depends on the facility's diameter or width, water height, and inlet momentum (Grayman et al. 2000). For a tank operating in a fill-and-draw mode, mixing occurs primarily during the fill cycle, and if the tank is relatively well mixed at the end of the fill cycle, mixing problems are unlikely. Fill time must exceed theoretical mixing time. This can be calculated using the following equation:

$$\Delta V/V > 9.03\, d/V^{1/3}$$

where ΔV is the change in water volume during the fill period, V is the volume of water at the start of the fill period, and d is the inlet diameter (Grayman et al. 2004). All units should be in consistent dimensions, typically either feet or meters.

In a study on steel and concrete tanks, the related technique of cycling (operating the reservoir to reduce water age) did not completely prevent nitrification from occurring, although the extent of mixing was not reported (Baribeau et al. 2001). In Wichita Falls, Tex., isolated pockets of nitrification were controlled by water cycling in storage tanks (Harms and Owen 2004). Tank samples taken at different depths found AOM in all liquid and biofilm samples. However, it is well known that poor cycling in storage facilities creates dead zones, which lead to impaired water quality, including reduced monochloramine

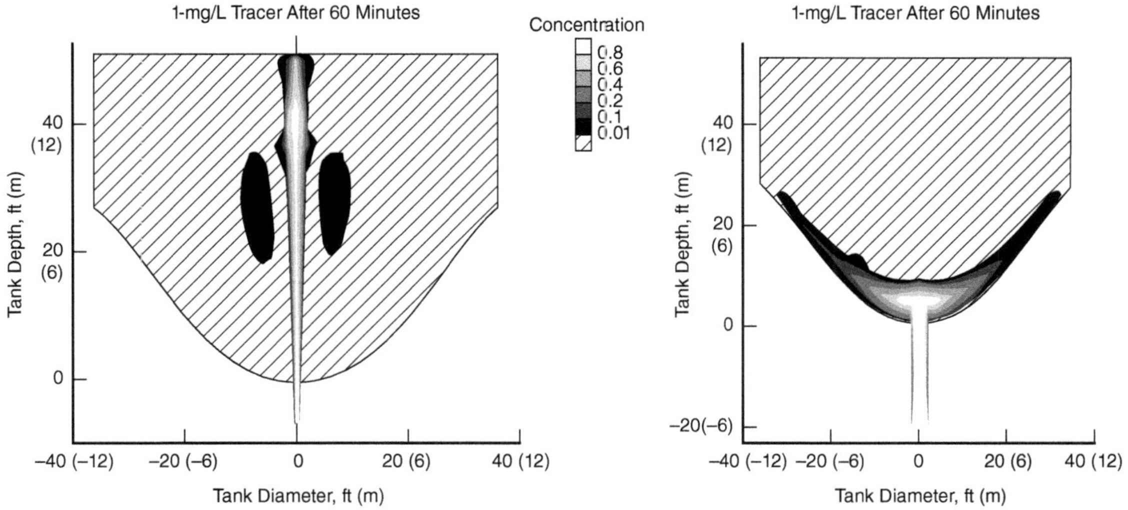

Inflow rate = 1,000 gpm (63 L/s); 16-in. (400-mm) vertical inlet; velocity = 1.6 fps (29 m/min); tank temperature = 68°F (20°C); inlet temperature = 64.4°F (18°C)

Inflow rate = 1,000 gpm (63 L/s); 48-in. (1,200-mm) vertical inlet; velocity = 0.2 fps (3.7 m/min); tank temperature = 68°F (20°C); inlet temperature = 64.4°F (18°C)

Source: Mahmood et al. 2005

Figure 7-10 The effect of inlet momentum on mixing characteristics of a 1 mil–gal elevated storage tank using computational fluid dynamics modeling

residual and increased bacterial growth. Therefore, improvements in cycling should reduce bacterial activity, including AOM (Grayman et al. 2000).

Increasing the fill time of a storage tank may be looked at as a long-term response to prevent nitrification. It is important to maintain sufficient inlet momentum during fill cycles to provide mixing. Simply reducing the fill rate to prolong the fill time will not improve mixing if the inlet pipe is too large. It may be necessary to reduce the size of the inlet to increase the inlet momentum or provide a combination of both increased fill time and increased the momentum by reducing the inlet pipe diameter. The increase in momentum is key for improving mixing in water tanks. Mahmood et al. (2005) demonstrated the effect of changing inlet momentum using computational fluid dynamics modeling for a 1 million–gal (MG) elevated tank. In this case, the velocity was altered by changing the inlet diameter.

As the inlet diameter was decreased, the velocity and momentum increased, resulting in a mixing zone that extended nearly to the surface of the elevated tank. Lower momentum (lower velocity and greater inlet diameter) showed a mixing pattern where the mixing zone only existed in the area surrounding the inlet (Figure 7-10).

If a storage facility has many short fill cycles in a day, changing tank operations to have fewer cycles or one longer fill cycle can also improve mixing. If a tank already has only one fill cycle per day, it may be possible to extend the duration of the fill cycle by draining the tank to a lower level before beginning the fill cycle.

Mixing may be further complicated by temperature. Temperature differences between the tank influent and the bulk water in the tank can lead to stratification and poor mixing (Mahmood et al. 2005). For example, in the summer, water temperature in a tank will usually be higher than the temperature of the water entering the tank. The cooler influent, being denser, will remain at the bottom of the tank, assuming the I/O line is at the bottom (Mahmood et al. 2005). Similar effects can occur in cold weather as well as where sunlight can increase the temperature of the stored water. It may be feasible to install a separate I/O line with appropriate check valves that force the tank to fill from the top and empty from the bottom and break up the thermal stratification.

The ability of a system to implement these changes will largely depend on system demand, hydraulics, and fire flow requirements. General factors that should be evaluated to improve mixing include

- increasing inlet momentum,
- increasing flow rate,
- decreasing inlet diameter (increase velocity),
- evaluating orientation and location of inlets, and
- monitoring temperature differentials and potential stratification.

In addition to the passive mixing/cycling described in the previous paragraphs, some utilities employ active mixing equipment. These devices require energy input—via line power, onsite solar, or compressed air—to circulate water within a tank or reservoir. Examples include mixers from PAX Scientific (San Rafael, Calif.), PulsAir Systems, Inc. (Bellevue, Wash.), and Big Wave Water Technologies (Oceanside, Calif.), as well as the SolarBee and GridBee by Ixom Water Care (Dickinson, N.D.).

Storage Capacity and Removal From Service

Distribution storage facilities have historically been sized with little consideration for water quality. Storage capacity has primarily been determined based on hydraulic issues, such as current and future system demand, pressures and flows, as well as fire flow requirements (AWWA 2008; Walski 2003; Theiss and Antoun 2000). Systems have generally sized storage facilities to provide uninterrupted service and fire flows.

A storage tank that is located below the system hydraulic grade line will become stagnant and will not contribute to the day-to-day operation and demand requirements unless pumped. Water stagnation can lead to monochloramine decay, increased microbiological activity, and nitrification.

As recommended by Ten States Standards (GLUMRB 2012), the "minimum storage capacity for systems not providing fire protection shall be equal to the average daily consumption. This requirement may be reduced when the source and treatment facilities have sufficient capacity with standby power to supplement peak demands on the system." The recommendations also state, "Excessive storage capacity should be avoided where water quality deterioration may occur." Storage design requirements may differ with different state regulators.

When a storage tank is oversized to address future system needs, it will cause inadequate volume turnover in the tank and excessive water age, which can potentially lead to nitrification. Proper sizing during design is important to balance water demand with water quality considerations. Dual-tank design, in which one tank can be valved out when not needed, is becoming more common.

Another means to decrease distribution system water age is to remove excess storage capacity from service, either temporarily or permanently. In particular, a combination of removal from service with regular cleaning and removal of deposits in affected facilities may reduce the occurrence of nitrification. It may be beneficial to remove facilities from service only seasonally when a utility has demonstrated seasonal vulnerability to nitrification (Ike et al. 1988). Permanent removal may be appropriate if the system has excess storage capacity. When repeated nitrification events are occurring within a storage facility that is not being used daily, an analysis of necessary versus available storage may determine a storage tank is not needed.

Decommissioning excess storage can provide a long-term solution to nitrification problems. However, decommissioning of any storage facility should only be considered after a thorough analysis of distribution system demand and capacity to make sure that the facility is not needed for equalization storage, fire flow, or emergency conditions like

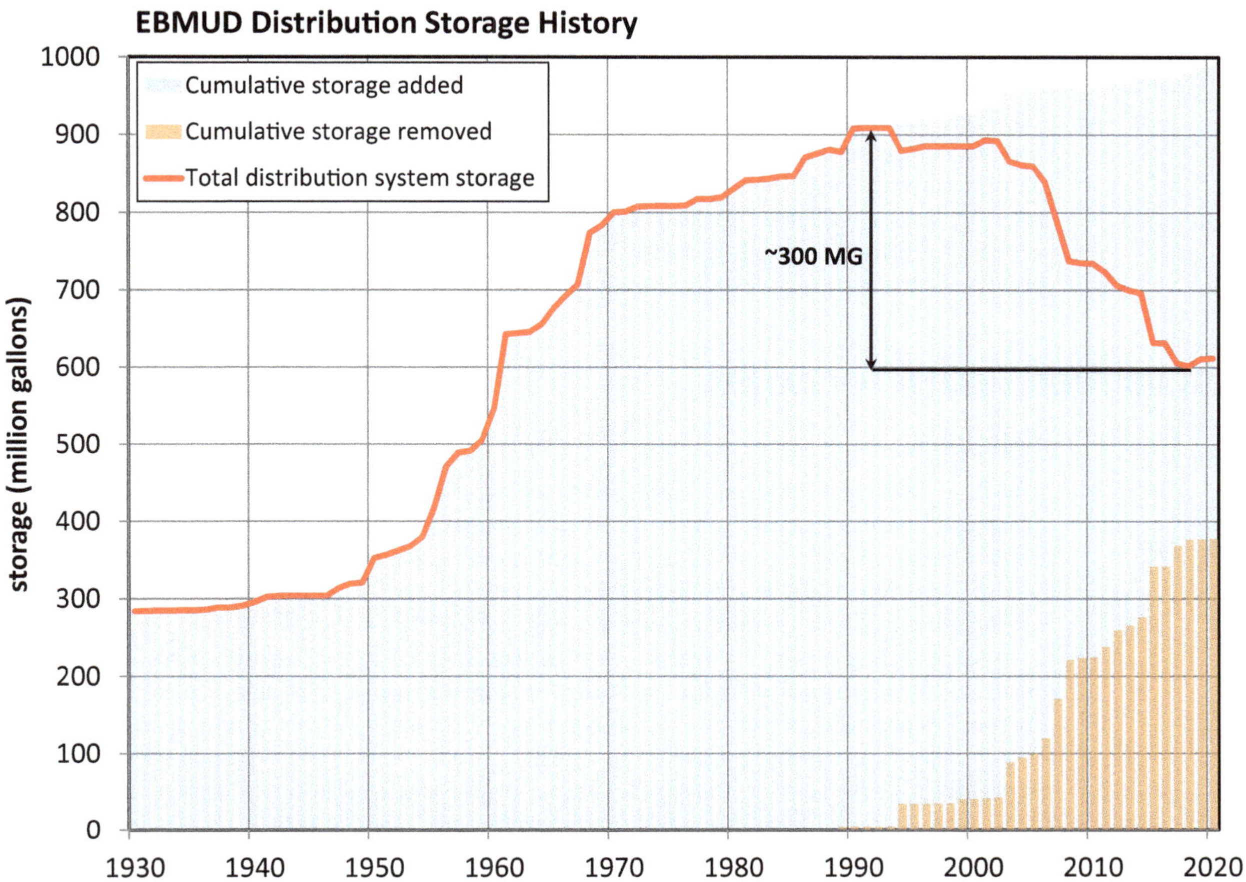

Source: East Bay Municipal Utility District
MG = million gal

Figure 7-11 Cumulative water storage, East Bay Municipal Utility District (EBMUD)

main breaks or treatment plant shutdowns. Reservoir decommissioning and replacement with smaller reservoirs is ongoing at EBMUD after its distribution system was converted to monochloramine in 1998. Since then, several water storage facilities have been taken out of service for water quality reasons, and others have been replaced with smaller tanks (Figure 7-11). Water quality is not the only driver for reservoir decommissioning; many of the large open-cut reservoirs have approached the end of their useful life and would require substantial upgrades to meet current dam safety standards. Replacing large, open-cut reservoirs with smaller tanks met both objectives. Based on the required capacity projected in 2030 for large EBMUD reservoirs and smaller tanks in the same pressure zones, the capacity in these storage facilities could be reduced significantly from around 900 MG to around 300 MG.

A creative solution to simultaneously decreasing storage while maintaining fire protection was previously practiced by the Newport News Waterworks, Va. This utility has a 5-MG tank that was kept full but isolated from the distribution system and only used for fires or emergencies. A rechlorination facility maintained the tank residual above 2.0 mg/L. The tank was periodically dumped and refilled with fresh water to minimize water quality degradation (Hotaling 2005). This practice has been discontinued in favor of retaining the tank in service (Hotaling 2011). The tank is now operated to turn the volume of the tank over every 3 days (i.e., 33 percent turnover daily, which is consistent with the recommendations previously discussed in this chapter). A smaller total maximum volume is now maintained in the tank (3 MG versus 5 MG). This operational protocol does have trade-offs in that the chlorine residual drops by about 1.0–1.5 mg/L while in the tank. However, it is no longer necessary to discharge 5 MG of water every time the water gets "old" (Hotaling 2011).

Master planning and hydraulic modeling of the distribution system are recommended to understand system dynamics and to properly size storage. If a storage facility is removed from service, it will need to be properly disinfected and tested before it is returned to service. A utility may wish to notify customers of these actions. For more discussion on return-to-service practices and customer notification procedures and issues, see Chapter 9.

DISTRIBUTION SYSTEM OPERATION

Hydraulics

Low-flow and high water age areas, such as dead ends, are the most likely nitrification locations. Therefore, it is important to understand distribution system hydraulics and identify areas of high water age. Hydraulic models can be used to estimate distribution system water age and identify high water age areas. There are numerous computer models for assessing age and related impacts on water quality.

Approaches such as tracer studies conducted using conservative substances (substances that do not degrade or change in concentration) or water quality monitoring approaches are other methods for determining water age and water quality. Tracer studies are often performed using fluoride or another similar substance whose feed rate can be modified (increased or decreased) and then monitored throughout the distribution system (Vasconcelos et al. 1996). By tracking the change in concentration of a substance, the residence time at different points can be estimated.

Many different tracer substances or chemical parameters have been successfully used, such as UV 254, conductivity, and calcium chloride (Grayman et al. 2000). It is important to use a tracer substance, like fluoride, that can be closely controlled, is conservative, has no adverse health effects, and has no taste or odor impact. In the absence of a tracer study, water quality parameters that degrade with water age can be monitored and assessed, like disinfectant residual. The selection of monitoring points is crucial to properly determine where the older water exists.

The expertise of the distribution system operators is very important both for selecting sample points for tracer studies and for verifying the results of any hydraulic modeling. Field and operations staff typically has extensive knowledge of distribution system operations and can identify areas of excessive water age based on their experience. Both modeling and system knowledge should be used to identify and pursue actions to reduce water age.

Systematic Distribution System Flushing

Routine flushing is an established practice for maintaining distribution system water quality. Dead ends and other areas with long detention times are particularly noted for nitrification incidents (Friedman 2010; Smith et al. 1996; Kirmeyer 1995; Skadsen 1993; Davis 1990; Negrin et al. 1990; Barrios and Stone 1989; Wolfe et al. 1988). By reducing distribution system water age and removing sediments and biofilm, it may be possible to reduce the frequency of nitrification. There is limited literature documenting the actual measured effectiveness of this approach (Kirmeyer et al. 1995). In a WRF survey of utilities, hydrant flushing was rated as one of the most important perceived methods for prevention of nitrification (Kirmeyer et al. 2004). It should be noted that once nitrification occurs, flushing is only sometimes effective at eliminating nitrification even though it may temporarily restore the monochloramine residual.

Types of flushing. There are two primary types of system-wide flushing: conventional and unidirectional flushing (UDF). The differences in each approach are discussed in the following sections. A third alternative, spot flushing, is discussed in Chapter 8 as a response to nitrification. Flushing velocities are an important consideration. Conventional flushing typically uses lower flow and therefore is less likely to remove biofilm and scale, although it may remove easily dislodged corrosion products. UDF is typically performed at velocities of 5–6 ft/s, which will scour the distribution piping and therefore is more effective at reducing scale and biofilm. In addition, quasi-UDF universal density filtration or a hybrid version of conventional and UDF is sometimes practiced, where only specific mains are valved to achieve higher flow rates but without flushing the entire distribution system. The effectiveness of quasi-UDF is dependent on the quality of the water used to flush the affected area.

Conventional flushing. Conventional flushing is conducted by opening one or more fire hydrants or flush outs and is primarily utilized to react quickly to water quality problems or as a routine practice. In this discussion, conventional flushing refers to flushing done anywhere in the distribution system (as for fire-flow testing) but not in any particular sequential order. Conventional flushing was defined by Antoun et al. (1999) as "the opening of hydrants in a specific area of the distribution system until pre-selected water quality criteria are met." This type of flushing does not include directing the flow with valves from one part of the system to another. Conventional flushing may also be used as a preventive action to maintain distribution water quality.

With conventional flushing, it is difficult to control the quality of water entering the main being flushed, and it is possible that the quality of this water may not be improved. Conventional flushing of distribution system pipes can sometimes even result in more water quality problems because of changes in flow directions and velocities, which can detach biofilms and stir up sediments. In addition, conventional flushing may be less than optimal in controlling other factors that can contribute to nitrification because, in most pipes, the velocity of at least 5–6 ft/s required to scour pipe walls is not achieved.

Conventional flushing requires minimum planning by the water utility. A small crew can perform conventional flushing with no special equipment (aside from standard dechlorination and erosion/sediment control equipment and routine water quality monitoring devices). Therefore, this is a simpler and less expensive approach, but the long-term benefits for nitrification prevention are questionable.

Unidirectional flushing. UDF is generally a more effective approach to maintaining water quality and preventing nitrification because of the planning involved and data collected. It is conducted in a systematic manner, directing the flow to enhance the flushing of the desired main by isolating particular sections of pipes. A properly designed and implemented UDF program can achieve water velocities greater than 5 ft/s and can scour the pipe.

For a successful UDF program, the order in which pipes are flushed, the hydrants that must be opened, and the valves that must be closed or opened are carefully planned to flush the entire system beginning at points of entry (e.g., water treatment plants) and moving outward toward system extremities (Freidman 2010; Hasit et al. 2004). Water that enters the main being flushed flows from other sections that have already been cleaned. UDF should be configured to maximize water velocity when a hydrant is opened while minimizing the chance of dirty water reaching customers. Usually, this requires that flushing start at a source of supply and work outward in the distribution system. Accurate maps of the distribution system, hydraulic models, and a complete database of valves and hydrants facilitate planning and execution of UDF programs. Throughout the process,

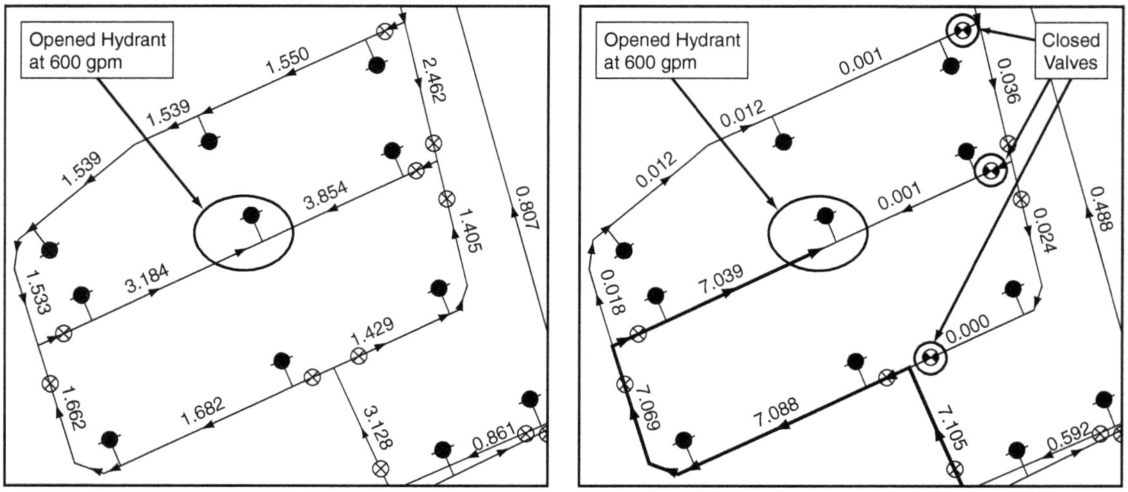

Source: InfraMetrix, LLC
gpm = gallons per minute

Figure 7-12 Difference in flow pattern and pipe velocity with conventional and unidirectional flushing: The numbers on the charts are water velocities in ft/s.

data, such as flows, pressures, and water quality parameters like chlorine residuals and turbidities, are gathered to document the work and improve future efforts.

Comparisons of the differences between conventional and UDF are shown in Figure 7-12. These models demonstrate the differences in flow patterns, velocities, and affected areas achieved by the two different approaches in a distribution system.

Before starting any flushing program, the benefits of flushing will need to be weighed against the drawbacks. In some cases, such as locations with heavily tuberculated unlined cast-iron water mains, UDF may cause red water issues and may not be appropriate. Water from flushing must be properly dechloraminated before disposal to the environment, thus necessitating treatment. In addition, depending on the extent of flushing, there can be a significant loss of water and an associated increase in cost to the utility. The tradeoff between water demand and conservation should be considered. Equipment is available (e.g., neutral output discharge elimination system [NO-DES]) that connects to water mains via hydrants, pumps the water through filters, and puts it back into the system, eliminating discharges to the environment. Such equipment can also add chlorine and continuously monitor water quality to optimize operation. Use of this type of no-discharge equipment helps alleviate customer concerns about water conservation that are raised when customers observe water intentionally flowing into storm drains.

Flushing needs to be conducted according to procedures that maintain required pressures in the system. The frequency of flushing will depend on the effectiveness and duration of the water quality improvement as well as season (temperature). In Ann Arbor, weekly conventional flushing of areas with high water age is performed. Flushing is continued until the target total chlorine residual is restored (>2.0 mg/L when the finished water leaves the treatment plant at 3.0 mg/L). These flushing programs vary with location and season (temperature).

Using flushing, it is possible to create artificial water demand, thereby reducing water age in storage facilities and the distribution system. The resulting reduction in water age may prevent nitrification. This can be done by continuous hydrant flushing, continuous blow-off, or programmable blow-off. These practices provide continuous water flow in dead-end pipes or other low-flow areas where water age is high. In Fort Eustis, Va., and

Figure 7-13 An example of programmable automatic flush device

San Bruno, Calif., automatic flushing has been implemented on a routine or adjustable basis (Harms and Owen 2004). An example of a programmable blow-off used by Ann Arbor is shown in Figure 7-13. The resulting reduction in water age will have the potential to decrease corrosion, decrease biofilm, and decrease nitrification.

Periodic Switch to Free Chlorine

The periodic switching to free chlorine is practiced by some water utilities as a nitrification prevention strategy. Free chlorine will inactivate AOM in the water and remove the free ammonia, thus preventing new growth. AOM are harder to remove from biofilm and pipe surfaces, particularly where tuberculation and corrosion are present, due to the protection afforded by these surfaces. Lee et al. (2011) compared free chlorine and monochloramine biofilm penetration; their study demonstrated monochloramine's faster and more effective penetration. However, free chlorine inactivated the microorganisms more effectively. In a WRF survey, 35 percent of respondents from chloraminated utilities indicated that they periodically use free chlorine; most use it once per year (67 percent), some seasonally (11 percent), and others as needed (22 percent) (Kirmeyer et al. 2004). Most respondents indicated that they use free chlorinate for 1 month (Kirmeyer et al. 2004). A 2017 survey indicated that 44 percent periodically switch to free chlorine.

Routine free chlorination of the distribution system is more common in warmer climates, such as Florida and Texas (Harms and Owen 2004). In a 2008 disinfection survey, 23 percent of utilities reported annual conversion to free chlorine for some duration annually (AWWA Disinfection Systems Committee 2008). As DBP regulations have become more stringent and public awareness has increased, many utilities are opting to not routinely free chlorinate their system. It is the recommendation of the authors of this manual that free chlorination of the entire distribution system be considered as a last resort for prevention of nitrification. If practiced, the impact of free chlorination on water quality and subsequent nitrification should be evaluated. Its effectiveness has generally been shown to be short lived.

In Raleigh, N.C., nitrification recurred within 3 months following free chlorination (DiGiano et al. 2002). In a study of three utilities, the effectiveness of a 1-month use of free chlorine for control of nitrification was assessed (Carrico et al. 2008). In two of these utilities, nitrification recurred within months of the return to chloramination. In one utility, 4 months of free chlorination was applied, but nitrification recurred within 1 month of the return to monochloramine.

In some cases, free chlorination of only one area of the distribution system where nitrification typically occurs is conducted as a preventive measure—typically a storage facility. In this situation, the utility breakpoint chlorinates only a storage reservoir or a low-flow area of the distribution system. The practice of using free chlorine in an isolated area of the distribution system minimizes the disruption of the entire system and the impact on customers and focuses on the affected area. However, this practice requires remote chlorination operations, flushing of the affected area, and possibly targeted public notification.

Contra Costa Water District in California routinely practices free chlorination of distribution system storage tanks in lieu of booster chloramination (Guistino 2004). At EBMUD, free chlorination of remote reservoirs is routinely used but has not been effective at long-term prevention of nitrification (Song et al. 1999). Guistino (2004) has described several methods for breakpoint chlorinating treated water storage facilities. The most important factor is that the sodium hypochlorite be mixed within the tank. Because sodium hypochlorite is denser than water, slug dosages can settle on the floor of the storage facility. Therefore, a mixer may be needed inside the reservoir to lift the slug of hypochlorite from the floor and disperse it throughout the reservoir.

Free chlorination has some undesirable effects that should be addressed before implementing a routine program for prevention of nitrification. When switching disinfectant types, the need for public notification of dialysis units and other critical facilities/customers is important. Short-term increases in trihalomethanes (THMs) and haloacetic acids (HAAs), plus chlorinous taste and odor, are some of the important concerns. Because most utilities practicing chloramination have done so to limit formation of THMs and HAAs, it is likely that high levels of these chlorine DBPs will form, which may threaten regulatory compliance and raise public concerns. The free chlorine dose will be important for minimizing tastes and odors and the formation of DBPs. The dose must be sufficient to produce a distribution system residual or noncompliance with the SWTR and insufficient microbial control may be experienced.

Impacts of free chlorination on the microbial population are mixed. Temporary changes from monochloramine to free chlorine may result in biofilm sloughing and thereby create increased levels of some microorganisms in the water column (Vikesland et al. 2007). Increases in total coliform positives have been observed during some periods of free chlorination (Odell et al. 1996; Wilczak et al. 1996; Skadsen 1991). Free chlorination following 11 months of chloramination at two utilities demonstrated decreased cell counts in samples taken from the water columns using flow cytometry. The cell counts increased after the return to monochloramine (Rosenfeldt et al. 2009). These studies did not address the biofilm. The use of fluorescent in situ hybridization has demonstrated that nitrifying bacteria are present in biofilms both before and during free chlorination (Vikesland et al. 2007). Carrico et al. (2008) also found that AOM survived in the biofilm despite 1 month of free chlorination. Vikesland et al. (2007) concluded that the efficacy of disinfectant switching practices is not established and that nitrifier levels are not affected long term. Therefore, it is questionable whether the benefits of this practice are sufficient to outweigh the demonstrated negative impacts.

If free chlorination is practiced as a nitrification prevention measure, it is recommended that the utility examine the impact on nitrification. The timing of free chlorine application may be critical if this approach is to have any success. In general, application at time of likely nitrification may be appropriate. However, many utilities free chlorinate during the winter to minimize the formation of DBPs. Because nitrification typically recurs after free chlorination, the application of free chlorine in cold weather may not be providing any benefit for nitrification prevention.

Therefore, it is often desirable to keep the frequency and duration of system-wide free chlorination to a minimum. Given the negative impacts on water quality and the potential associated regulatory compliance issues, this practice has limited benefit and is recommended as a last resort. At best, the frequency of free chlorination needed to prevent nitrification depends on the utility based on water quality, temperature, pipe material, demand, and other utility-specific factors.

Booster Chloramination

Booster chloramination is a method for raising the distribution system total chlorine residual in areas where it has degraded. The boosting principle remains the same whether boosting free chlorine or monochloramine. However, booster chloramination is much less commonly practiced than booster chlorination. There are two basic approaches to booster chloramination: (1) both ammonia and chlorine may be added in the distribution system, or (2) chlorine alone may be added. If the utility can assess and manage the ammonia and chlorine concentrations at a given location, the elevated monochloramine residual achieved through boosting may help prevent the occurrence of downstream nitrification. As when forming monochloramine at the treatment plant, it is still critical to maintain the proper chlorine-to-ammonia-N ratio and minimize excess free ammonia.

When boosting both chlorine and ammonia, close attention must be given to the total chlorine residual, total ammonia, and free ammonia levels. The concentration of these parameters may vary daily at the influent to the booster station due to varying monochloramine decay rates in the system. Therefore, frequent water quality measurements, either by grab samples or online instruments, and adjustments of chemical pumps are strongly recommended. If not properly controlled, this approach to booster chloramination could inadvertently enhance nitrification by overfeeding ammonia. Alternatively, inadvertent breakpoint chlorination, producing a low total residual of free chlorine, could occur. Booster chloramination must be carefully monitored and controlled.

Booster chloramination at reservoir sites, as practiced by the Golden State Water Company in California, can be particularly beneficial when attempting to reduce heterotrophic plate count (HPC) levels and maintain a minimum residual in remote ends of the distribution system (Figure 7-13) (Cohen 1998). Trimming, which refers to the practice of adding chlorine only to bind with any remaining free ammonia, is becoming a common practice. The Florida Keys Aqueduct Authority practices booster chlorine and ammonia application at their storage and pump stations. Boosting the residual from 3.0–4.5 mg/L at a pump station resulted in a significant decrease in nitrite concentration (Figures 7-14 and 7-15, see case study). Booster chloramination has proven effective at preventing nitrification.

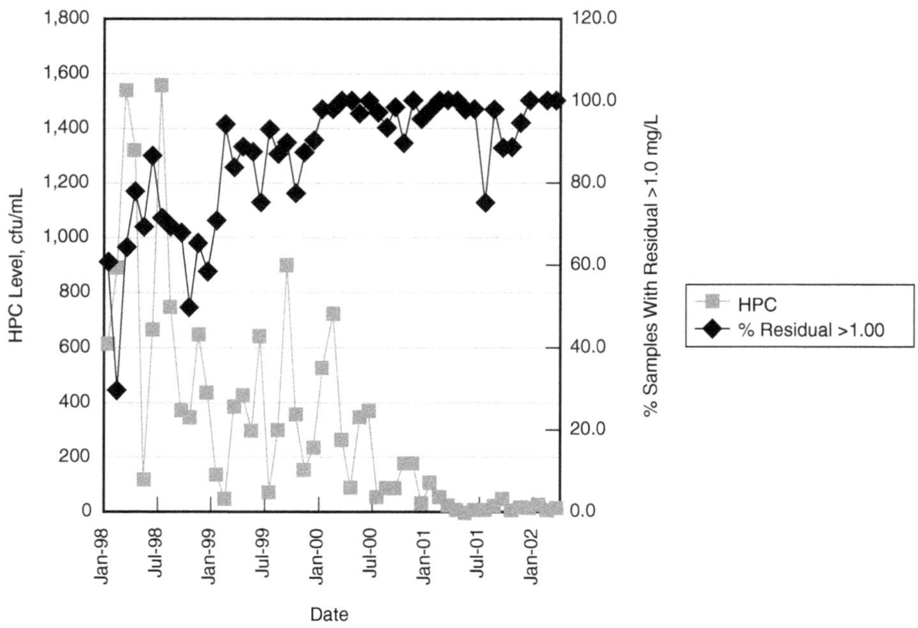

Source: Cohen 1998

Figure 7-14 Relationships between total chlorine residual and heterotrophic plate count (HPC) levels with booster chloramination at Golden State Water Company

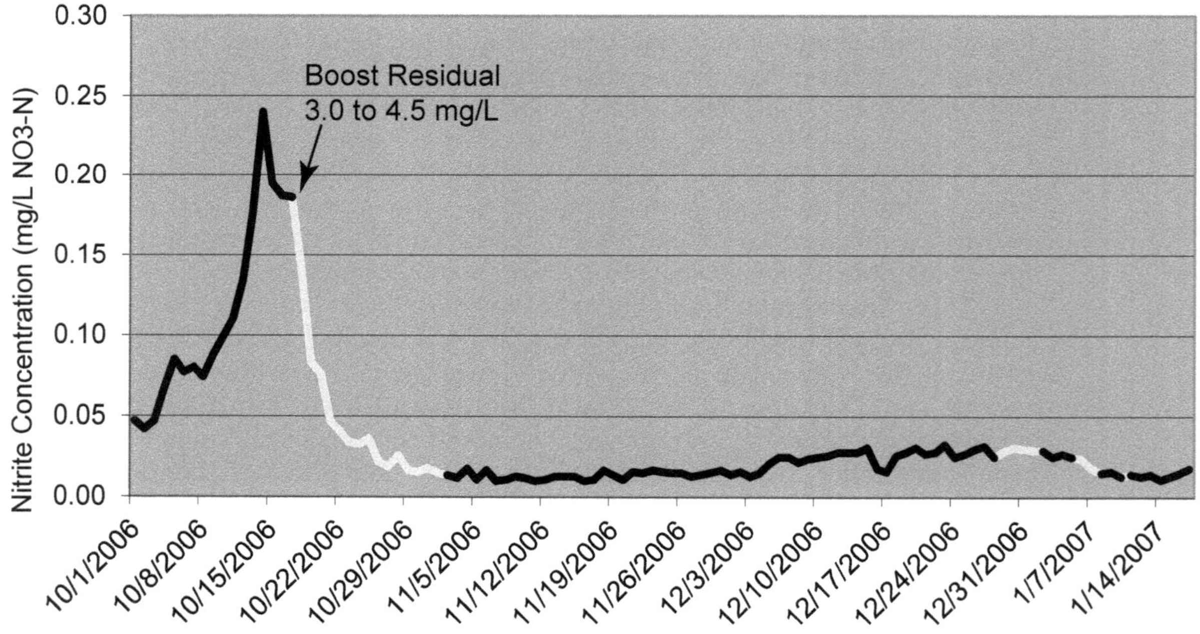

Source: Coley et al. 2010

Figure 7-15 Impact of booster chloramination on nitrification

In order for trimming to be successful, sufficient ammonia must be present to recombine with the newly added chlorine. If sufficient ammonia, at least 0.1 mg/L, is not present for the proper formation of more monochloramine, application of chlorine alone will shift disinfection past breakpoint, resulting in a decrease in total chlorine residual, potential free chlorine formation, and potential dichloramine formation. It is important when using chlorine only for booster operations that careful monitoring and

Case study—nitrification prevention strategies. The Florida Keys Aqueduct Authority (FKAA) transmission system extends approximately 130 mi from the 30 million gal per day J. Robert Dean Water Treatment Plant in Florida City on Florida's mainland to Key West. The system has 182 mi of transmission mains, five transmission booster pump stations, 650 mi of distribution mains, 14 distribution pump stations, and 45 million gal of treated water storage capacity. Because of the length and uniqueness of its water system, transit times for finished water can exceed 7–10 days at the extremities of the system in Key West. FKAA uses free chlorine for primary disinfection, but because of the high transit times to end users, FKAA uses monochloramine for residual disinfection.

Nitrification has historically been problematic for FKAA. Nitrification occurs in the FKAA system because of large system storage volumes, high transit times through transmission and distribution lines, and warm water temperatures inherent in most south Florida water sources. Chlorine and ammonia were historically added at a ratio of 4:1 at the water treatment plant, which resulted in some minimal level of free ammonia entering the distribution system. Additional free ammonia due to monochloramine decay enhanced the conditions for nitrification, which historically had resulted in frequent nitrification outbreaks, followed by free chlorine burnouts, frequent flushing, and draining and disposal of potable water in transmission storage tanks.

Because of the nature of the FKAA system and the need to maintain adequate distribution storage, proper chlorine-to-ammonia-N dose control and minimization of free ammonia in the distribution system was determined to be the only way to control nitrification. FKAA's Florida City Water Plant now doses ammonia to achieve nondetectable ammonia levels and maximum monochloramine residual (i.e., approximately 5:1 Cl_2:NH_3-N ratio and a 4.0 mg/L total chlorine residual). FKAA also now chlorinates at a midway point of the transmission system to reform monochloramine and bind any free ammonia resulting from chloramine decay or chlorine demand during transmission. As a result, there have been no nitrification events and no wasted stored water since the change in operation in 2007. Additionally, FKAA has eliminated plans to perform annual free chlorine conversions to control nitrification.

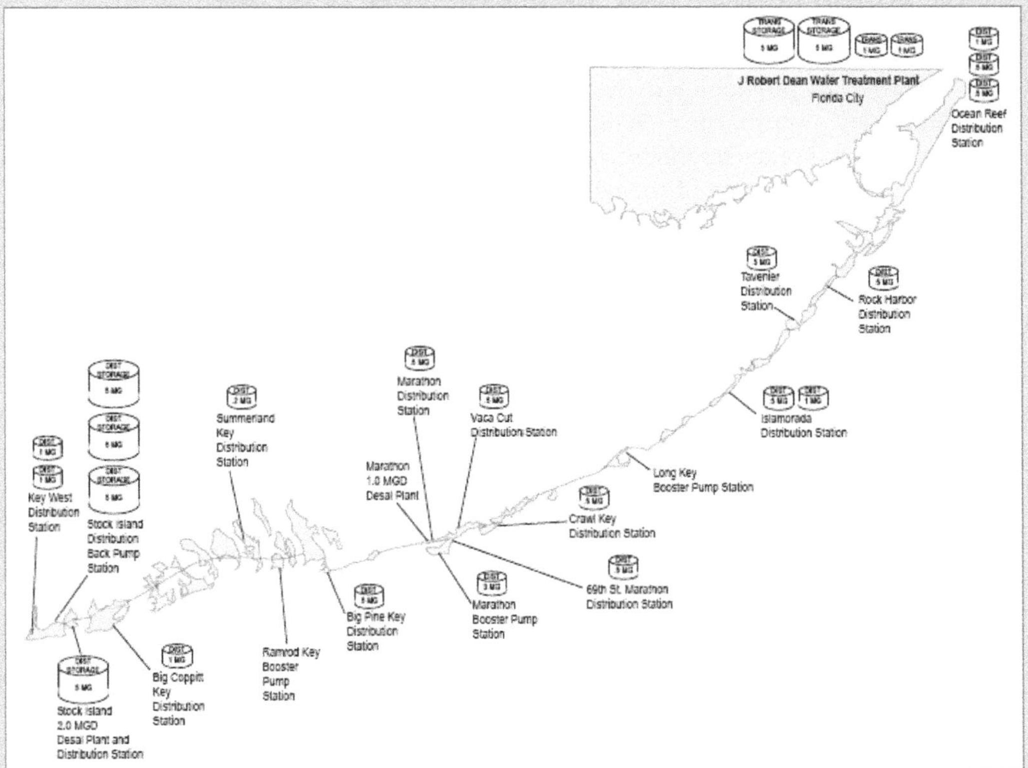

Figure 7-16 Florida Keys Aqueduct Authority transmission system

process control are performed to avoid breakpoint reactions while achieving the goal of recombining the free ammonia. This system adds sodium hypochlorite to large reservoirs during the fill portion of the cycle, and the reservoirs are equipped with mixers and online chlorine analyzers.

Glendale, Calif., also practices trimming (Sykes 2003). The chlorine dose is calculated based on the free ammonia concentration and the target chlorine ratio. The amount of ammonia available for the reaction is monitored and controlled to achieve a ratio of 4.75:1 chlorine to ammonia-N at the chlorine boosting station. This strategy may be useful for reservoirs with long detention times or reservoirs in series and where chlorine boosting facilities already exist (Palacios and Smith 2003). In bench-scale tests, this strategy demonstrated an initial increase in total monochloramine residual, but the "recombined" residual appeared to decay at a slightly faster rate than the nonboosted sample. This observation was not verified in field data as the system nitrified and the free ammonia from monochloramine decay was consumed (Palacios and Smith 2003).

In areas of high water age, there may be sufficient ammonia to make trimming an effective strategy. Experiments in 2006 and 2007 at over 50 storage facilities indicated that increased nitrite concentrations and decreased monochloramine residual could be at least partially mitigated when booster chloramination was performed in a timely manner (Bleth 2008). This report noted that good mixing is an essential component of the response plan to avoid localized breakpoint reactions.

Booster chloramination must be carefully monitored and controlled to ensure that the stoichiometric amounts of chlorine and ammonia-N react. Boosting monochloramine residual as a practice requiring engineering improvements and capital expenditures is discussed in detail in Chapter 9.

ALTERNATIVE APPROACHES TO NITRIFICATION PREVENTION

Because prevention of nitrification is not always successful using the traditional approaches previously discussed, there have been continued interest and investigation into new approaches. While the previous recommendations reflect general good operational practices that should be routinely practiced, there may be times when they are not sufficient. Some of the newer approaches that are still being evaluated are elevation of pH, use of chlorite, blending of different water qualities, booster chloramination, and use of ultraviolet (UV) light. These actions are supplemental to the practices previously recommended as traditional approaches to nitrification prevention.

Blending of Distribution Waters or Raw Waters

Blending may create a different water quality matrix that may decrease the potential for nitrification. Blending can be performed with raw waters if multiple sources are available and/or in distribution systems where multiple treated waters are present.

Blending of different chloraminated or chloraminated/chlorinated distribution waters will alter the chemistry and microbiology of those waters, thus affecting the potential for nitrification. Such blending must be carefully evaluated, as the blending can be either beneficial or detrimental regarding nitrification and water quality in general.

An example of the detrimental effect of distribution water blending on disinfectant residual was reported by Sykes (2003). When combining well water that had been disinfected with free chlorine with purchased chloraminated water, the free chlorine reacted with the monochloramine and reduced the final total chlorine residual. This occurred in the Glendale, Calif., when their wholesale supplier switched disinfection from free chlorine to monochloramine.

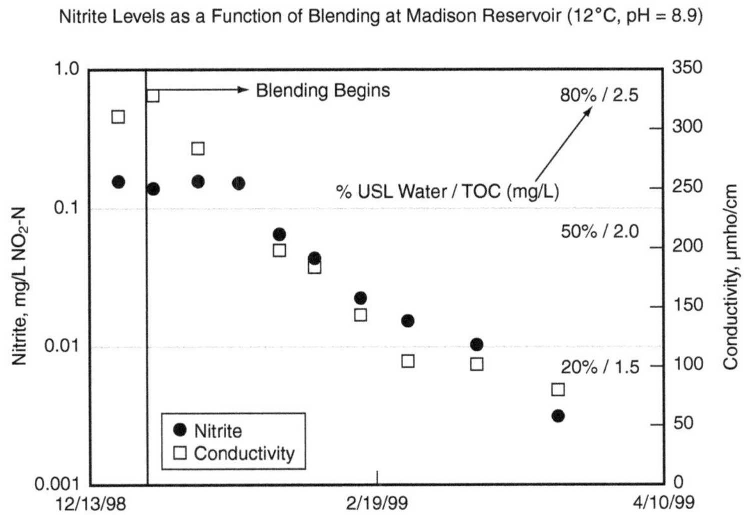

Source: Kirmeyer et al. 2004

TOC = total organic carbon; USL = upper San Leandro

Figure 7-17 Reduction in nitrite levels following water blending

A very thorough understanding of distribution system behavior is necessary when blending different waters. It is best if the chlorine-to-ammonia-N ratio is maintained during blending. The issues with blending distribution waters are similar to the issues with booster chloramination.

The positive benefits of blending of raw waters have been demonstrated at EBMUD. The blending of two different waters (Mokelumne River water with other East Bay area water reservoirs with differing TOC concentrations) increased the pH, lowered the TOC of the blended product, and thereby improved the monochloramine stability in the distribution system (Song et al. 1999). An example of nitrite reduction following blending at the distribution storage reservoir is shown in Figure 7-17. In this case, upper San Leadro (USL) water was blended with Mokelumne River water, resulting in lower TOC. As the percent of USL water decreased and the percent of Mokelumne River water increased, the resulting blend of water resulted in a reduction in nitrite and conductivity. Nitrite concentrations dropped significantly.

pH and Nitrification Prevention

The use of pH control to prevent nitrification has been explored by some utilities and has met with mixed success. The optimal pH for AOB growth is 6.5–8.5 (Kirmeyer et al. 1995; Odell et al. 1994; Jolley et al. 1984). Theoretically, either raising or lowering the pH outside the optimum growth range should reduce the growth of AOB and thereby limit nitrification. Also, increasing the pH will slow the nitrite and monochloramine reaction and slow the decay of monochloramine.

AOB have been shown to be highly sensitive to pH. Growth of AOB stops at pH ≤6.5 (Keen and Prosser 1987). However, because utilities are concerned with corrosion control, the option of reducing the pH is less feasible than raising the pH. Optimal AOB growth is at pH 8.0. Growth of AOB at pH 7 and 9 exhibits an 80 percent decrease in growth rate compared to pH 8. Harrington et al. (2002) observed that the time to onset of nitrification was fastest around pH 8.5.

In one distribution system study, the use of a finished water of pH 9.3 versus pH 8.5 prevented nitrification (Skadsen 1993). An inverse correlation between pH and nitrite concentrations in the distribution system was observed over the pH range studied and is depicted in Figure 8-16. Long-term studies at this utility found the elevated pH 9.3 to be effective at eliminating the occurrence of nitrification for 7 out of 8 years (Skadsen 2002). After 8 years, significant increases in nitrite concentrations at distribution locations with high water age were again routinely observed. In 2011, nitrite values of 350 ug/L were observed at one point in the distribution system (Figure 8-17). However, other parameters, such as a decrease in monochloramine residual, were not observed to the extent typical of past nitrification incidents. Total chlorine at the same location never decreased below 1.5 mg/L. HPC (R2A agar, pour plate, 7-day incubation at 22°C) increased to >500 cfu/mL, but the results were lower than previously observed during nitrification.. It is possible that either the species of nitrifying bacteria had changed or the existing species had become more tolerant of high pH.

In another study comparing pH 9.5 versus pH 8.5, it was observed that the utility with the lower pH was more likely to nitrify (Schrempp et al. 1994). In the Florida Keys, lowering the pH from 9.5 to 7.8 coincided with nitrification in storage tanks (Cates and Lavinder 1996). Further experiments demonstrated increased monochloramine stability and lower free ammonia as pH increased (Figures 7-18 through 7-21).

In a limited experiment, Song et al. (1999) found that raising the pH to 9 appeared to prevent nitrification in one distribution system but not in another when using system blending. Increasing the pH from 8.6 to 9.0 had mixed success on reduction of nitrification incidents. Wichita Falls, Tex., saw a 50 percent reduction in nitrite concentration in 24 h during a nitrification episode when the pH was raised from 8.5 to 9.2 (Harms and Owen 2004). Operation of a pilot system at pH >9 during the summer was unable to induce nitrification, even after dropping the total chlorine residual to 0.5 mg/L (Broad 2005).

Not all experiences have supported this pH control hypothesis. Nitrification in distribution systems has been observed over a wide range of pH, from 6.6 to 9.8 (Wilczak et al. 2003a; Odell et al. 1996; Kirmeyer et al. 1995). Laboratory inactivation experiments

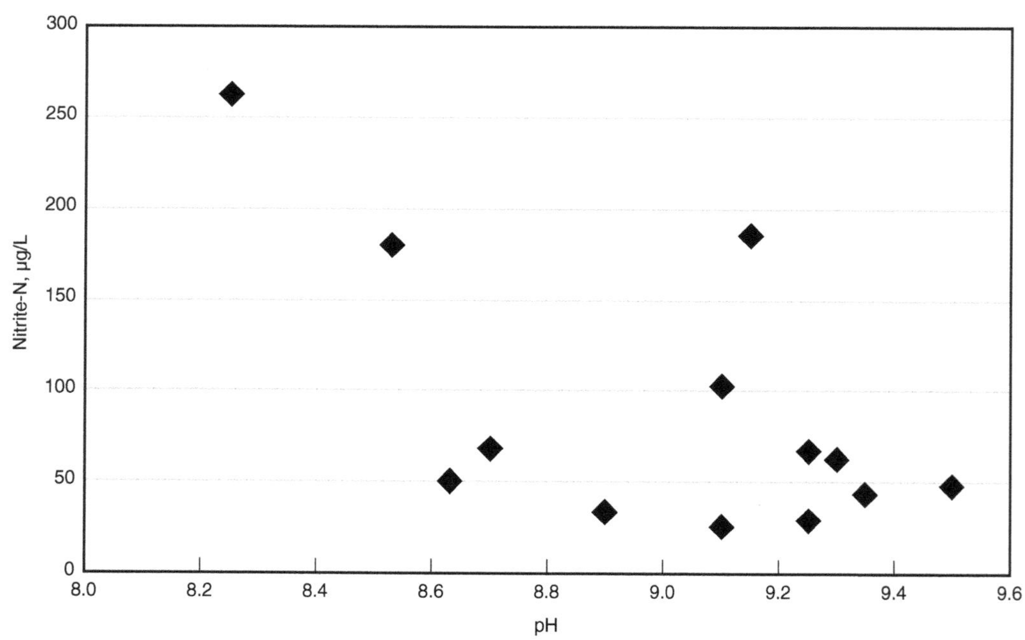

Source: Skadsen 2002

Figure 7-18 Correlation of pH and nitrite at a distribution system sampling location

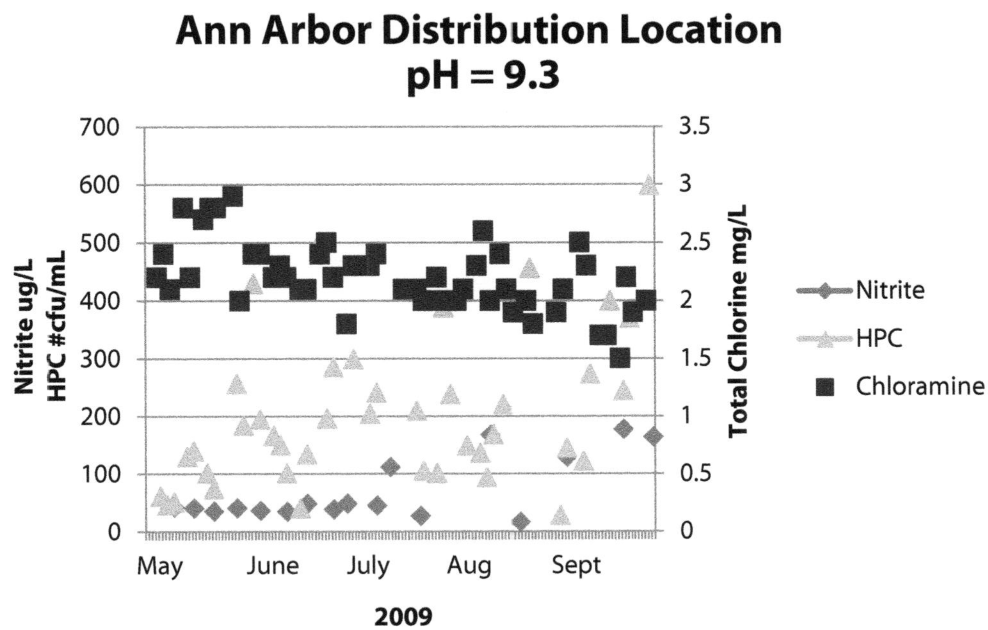

Source: Ann Arbor (Mich.) Water Plant

HPC = heterotrophic plate count

Figure 7-19 Limited nitrification in Ann Arbor (Mich.) water system after 15 years of operation at pH 9.3

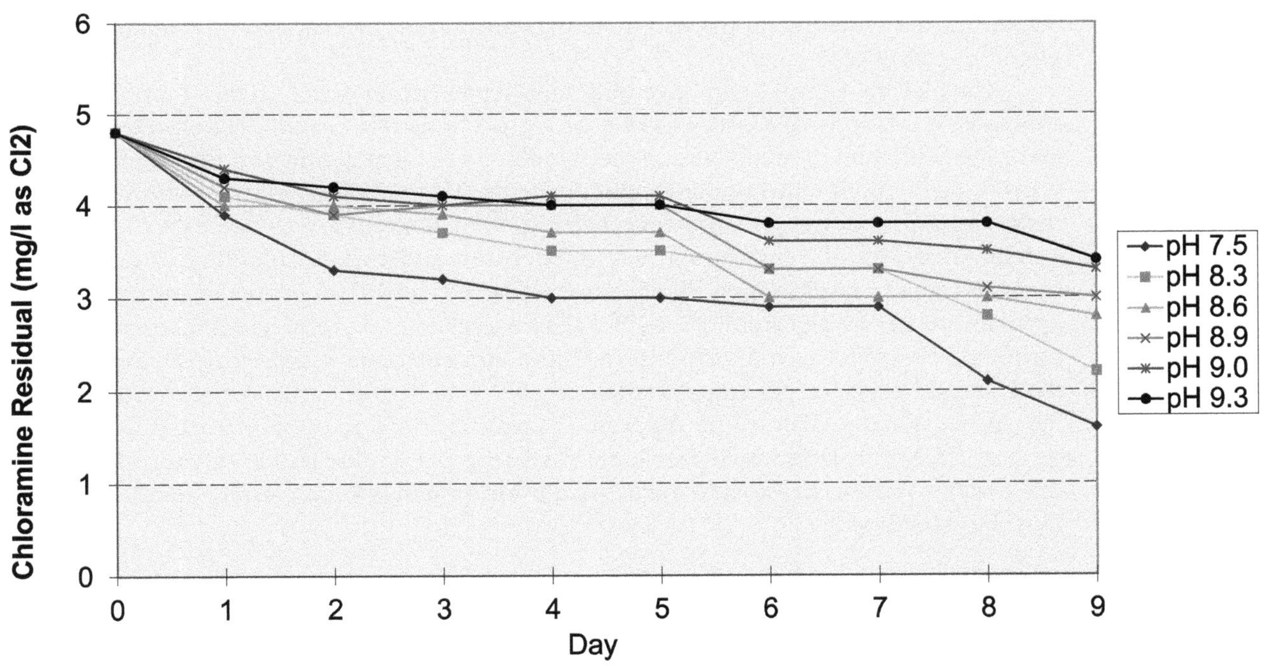

Source: Coley et al. 2010

Figure 7-20 Monochloramine stability and pH

have found that AOM were more easily inactivated with monochloramine at pH 7 versus pH 9 (Noguera et al. 1999). Oldenburg et al. (2002) found that the rate of inactivation decreased as pH increased. However, this study noted that the inactivation rates were dependent on the method used, with the most probable number rates being significantly higher. Harrington et al. (2002) found that a biocide-to-food ratio of ≥1.9 mg Cl_2/mg N was

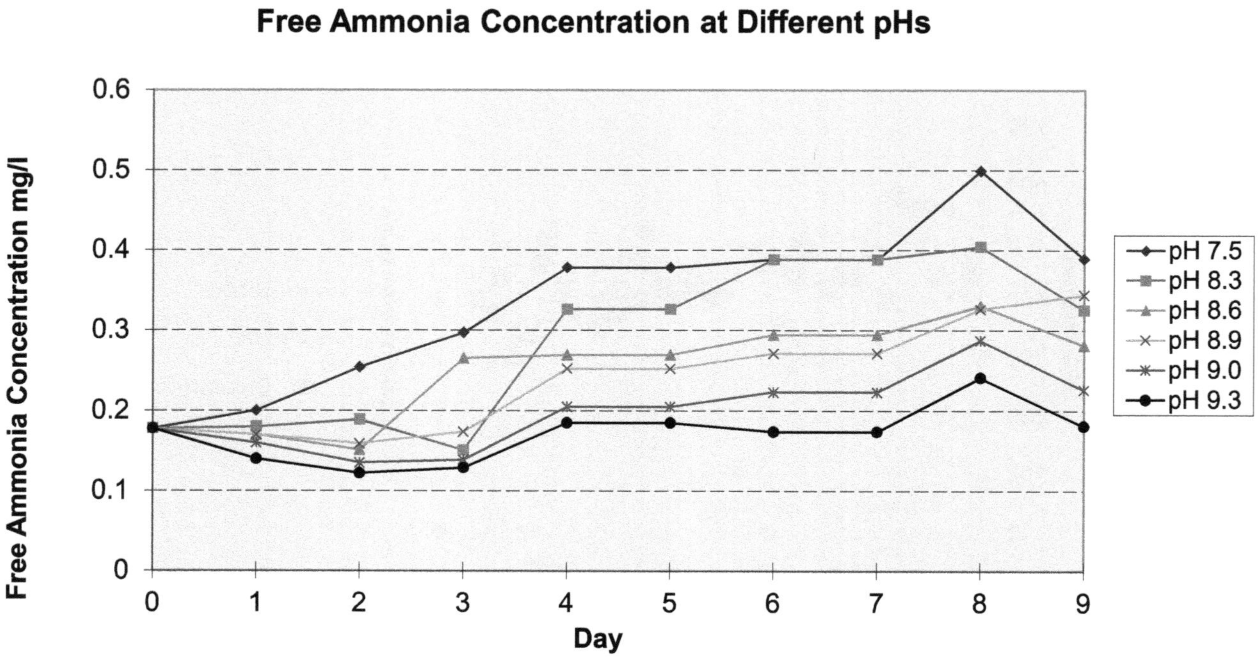

Source: Coley et al. 2010

Figure 7-21 Free ammonia and pH

necessary to inhibit nitrification. The balance between growth and inactivation should be considered when using pH as a potential prevention mechanism (Speitel et al. 2011, Fleming et al. 2005).

The elevation of pH will have other consequences on water quality that need to be evaluated if this approach is implemented for preventing nitrification. The elevated pH will favor the formation of monochloramine and eliminate the potential for dichloramine, thus preventing dichloramine taste-and-odor incidents. Also, the decay rate of monochloramine decreases as pH is raised, creating a more stable monochloramine residual in the distribution system. However, the higher pH may impact lead and copper solubility or have other corrosion control implications. Phosphate-based corrosion inhibitors, for example, perform optimally at pH ranges around 7.6–7.8. Excessive scaling on pipes, resulting from calcium carbonate precipitates, may affect water flow and customer satisfaction. Depending on the chemical used for pH adjustment, increases in both cost and sodium concentrations may occur. Finally, disinfection byproduct concentrations may be impacted as well. In general, THM formation increases with increasing pH while HAA formation slows, but this is water specific. Before making a distribution system pH change, all potential impacts must be considered.

Chlorite Ion Effects

The addition of chlorite ion has been proposed as a means to prevent nitrification but has met with mixed success. This theory was originally developed when it was observed that the Gulf Coast Water Authority in Texas used chlorine dioxide as a primary disinfectant (chlorite concentrations in their distribution system were 0.25–0.35 mg/L) and monochloramine as a final disinfectant but had not noticed nitrification in their distribution system (McGuire et al. 1999). A comparison of two utilities in Texas supported this observation: high concentrations of nitrite were observed at sample points throughout the distribution system of the chloraminated utility, but much lower nitrite levels were concurrently observed in the utility using chlorine dioxide/monochloramine.

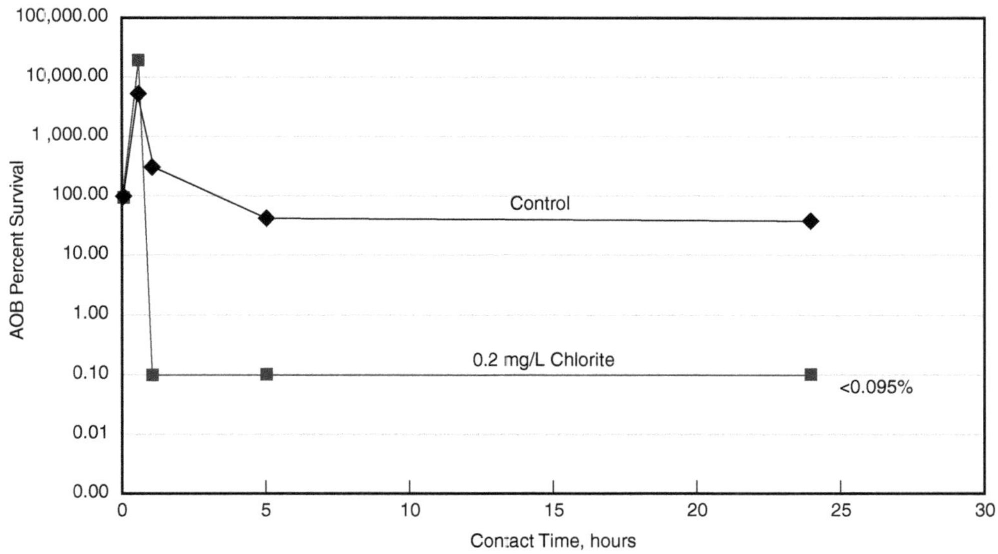

Source: Arizona Water Company

Figure 7-22 Survival of ammonia-oxidizing bacteria as affected by chlorite ion

Based on this observation, laboratory tests were conducted using chlorite ion concentrations of 0, 0.2, 0.5, and 1 mg/L to assess the effects on AOB. All concentrations were found to reduce the survival of AOB. As shown in Figure 7-20, the percentage of survival of AOB cultures was significantly reduced in the presence of 0.2 mg/L chlorite (McGuire et al. 1999). Field studies also found that the presence of chlorite delayed the development of nitrification in distribution systems in warm climates at two of the three utilities. One system noted nitrification still occurred. In that instance, high concentrations of ammonia-N may have limited the effectiveness of the chlorite (McGuire et al. 1999).

Experimentation using plug-flow reactors at Tucson Water has found that continuous chlorite feed as low as 0.1 mg/L prevented nitrification from becoming established (McGuire et al. 2004). In this study, it appears that nitrification can be prevented by short-term applications of 0.2 mg/L chlorite ion (McGuire et al. 2004). However, intermittent chlorite application may not be effective for long-term control if nitrification and AOB colonies are well established (McGuire et al. 2006).

Full scale experiments were conducted at Glendale, Calif. When 0.6 mg/L of chlorite was applied in the distribution system, it was concluded that chlorite was effective in preventing nitrification in systems where it commonly occurred but that it was not effective alone in reversing nitrification that had already started. A period of free chlorination was recommended prior to addition of chlorite (McGuire et al. 2007). Further studies at the same utility with 0.6 mg/L chlorite met with mixed success. In some cases, it proved effective (Figure 7-21), and in others, nitrification recurred (Figure 7-22). This study also examined the impact of a free chlorination period prior to application of chlorite. Nitrification was not observed for the 76 days where data were collected and analyzed following this free chlorination. This suggests that the free chlorination was effective at inactivating AOM and subsequently helping to prevent nitrification for at least a limited period; the use of chlorite following free chlorination is likely to be more effective in preventing nitrification than addition of chlorite to a system in which AOM have already colonized.

Addition of sodium chlorite at certain areas of its distribution system has been used at the Louisville Water Company for over 10 years with support and approval from Kentucky Division of Water (KDOW). An effective application strategy was developed based on extensive kinetic and full-scale studies (Rungvetvuthivitaya et al. 2014; Zhu et al. 2010). The kinetic studies demonstrated potential nitrification prevention at doses as

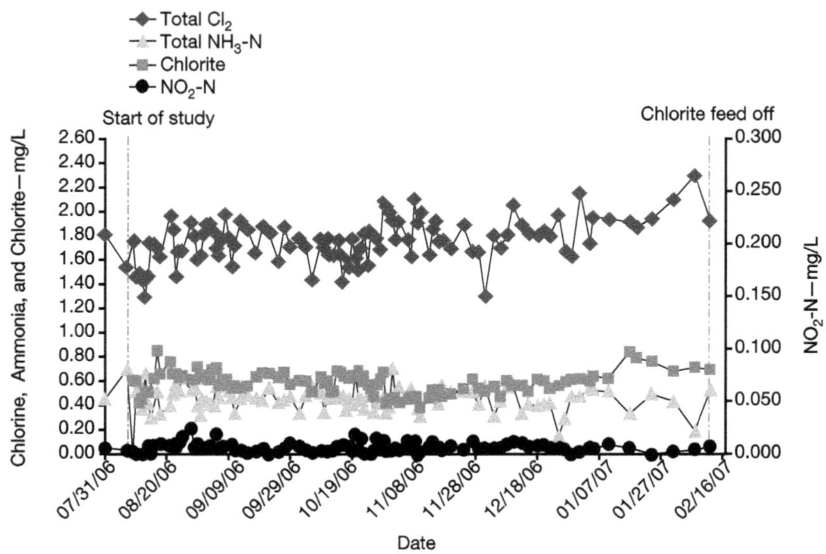

NO_2-N—nitrite-nitrogen, total Cl_2—total chlorine, total NH_3-N—total ammonia-nitrogen

Source: McGuire et al. 2009

Figure 7-23 Control of nitrification by use of chlorite ion

low as 0.02 mg/L. A 2-year full-scale demonstration study showed that direct application of chlorite at a dosage level of 0.3–0.5 mg/L was generally effective in preventing and controlling nitrification (Figure 7-23). However, it was noted that chlorite ion was less effective in areas where nitrification was already developed. During the first season of the study, chlorite was applied when nitrification had already become established. Chlorite feed alone was able to delay but not prevent the reestablishment of nitrification in certain dead-end areas. Chlorite application then started in early April during the following season when water temperature was still below 15°C and no signs of nitrification were observed. As a result, chloramine residual was maintained throughout the chlorite feeding area over the 8-month study period with no need for additional flushing. Louisville Water also developed a range of protective measures, including online monitoring, to ensure there was no chlorite overfeed. Since the implementation of this chlorite application strategy in 2009, the occurrence of nitrification in the treated areas has been virtually eliminated. Rungvetvuthivitaya et al. (2019) also developed a kinetic model to simulate the chlorite decay under various chloramine concentrations and other environmental conditions, which can provide guidance to water utilities for nitrification prevention and control via chlorite treatment.

As an alternative to direct chlorite application, indirect addition of chlorite ion through chlorine dioxide treatment has been practiced in several locations. Its effectiveness in nitrification prevention has been reported at several Texas and California water systems (Santos 2018; Holden 2017; McGuire et al. 1999). For example, one of the three San Diego water treatment plants (the Otay Water Treatment Plant) upgraded its process with chlorine dioxide application in 2011 for DBP formation control. As a result, 0.2–0.3 mg/L of chlorite residual was maintained in the distribution system from the Otay facility. Since the implementation of the chlorine dioxide treatment, no major nitrification issues have been observed in the part of the city's distribution system served by the Otay plant (Holden 2017). In addition to nitrification control, chlorine dioxide treatment provides an additional barrier for microbial disinfection, less THM formation, and lower chlorine/ammonia demand. On the other hand, the challenges of this indirect chlorite addition for nitrification prevention involve appropriate process engineering to maintain proper

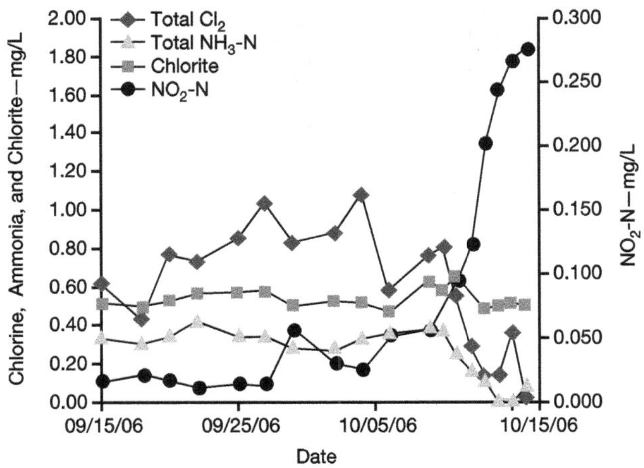

Source: McGuire et al. 2009

Figure 7-24 Loss of nitrification control in the presence of chlorite ion

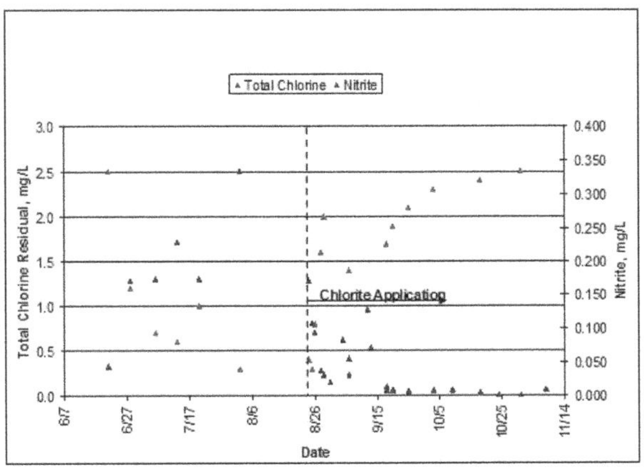

Source: Zhu et al. 2010

Figure 7-25 Nitrification control by use of chlorite ion in Louisville (Ky.) Water System

chlorite concentrations in the distribution systems and to minimize the formation of unregulated DBP in chlorate (see Figures 7-22 through 7-25).

It is important to consider that chlorite concentration is regulated under the stage 1 D/DBP Rule, with a maximum contaminant level of 1.0 mg/L at the entry point to the distribution system and within the distribution system. The maximum contaminant level goal for chlorite is 0.8 mg/L. Therefore, any use of chlorite, either through direct addition of sodium chlorite or the use of chlorine dioxide, for the prevention of nitrification would need to be strictly controlled. The issue of degradation of chlorine dioxide into chlorite over time has the potential to increase chlorite concentrations within the distribution system. Chlorite ion concentration has been shown to change within a distribution system (Baribeau et al. 2002), so it would be important to monitor for chlorite in different distribution areas with different residence times to determine the exposure and effectiveness of chlorite at inhibiting nitrification. Note that the regulatory standard for chlorite, unlike other DBPs, is based on an acute health impact, not a chronic one. Therefore, immediately public notification is required if the regulatory standard is exceeded.

The issue of adding a regulated substance to the drinking water for prevention of nitrification is a debatable subject that needs to be further evaluated by a utility, its regulators, and its customers. Customer acceptance of the practice needs to be carefully considered before implementation. The appropriate state agency should also be consulted before this approach is used.

If chlorine dioxide feed is used to obtain both disinfection and chlorite, the hazards associated with chlorine dioxide formation should be considered. The use of chlorine dioxide can provide an additional benefit of providing disinfection credits, but the control of chlorite concentration will be limited compared to direct chlorite addition. As with all preventive measures, this approach should be combined with other good operational practices, such as the limitation of excess free ammonia and minimization of detention times.

Ultraviolet Light

It is known that AOM are inactivated in the presence of sunlight (Kirmeyer et al. 2004). Therefore, applying a UV dose to a water storage facility should theoretically inactivate at least the organisms in the water column. Investigation on the use of blacklight (ultraviolet A [UVA]) for control and prevention of nitrification was reported (White and Adams 2010). In that study, the Los Angeles Department of Water and Power (LADWP) performed a pilot study in July 2009 wherein they added near-UV lights in covered storage facilities (Figure 7-26). Though the ability to penetrate the biofilm was limited, initial results were promising based on the reduction of the presence of nitrite-N as an indicator of nitrification (Figure 7-27). LADWP conducted follow-up studies to generate UV dose–response curves using their native AOB (Sarkisian et al. 2020) and various types of light (medium-pressure [MP] UV, low-pressure [LP] UV, and near-UV). The study also looked at dark repair and the impact on monochloramine degradation. They found that both MP- and LP-UV lights at cost-effective doses can be used for 3-log and 4-log inactivation of AOB in LADWP water. For the near-UV lamps (365 and 395 nm), they found that hours of contact time would be needed to achieve significant inactivation. A full-scale installation is planned for their Temescal Valley tank with an external recirculating MP-UV reactor (So 2020).

A full-scale pilot was conducted by EBMUD starting in 2015 at one distribution reservoir. In this case, water was pumped out of the reservoir, treated through the UV reactor, and discharged back into the reservoir. Data from 4 years of operation indicate that this reservoir nitrified significantly less frequently when UV treatment was employed (once per year compared to 4–5 times per year).

ASSESSMENT OF EFFECTIVENESS OF PREVENTIVE OPERATIONAL MEASURES

An important part of nitrification prevention will include continuous assessment of water quality and the effectiveness of the nitrification prevention measures. Indicators of nitrification in the distribution system include routine testing and assessment of relevant parameters at areas of high water age (Odell et al. 1996; Wilczak et al. 1996). Following are these parameters:

- low total chlorine residual,
- high nitrite concentration,
- low free ammonia concentration (indicating ammonia consumption),
- increased nitrate concentration,
- high HPC R2A counts, and
- decrease in pH.

These parameters should be routinely evaluated for changes that might indicate developing nitrification. Utilities should track background concentrations and assess changes and trends. An action plan and trigger levels should be developed. It is important

Source: White and Adams 2010

Figure 7-26 Photo of full-scale ultraviolet light application in storage facility

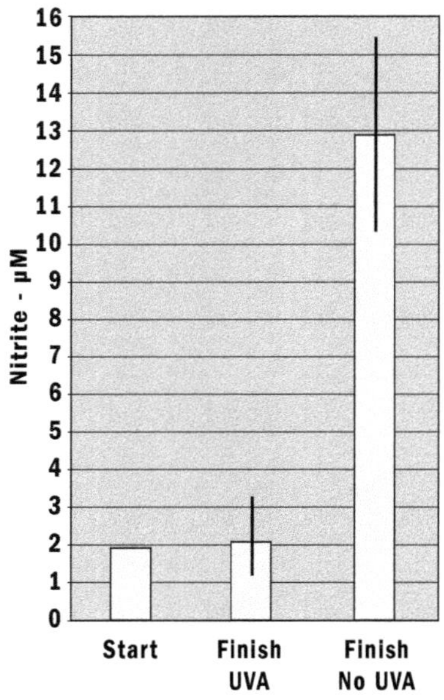

Source: White and Adams 2010

Figure 7-27 Low-intensity ultraviolet A (UVA) radiation installed in storage facility for inhibition of nitrification in Los Angeles

to assess the normal background levels of these parameters and monitor for any change in relative concentration.

KEY TAKEAWAYS

There are many approaches and operational practices that can be used to prevent nitrification in the distribution system. At most utilities, the following practices are recommended.

Always

- Minimize free ammonia and control the chlorine-to-ammonia-N ratio to 4.5–5:1
- Maintain a sufficient monochloramine residual, >2.0 mg/L, leaving the treatment plant and >1.0 mg/L or higher in the distribution system depending on site-specific requirements
- Routinely clean storage facilities and distribution system piping
- Practice systematic distribution system flushing
- Provide good corrosion control

Optimization of practices

- Minimize distribution system water age
- Maximize storage facility turnover and cycling
- Maximize hydraulic mixing
- Optimize storage volume
- Install online chlorine residual analyzers
- Evaluate booster chloramination

Emerging/nontraditional approaches

- Blend waters to improve quality

- Evaluate increased pH
- Evaluate use of chlorite ion
- Evaluate use of UVA

Last resort

- Use free chlorination: Evaluate effectiveness on nitrification as well as unintended consequences and assess both beneficial and detrimental impacts on water quality.

However, most of these approaches have met with mixed success. Many of these approaches may be utility specific due to differences in water quality and distribution configurations and operations. Others need more investigation to determine how universally applicable and effective they are over time. Many of these approaches coincide with good water quality management practices and therefore will be a good starting point for any utility using or considering monochloramine. A utility should select those approaches that are most appropriate to their situation. The frequency and extent of nitrification episodes will help determine the type and extent of prevention measures considered and selected.

Despite extensive prevention measures, given to the complexity of distribution systems and constantly changing environmental factors that affect source quality, all chloraminated utilities should have a robust nitrification detection and response plan in place.

REFERENCES

American National Standards Institute/American Water Works Association (ANSI/AWWA). 2015. Distribution Systems Operation and Management. G200. Denver: AWWA.

Antoun, E.N., J.E. Dyksen, and D.J. Hiltebrand. 1999. "Unidirectional Flushing: A Powerful Tool." *Journal AWWA*, 91:7:62–71.

AWWA (American Water Works Association). 2008 (4th edition). Manual M31, *Distribution System Requirements for Fire Protection*. Denver: AWWA.

AWWA. 2018a. *2017 Water Utility Disinfection Survey Report*. Denver: AWWA.

AWWA. 2018b. *Self-Assessment Guide for Distribution System Optimization: Partnership for Safe Water*. Denver: AWWA.

AWWA Disinfection Systems Committee. 2008. "Committee Report: Disinfection Survey, Part 2—Alternatives, Experiences, and Future Plans." *Journal AWWA*, 100:11:110–124.

Baribeau, H., C.A. Kinner, R. De Leon, and P.A. Rochelle. 2001. Nitrification in Steel and Concrete Drinking Water Reservoirs: A Case Study. *In Proc. of the AWWA Annual Conference*. Denver: AWWA.

Baribeau, H., M. Prévost, R. Desjardins, P. Lafrance, and D. Gates. 2002. "Chlorite and Chlorate Ion Variability in Distribution Systems." *Journal AWWA*, 94:7:96–105.

Barrios, M.R., and P. Stone. 1989. "Chloramine Depletion in Covered Reservoirs." *In Proc. of the AWWA Water Quality Technology Conference*. Denver: AWWA.

Black & Veatch. 2010 (5th edition). *White's Handbook of Chlorination and Alternative Disinfectants.* New York: John Wiley & Sons.

Bleth, J. 2008. "Fast Response, Early Boost: Eliminate Ammonia-Oxidizing Bacteria." *Opflow*, 34:12:14–16.

Broad, personal communication, 2005.

Camper, A.K., W.L. Jones, and J.T. Hayes. 1996. "Effect of Growth Conditions and Substratum Composition on the Persistence of Coliforms in Mixed-Population Biofilms." *Applied and Environmental Microbiology*, 62:11:4014–4018.

Carrico, B.A., F.A. DiGiano, N.G. Love, P. Vikesland, K. Chandran, M. Fiss, A. Zaklikowski. "Effectiveness of Switching Disinfectants for Nitrification Control." *Journal AWWA*, 100:10:104–115.

Cates, J., and S. Lavinder. 1996. "Action Plan for Improving Chloramine Residuals and Minimizing the Occurrence of Nitrification in the Florida Keys Aqueduct Authority's Distribution System." *In Proc. of the AWWA Water Quality and Technology Conference.* Denver: AWWA.

Cohen, Y.K. 1998. "Forming Chloramine and Maintaining Residual." *Opflow*, 24:9:1–5.

Coley, R., J. Reynolds, and C. Johnson. 2010. "Nitrification Control in the Florida Keys." *In Proc. Of the AWWA Water Quality and Technology Conference.* Denver: AWWA.

Cunliffe, D.A. 1991. "Bacterial Nitrification in Chloraminated Water Supplies." *Applied and Environmental Microbiology*, 57:11:3399–3402.

Davis, M.K. 1990. "The Role of the Laboratory in Water Utility Decision Making." *In Proc. of the AWWA Water Quality and Technology Conference.* Denver: AWWA.

DiGiano, F.A., W. Zhang, A. Travaglia, D.E., Francisco, and M. Wood. 2002. *Occurrence of Bacterial Regrowth and Nitrification in the Raleigh Distribution System and Development of an EPANET Model for Future Assessments.* UNC-WRRI. 2002–338. Raleigh, NC: Water Resources Research Institute of the University of North Carolina.

Fleming, K.K., G. Harrington, and D.R. Noguera. 2005. "Nitrification Potential Curves: A New Strategy for Nitrification Prevention." *Journal AWWA*, 97:8:90–99

Friedman, M., G. Kirmeyer, J. Lemieux, M. LeChevallier, S. Seidl, and J. Routt. 2010. Criteria for Optimized Distribution Systems. Partnership for Safe Water. Denver: WRF.

GLUMRB (Great Lakes Upper Mississippi River Board of State Public Health and Environmental Managers). 2012. Interim Standard-Use of Chloramine Disinfectant for Public Water Supplies. Policy Statement Adopted April 1997; Re-Adopted as Interim Standard April 2003, Revised April 2007.

Grayman, W.M., R.A. Deininger, L.A. Rossman, C.D. Smith, C.N. Arnold, and J.F. Smith. 2004. "Mixing and Aging of Water in Distribution System Storage Facilities." *Journal AWWA*, 96:9:70–80.

Grayman, W.M., L.A. Rossman, C. Arnold, R.A. Deininger, C. Smith, J.F. Smith, and R. Schnipke. 2000. *Water Quality Modeling of Distribution System Storage Facilities*. Denver: WRF.

Guistino, J. 2004. "Breakpoint Chlorination of Reservoirs—A Nitrification Band-Aid." *Presented at the 2004 AWWA Cal-Nevada Section Spring Conference*. Denver: AWWA.

Harms, L.L., and C. Owen. 2004. *A Guide for the Implementation and Use of Chloramines*. Denver: WRF.

Harrington, G.W., D.R. Noguera, C.C. Bone, A.I. Kandou, P.S. Oldenburg, J.M. Regan, and D. Van Hoven. 2003. *Ammonia from Chloramine Decay: Effects on Distribution System Nitrification*. Denver: WRF.

Harrington, G.W., D.R. Noguera, A.I. Kandou, and D.J. Vanhoven. 2002. "Pilot-Scale Evaluation of Nitrification Control Strategies." *Journal AWWA*, 94:11:78–89.

Hasit, Y.J., A.J. DeNadai, H.M. Gorrill, S.B. McCammon, R.S. Raucher, and J. Witcomb. 2004. *Cost and Benefit Analysis of Flushing*. Denver: WRF.

Henrie, T., L.W. Rezania, D. Nagy, and D.A. Lytle. 2012. "Got Ammonia?" Opflow, 38:6:22–24.

Holden, G. W. 2017. "Chlorine Dioxide Preoxidation for DBP Reduction." *Journal AWWA*, 109:7:36–43.

Holt, D., R.D. Todd, A. Delanoue, and J.S. Colborne. 1995. "A Study of Nitrite Formation and Control in Chloraminated Distribution Systems." *In Proc. of the AWWA Water Quality Technology Conference*. Denver: AWWA.

Hotaling, personal communication, 2005.

Hotaling, personal communication, 2011.

Ike, N.R., R.L. Wolfe, and E.G. Means. 1988. "Nitrifying Bacteria in a Chloraminated Drinking Water System." *Water Science & Technology*, 20:11–12:441–444.

Ireland, C., and M. Knudson. 1998. "Portland's Experience with Chloramine Residual Management." *In Proc. of the AWWA Protecting Water Quality in the Distribution System*: What is the Role of Disinfect Residuals. Denver: AWWA.

Jolley, R.L., R.J. Bull, W.P. Davis, S. Katz, M.H. Roberts Jr., and V.A. Jacobs. 1984. *Water Chlorination: Environmental Impact and Health Effects*. Vol. 5. Chelsea, Mich.: Lewis Publishers.

Keen, G.A., and J.I. Prosser. 1987. "Interrelationship Between pH and Surface Growth of Nitrobacter." *Soil Biology and Biochemistry*, 19:6:665–672.

Kirmeyer, G., K. Martel, G. Thompson, L. Radder, W. Klement, M. LeChevallier, H. Baribeau, and A. Flores. 2004. *Optimizing Chloramine Treatment*, 2nd ed. Denver: WRF.

Kirmeyer, G.J., L.H. Odell, J.G. Jacangelo, A. Wilczak, and R.L. Wolfe. 1995. *Nitrification Occurrence and Control in Chloraminated Water Systems*. Denver: WRF, AWWA.

Krasner, S.W., W.A. Mitch, D.L. McCurry, D. Hanigan, and P. Westerhoff. 2013. "Formation, Precursors, Control, and Occurrence of Nitrosamines in Drinking Water: A Review." *Water Research*, 47:13:4433–4450

Lee, W.H., D.G. Wahman, P.L. Bishop, and J.G. Pressman. 2011. "Free Chlorine and Monochloramine Application to Nitrifying Biofilm: Comparison of Biofilm Penetration, Activity, and Viability." *Environmental Science & Technology*, 45:4:1412–1419.

Lipponen, M.T.T., M.H. Suutari, and P.J. Martikainen. 2002. "Occurrence of Nitrifying Bacteria and Nitrification in Finnish Drinking Water Distribution Systems." *Water Research*, 36:17:4319–4329.

Mahmood, F., J.G. Pimblett, N.O. Grace, and W.M. Grayman. 2005. "Evaluation of Water Mixing Characteristics in Distribution System Storage Tanks." *Journal AWWA*, 97:3:74–88.

Maier, W. 2003. "The Motor Wheel Disposal Site or: Regulatory Soup." Presented at the 2003 AWWA Michigan Section Groundwater Conference, Lansing, Mich.

Maier, W., personal communication, 2011.

McGuire, M.J., N.I. Lieu, and M.S. Pearthree. 1999. "Using Chlorite Ion to Control Nitrification." *Journal AWWA*, 91:10:52–61.

McGuire, M.J., M.S. Pearthree, N.K. Blute, T. Hoogerwerf, and K.F. Arnold. 2004. Tucson Nitrification Control Pilot Study: A Significant Advance in the State-of-the-Art of Implementing Chloramines in a Desert Environment. Arizona Water & Pollution Control Association Annual Conference.

McGuire, M.J., P.A. Hayes, G. Qin, N.K. Blute, and X. Wu. 2007. "Control of Nitrification Using Chlorite Ion in a Glendale, California Demonstration Area." *In Proc. For the AWWA Annual Conference*. Denver: AWWA.

McGuire, M.J., X. Wu, N.K. Blute, D. Askenazier, and G. Qin. 2009. "Prevention of Nitrification Using Chlorite Ion: Results of a Demonstration Project in Glendale, Calif." *Journal AWWA*, 101:10:47–59.

Negrin, J.M., W.M. Heyer, and S. Cheng. 1990. "Evaluation of Nitrification Control Strategies at the Los Angeles Department of Water and Power." *In Proc. of the AWWA Annual Conference*. Denver: AWWA.

Noguera, D.R., P.S. Oldenburg, C. Bone, and G.W. Harrington. 1999. "Inactivation of Ammonia-Oxidizing Bacteria by Chloramine: Implications for Prevention and Control of Nitrification in Distribution Systems." *In Proc. of the AWWA Water Quality Technology Conference*. Denver: AWWA.

Odell, L.H., G.J. Kirmeyer, A. Wilczak, J.G. Jacangelo, J.P. Marcinko, and R.L. Wolfe. 1996. "Controlling Nitrification in Chloraminated Systems." *Journal AWWA*, 88:7:86–98.

Odell, L.H., G. Kirmeyer, A. Wilczak, J.G. Jacangelo, J.P. Marcinko, and R. Wolfe. 1994. "The Control of Nitrification in Chloraminated Distribution Systems." *In Proc. of the AWWA Water Quality Technology Conference.* Denver: AWWA.

Oldenburg, P.S., J.M. Regan, G.W. Harrington, and D.R. Noguera. 2002. "Kinetics of Nitrosomonas europaea Inactivation by Chloramine." *Journal AWWA*, 94:10:100–110.

Palacios, B., and C.D. Smith. 2003. "The Use of Chlorine Decay Studies to Plan San Francisco's Conversion to Chloramines." *In Proc. of the Distribution and Plant Operations Symposium.* Denver: AWWA.

Pintar, K., R.M. Slawson, and P.M. Huck. 2000. "Investigation of Conditions and Control Strategies Influencing Nitrification in a Chloraminated Bench-Scale Distribution System." *In Proc. of the AWWA Water Quality Technology Conference.* Denver: AWWA.

Rezania, L.W. 2010. Minnesota Ammonia Study–Phase 1 Problem Assessment and Data Collection, AWWA Inorganics Contaminants Workshop, Denver, Feb. 28–Mar. 2, 2010.

Rittmann, B.E., and V.L. Snoeyink. 1984. "Achieving Biologically Stable Drinking Water." *Journal AWWA*, 76:10:106–114.

Rodriguez, J. 2011. "Sustainable Techniques Improve Water Quality, Lower Costs." *Opflow*, 37:8:16–18.

Rosenfeldt, E.J., C. Baeza, and D.R.U. Knappe. 2009. "Effect of Free Chlorine Application on Microbial Quality of Drinking Water in Chloraminated Distribution Systems." *Journal AWWA*, 101:10: 60–70.

Rungvetvuthivitaya, M., R. Song, M. Campbell, and C. Ray. 2014. "A Kinetic Study of Nitrification Inhibition in Water Distribution Systems Using Low Levels of Chlorite." *Journal of Water Supply: Research & Technology-AQUA*, 63:6:497–506.

Rungvetvuthivitaya, M., R. Song, M. Campbell, E. Zhu, T. Zhang, and C. Ray. 2019. "Decay Kinetics of Chlorite under Simulated Distribution System Conditions." *Journal of Environmental Engineering*, 145:4.

Santos, M. 2018. "Chlorine Dioxide Application for Water Quality Challenges: Experience from Texas Systems." *In Proc. for the AWWA Water Quality Technology Conference.* Denver: AWWA.

Sarkissian, A., C. Munis, M. So, J. Malley, and C. Cotton. 2020. "Non-chemical Inhibition of Nitrification: A Bench and Full Scale Pilot of UV Reactors at the Los Angeles Dept of Water and Power." Presented at IUVA Americas Conference, 2020

Schrantz K.A., J.G. Pressman, and D.G. Wahman. 2013. "Simulated Distribution Nitrification: Nitrification Index Evaluation and Viable AOM." *Journal AWWA*, 105:5:E242–E254.

Schrempp, T., R. Goold, B. Kwan, D. Sarai, and M. Horsley. 1994. "Effect of Mechanical Cleaning and Free Residual Chlorine on the Chlorine Demand and Nitrification

Process in a Chloraminated System." *In Proc. of the AWWA Water Quality Technology Conference*. Denver: AWWA.

Skadsen, J. 2002. "Effectiveness of High pH for Control of Nitrification." *Journal AWWA*, 94:7:73–83.

Skadsen, J.M. 1991. "Nitrification in a Distribution System and the Effects of Changing Disinfectants on Bacterial Populations." *Proc. AWWA Water Quality Technol. Conf.* p. 773 and presentation at Water Quality Technology Conference. Denver: AWWA.

Skadsen, J.M. 1993. "Nitrification in a Distribution System." *Journal AWWA*, 85:7:95–103.

Skadsen, J., R. Janke, W. Grayman, W. Samuels, M. Tenbroek, B. Steglitz, and S. Bahl. 2008. "Distribution System On-line Monitoring for Detecting Contamination and Water Quality Changes." *Journal AWWA*. 100:7:81–94.

Smith, C.D., J.F. Smith, and B. Milosky. 1996. "Excessive Loss of Chloramine Residual: A Case Study." *In Proc. of the AWWA Annual Conference*." Denver: AWWA.

So, personal communication, 2020.

Song, D.J., A. Sheikholeslami, L.L. Hoover, K.A. Turner, H.H. Lai, and A. Wilczak. 1999. "Improvement of Chloramine Stability Through pH Control, TOC Reduction, and Blending at EBMUD, California." *In Proc. of the AWWA Annual Conference*. Denver: AWWA.

Speitel, G.R. Jr., R. Kannappan, and B.M. Bayer. 2011. "The Nitrification Index: A Unified Concept for Quantifying the Risk of Distribution System Nitrification." *Journal AWWA*, 103:1:69–80.

Stewart, M.H., and N.I. Lieu. 1997. "Nitrification in Chloraminated Drinking Water and Its Association with Biofilms." *In Proc. of the AWWA Water Quality Technology Conference*. Denver: AWWA.

Sykes, M. 2003. "Issues Related to Mixing Free Chlorine and Monochloramine Residuals in the Distribution System." *In Proc. of the AWWA Water Quality Technology Conference*. Denver: AWWA.

Theiss, R.J., and E.N. Antoun. 2000. "Excess Storage: Utilities' Achilles' Heel." *In Proc. of the AWWA Distribution System Symposium*. Denver: AWWA.

Uhl, W., and R. Gimbel. 2000. "Dynamic Modeling of Ammonia Removal at Low Temperatures in Drinking Water Rapid Filters." *Water Science & Technology*, 41:4–5:199–206.

USEPA (US Environmental Protection Agency). 2002. *Finished Water Storage Facilities*. Washington: USEPA.

Vasconcelos, J.J., P.F. Boulos, W.M. Grayman, L. Kiene, O. Wable, P. Biswas, A. Bhari, L.A. Rossman, R.M. Clark, and J.A. Goodrich. 1996. *Characterization and Modeling of Chlorine Decay in Distribution Systems*. Denver: WRF.

Vestal and A. Wilczak, personal communication, 2005.

Vikesland, P., N. Love, M. Fiss, A. Zaklikowski, K. Chandran, F. DiGiano, and B. Carrico. 2007. "Water Quality Impacts of Disinfectant Switching Practices at Pilot and Full Scale." *In Proc. for the AWWA Annual Conference*. Denver: AWWA.

Vokes, C.M. 2007. "Impact of Ozone and Biological Filtration on Water Quality Parameters in Arlington, Texas." *Ozone Science & Engineering*, 29:4:261–271.

White, B., and M. Adams. 2010. "Battling Nitrification With Blacklights." *Opflow*, 36:2:10–13.

Wilczak, A., personal communication, 2004.

Wilczak, A., L.L. Hoover, and H.H. Lai. 2003a. "Effects of Treatment Changes on Chloramine Demand and Decay." *Journal AWWA*, 95:7:94–106.

Wilczak, A., J.G. Jacangelo, J.P. Marcinko, L.H. Odell, and G.J. Kirmeyer. 1996. "Occurrence of Nitrification in Chloraminated Water Distribution Systems." *Journal AWWA*, 88:7:74–85.

Wilczak, A., C.D. Smith, K.Y. Cohen, and P. Martin. 2003b. "Strategies for Combining Free Ammonia and Boosting Chloramines—Survey of Utility Practice." *In Proc. of the AWWA Water Quality Technology Conference*. Denver: AWWA.

Wolfe, R.L., and N.I. Lieu. 2003. "Nitrifying Bacteria in Drinking Water." *Encyclopedia of Environmental Microbiology*. G. Britton, editor. Hoboken, NJ: Wiley.

Wolfe, R.L., N.I. Lieu, G. Izaguirre, and E.G. Means III. 1990. "Ammonia-Oxidizing Bacteria in a Chloraminated Distribution System: Seasonal Occurrence, Distribution, and Disinfection Resistance." *Applied and Environmental Microbiology*, 56:2:451–462.

Wolfe, R.L., E.G. Means III, M.K. Davis, and S.E. Barrett. 1988. "Biological Nitrification in Covered Reservoirs Containing Chloraminated Water." *Journal AWWA*, 80:9:109–114.

Work, L., H. Lai, J. Lau, C. Cheok, and S. Teefy. 2019. "Impact of Chlorine to Nitrogen Ratio on Disinfection Byproduct Formation and Nitrification in the Distribution System" presented at AWWA Annual Conference and Exhibition, June 2019.

Zhu, E., V. Monks, and R. Song. 2010. "Effectiveness of Early Application of Sodium Chlorite in Nitrification Prevention–A Demonstration-Scale Study." *In Proc. for the AWWA Water Quality Technology Conference*. Denver: AWWA.

This page intentionally blank.

Chapter 8

Assessment and Operational Responses to Nitrification Episodes

Christopher P. Hill, Sabine Arweiler, Susan Teefy

INTRODUCTION

This chapter provides guidelines for conducting a nitrification assessment to determine when nitrification is occurring and discusses water quality indicators of nitrification in drinking water distribution systems. This chapter also discusses development of a nitrification response plan and operational responses to nitrification in the distribution system and finished water storage facilities.

When loss of disinfectant residual or any unusual change in water quality occurs in a chloraminated distribution system, the utility must determine if the cause is nitrification to be able to respond in an effective manner. It is important to learn and understand the impacts of system operation on nitrifications events and consequently to be able to prevent nitrification in the future. Table 8-1 presents a summary of the key points presented in this chapter.

NITRIFICATION ASSESSMENT

This section discusses the impacts of nitrification on finished water quality and how to use water quality data to determine when nitrification is occurring. It first identifies common water quality characteristics of nitrification. It then provides examples of how water quality data can be used to identify nitrification in the distribution system and storage facilities.

Table 8-1 Key points from Chapter 8

Nitrification assessment	• A nitrification monitoring plan is needed to assess distribution system water quality and determine when nitrification is occurring. • Low total chlorine residual concentrations compared to baseline concentrations can be indicative of excessive chloramine degradation or nitrification. • Progressively higher nitrite concentrations indicate that nitrification is occurring. High nitrite and/or nitrate concentrations are among the best indicators of nitrification. • Higher-than-normal free ammonia concentrations may promote growth of ammonia-oxidizing microorganisms. Lower-than-normal free ammonia levels can also indicate that nitrification is already occurring. • Nitrification is often accompanied by increased heterotrophic plate count (HPC) R2A counts. Because HPC results take days to be available, this is a less useful indicator. • Nitrification can result in a depression in pH in low alkalinity waters. As such, localized reductions in pH, coupled with any of the other indicators, may be an indicator of nitrification. • After it has been determined that nitrification is occurring, it is important to respond in a timely and effective manner. To prevent recurring nitrification, it is important to identify the cause of the nitrification. • Determining the cause of nitrification requires utilities to assess the results from distribution monitoring, and impacts of finished water quality, treatment plant operation, distribution system and storage tank operation, and distribution piping on nitrification.
Responses to distribution system nitrification episodes	• Breakpoint chlorination can be an effective short-term response to nitrification but is not a long-term solution. The goals of breakpoint chlorination relative to nitrification response are to oxidize any free ammonia in the distribution system, depriving nitrifying bacteria of a nutrient source; inactivate the nitrifying bacteria in the system; and oxidize nitrite, which exerts an oxidant demand and increases chloramine degradation. • Flushing can be an effective emergency response to both nitrification and indicators of the potential for nitrification. When properly designed, flushing can remove distribution system biofilms and sediments, reduce disinfectant demand, and bring fresher water into areas of the distribution system with low flow. • Increasing chloramine residual is primarily used to prevent nitrification and has never been demonstrated to be effective as a response measure, even during the early stages of nitrification.
Responses to nitrification in distribution system storage facilities	• Breakpoint chlorination of storage tanks can be an effective response to nitrification. • Deep cycling of storage facilities can be an effective means to prevent or respond to nitrification but is only effective in well-mixed tanks. The ability to implement deep cycling is dependent on system pressure and emergency storage requirements. • Draining and disinfection of storage facilities may be practical and effective in systems in which operations can be adapted to having storage off-line for an extended period. Before taking the facility off line, it is important to assure the ability to maintain system pressure and fire-flow requirements with the facility out of service.

Finally, this section discusses the importance of identifying the cause of a nitrification episode and to effectively respond and prevent future nitrification events.

Water Quality Indicators of Nitrification

A water quality monitoring plan is needed to assess distribution system water quality and determine when nitrification is occurring. Monitoring plans are typically identified in a nitrification action plan, which not only identifies monitoring requirements (i.e., which parameters, where, and when to monitor) but also identifies alert levels and action levels for the parameters to be monitored. Alert levels are indicators that conditions are right for nitrification to occur. Action levels generally indicate that nitrification is occurring, and some level of response is necessary.

Indicators of nitrification in the distribution system can include the following compared with upstream or baseline water quality in the system (Odell et al. 1996; Wilczak et al. 1996):

- Lower or depleted total chlorine residual
- Elevated nitrite concentration
- Elevated or depleted free ammonia concentration
- Increased nitrate concentration
- Elevated HPC R2A counts
- Low pH
- Increased temperature (not a direct indicator of bacterial activity, but some systems may use it as an indicator of nitrification potential and alter operations, e.g., summer versus winter operation)

Chloramine degradation over time releases free ammonia, which is a nutrient source for ammonia-oxidizing microorganisms (AOM) and can cause nitrification. Disproportionately low total chlorine residual concentrations compared to baseline concentrations (entering the system or upstream of a localized area) can be indicative of nitrification. Rapid and/or disproportionate chloramine degradation may also be due to the presence of nitrite, a product of nitrification, which reacts with chloramine and can result in rapid chloramine degradation and the further release of ammonia. The result is a complete destruction of chloramine residual and rapid residual decline even after the area has been flushed with "fresh" water containing a chloramine residual.

Nitrite and nitrate are products of nitrification, and elevated nitrite concentrations are among the best indicators of nitrification. Though nitrification can result in elevated concentrations of nitrate as well, oxidation of nitrite to nitrate typically occurs much later. Therefore, elevated nitrate concentrations can be indicative of nitrification, but nitrate may not be useful as a monitoring parameter because it occurs after nitrification is well established. Higher-than-usual free ammonia concentrations, particularly in areas of high water age and low disinfectant residual, indicate there are sufficient ammonia levels present to promote AOM growth. When ammonia levels are lower than normal or fully depleted, nitrification may already be occurring. Nitrification is often accompanied by increased heterotrophic plate count (HPC) R2A counts. The increase in HPC R2A bacterial densities can result from both the low monochloramine residual (insufficient disinfectant) and the presence of organic-rich metabolic byproducts released by AOM, which provide a nutrient source for HPC bacteria. (It is important to note that the HPC analytical methods do not capture nitrifying microorganisms and are only an indicator of increased microbiological activity.) Nitrification may result in a depression in pH in low-alkalinity waters. As such, localized reductions in pH coupled with any of the other indicators may also be an indicator of nitrification.

Nitrification monitoring plans typically include some or all of the previously mentioned water quality parameters. Temperature should also be monitored simultaneously with the previous parameters. AOM and nitrite-oxidizing bacteria (NOB) activity and growth increase with temperature, as discussed in Chapters 2, 3, and 5. Therefore, utilities can expect some or all of the water quality trends previously listed to coincide with rising water temperature and the reverse with decreasing temperature.

Because several related phenomena can occur simultaneously or sequentially, for best diagnostics, all related water quality indicators should be reviewed jointly along with the assessment of system operations. Note that AOM have not previously been considered useful operational indicators of nitrification because of the turnaround time (up to 1 month)

Table 8-2 Examples of water quality characteristics* of various stages of nitrification

	Background (No Nitrification)	Beginning of Nitrification	Incomplete Nitrification (NOB not active)	Complete Nitrification (NOB active)
Total chlorine (mg/L Cl_2)	>1.5	0.5–1.5	<0.5	0
Nitrite (mg/L N)	0–0.010	0.010–0.050	>0.050	0
Free ammonia (mg/L N)	0–0.1	>0.1	0	0
HPC R2A (cfu/mL)	<500	1,000	>1,000	>1,000
Temperature (°C)	<15	>15	>15	>20
pH (units)	Normal	Normal	Less than target	Less than target

* Levels will vary depending on system's background levels.
HPC, heterotrophic plate count; NOB, nitrite-oxidizing bacteria.

and complexity of the methods required to assay these organisms Advances in molecular methods have resulted in significant reductions in the time associated with AOM analyses (1–2 days); however, these methods remain complicated. Correlations between nitrite and AOM and HPC R2A and AOM are tentative at best and system specific. Although it is more time consuming to monitor HPC R2A compared to nitrite, HPC R2A can be another good indicator of microbiological water quality. Increased HPC R2A counts can occur for many reasons, but consistently high and/or unstable HPC R2A counts in the absence of high nitrite can indicate that conditions are ideal for nitrification and an appropriate preventive response is warranted. If less time-consuming AOM detection and enumeration methods are available, AOM monitoring may become the primary water quality indicator. Until then, surrogate measurements for AOM will continue to serve as the main tools primarily available to operations.

Because several related phenomena can occur simultaneously or sequentially, for best diagnostics, all related water quality indicators should be reviewed jointly along with the assessment of system operations. Table 8-2 presents an example of how these indicators might be used to determine if nitrification is occurring. It should be noted that the concentrations presented in Table 8-2 will be highly system specific and as such may not be applicable to all systems: e.g., background HPC R2A levels will depend on the concentration of organic carbon in the water. Chapter 6 discusses how a utility can develop their own water quality monitoring plan to identify when nitrification is occurring in their water system.

Identifying Nitrification in Drinking Water Distribution Systems

All water systems utilizing chloramines should have a nitrification monitoring plan. However, if no nitrification monitoring plan exists but nitrification is suspected, a utility should monitor the following minimum number of parameters in the area(s) of suspected nitrification: total chlorine residual, nitrite, and free ammonia. A nitrification monitoring plan should then be developed as soon as possible. Figure 8-1 presents a flowchart to determine if nitrification is occurring. It is based on the use of monitoring parameters recommended in Chapter 6 and is prioritized based on their usefulness in assessing whether or not nitrification is occurring.

Low total chlorine residual is the primary screening parameter for identifying areas in which nitrification might be occurring. Although nitrification has been shown to occur in distribution systems in the presence of chloramine residuals 2 mg/L or greater, it generally requires the presence of sufficient free ammonia such that the rate of AOM growth exceeds the rate of AOM inactivation (Harrington et al. 2003; Wolfe et al. 1990,1988).

Increasing nitrite and/or nitrate concentrations, particularly in areas with low total chlorine residuals, are indicative of nitrification; unless high nitrite or nitrate levels are observed, it is not likely that nitrification is occurring. Low total chlorine residuals

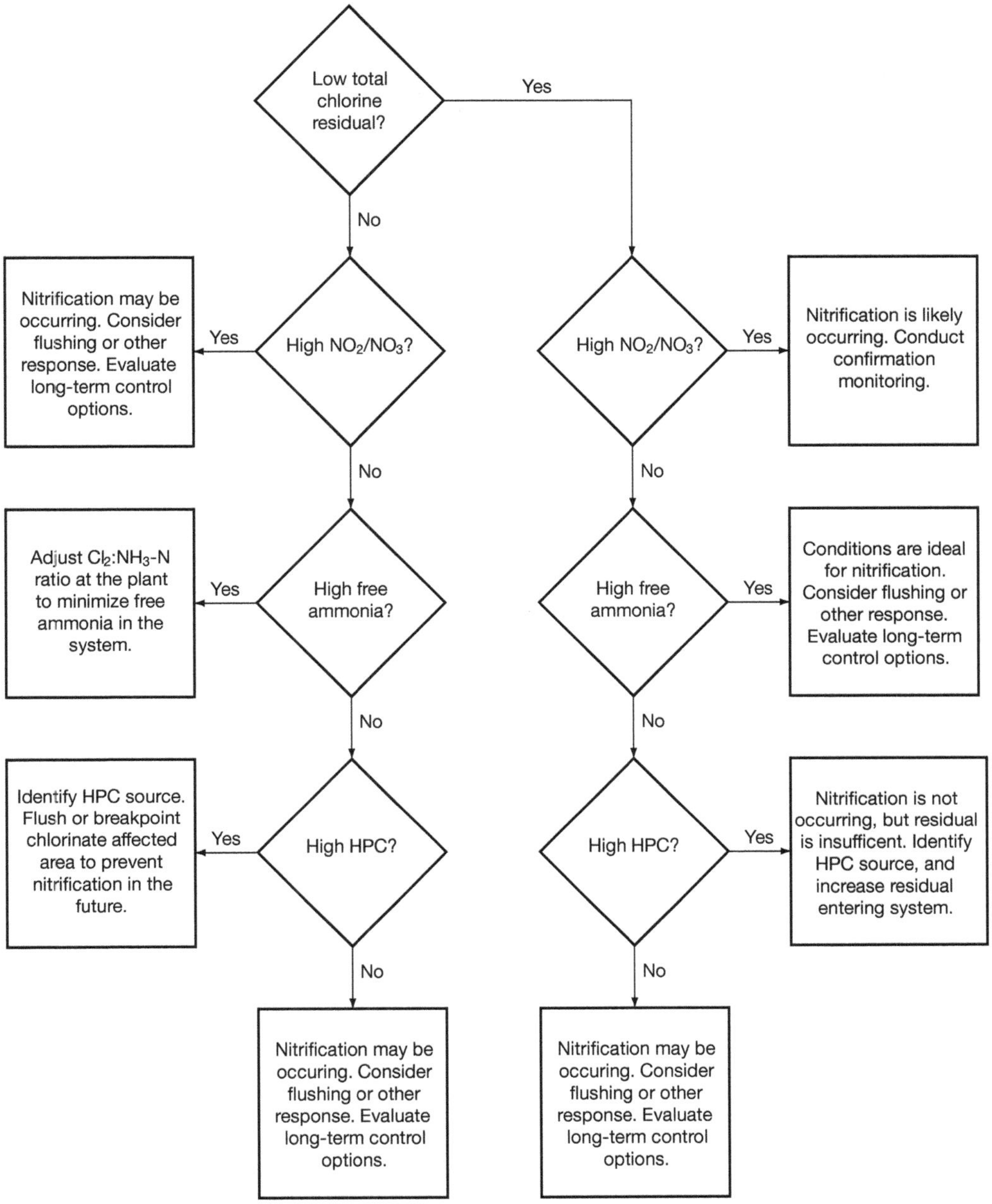

Figure 8-1 Nitrification assessment flowchart

accompanied by high free ammonia concentrations indicate conditions are ideal for nitrification. Low total chlorine residuals, high nitrite or nitrate concentrations, and low free ammonia concentrations indicate that nitrification is occurring.

Figure 8-3 presents an example of how water quality changes during nitrification. In this example, total chlorine, nitrite, pH, temperature, and HPC R2A are presented for one distribution system monitoring location. This location was characterized by initial total chlorine concentrations of approximately 2.7 mg/L Cl_2, low nitrite (<0.05 mg/L N), and low HPC R2A (<100 cfu/mL). As nitrification began to occur in early September, total chlorine levels began to drop significantly at this monitoring location before reaching a low of 0.2 mg/L Cl_2 in mid-October. The drop in residual was accompanied by significant increases in nitrite (from less than 0.05 to 0.26 mg/L N) and HPC (from less than 10 cfu/mL to greater than 2,000 cfu/mL). Nitrification also resulted in a drop in finished water pH from 9.5 to 8.2. In this case, as the system water temperatures decreased and total chlorine residuals returned to baseline conditions, the nitrification subsided, and water quality parameters returned to their baseline concentrations without implementing any other mitigation measures.

Case Study—Storage tank nitrification and response assessment

This case study demonstrates how monitoring results can be used to identify when nitrification is occurring in a distribution system storage facility and to evaluate the effectiveness of the response action taken. As shown in the figures below, water quality in a 3-mil gal steel distribution system reservoir located low in a pressure zone (i.e., not turning over very well) was initially characterized by low total chlorine residual (≤0.1 mg/L Cl_2), high nitrite concentrations (0.2 mg/L N), nondetectable levels of free ammonia, pH of 8.3 (about 0.5 unit less than normal), and temperature of 17°C. Based on these data, it was determined that the tank was nitrifying.

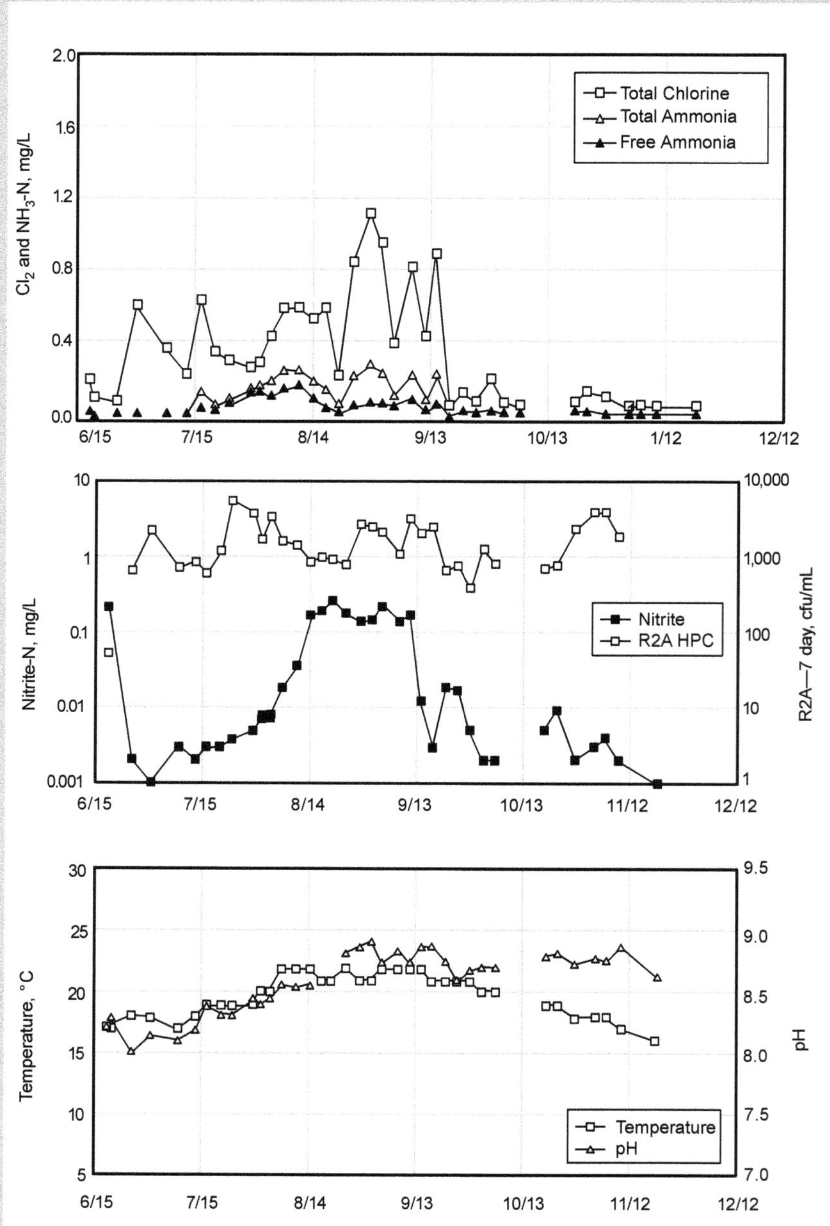

Figure 8-2 An example of storage tank nitrification and response assessment

A decision was made to breakpoint-chlorinate the tank on June 22, 1999, in an attempt to eliminate nitrification and improve water quality. Sodium hypochlorite was added to the tank inlet at the beginning of the fill cycle to achieve a free chlorine residual of approximately 1 mg/L Cl_2. Following breakpoint chlorination, the total chlorine residual increased to approximately 0.5 mg/L Cl_2, and nitrite was completely eliminated.

Concurrent with the free chlorination, attempts were made to improve routine tank turnover, including modification of fill and draw operations to improve mixing. Approximately 1 month after breakpoint chlorination, nitrite levels began to increase. The total chlorine residual trend was characteristically "jagged," displaying a pattern of unstable water quality. Traces of free ammonia had historically been detected in the stored water in July, but by mid-August all of the free ammonia was gone, and the nitrite had returned. Over the entire study period, the R2A HPC bacterial counts in stored water were high (approximately 1,000 colony-forming units [cfu]/mL and greater), indicating bacterial growth in the storage tank had not been arrested by operating the tank with free chlorine. This is an example of an insufficient nitrification response that did not resolve the root cause of the nitrification. It was surmised that the root cause was most likely due to an insufficient free chlorine dose during breakpoint chlorination and/or inadequate distribution of the free residual throughout the tank due to inadequate tank cycling. This is also an example of the iterative approach that may be necessary to mitigate a nitrification event.

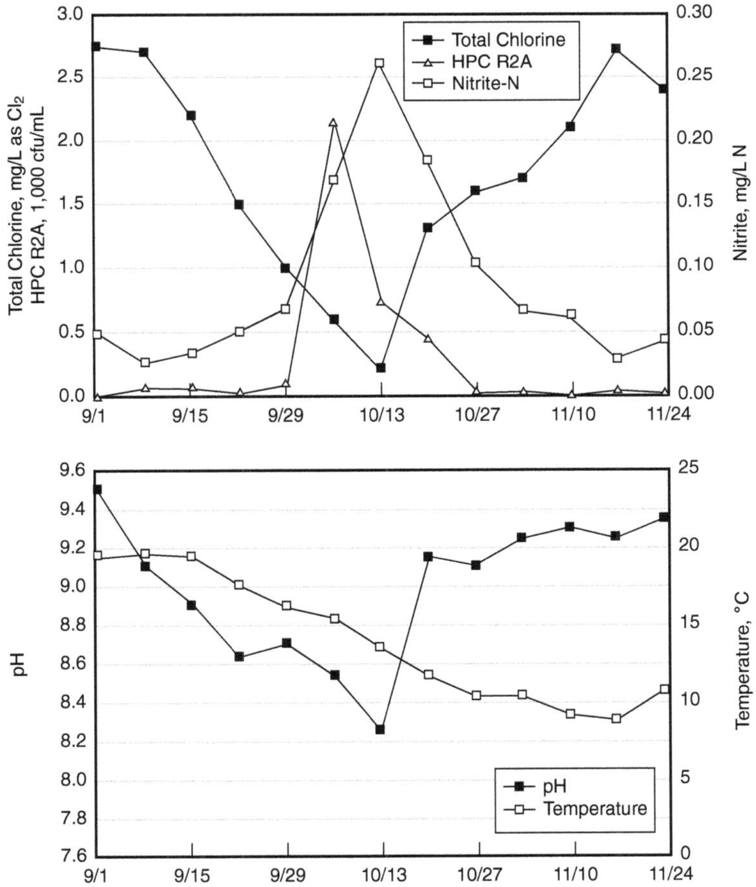

Source: Janice Skadsen

Figure 8-3 An example of distribution system nitrification assessment

Determining the Cause of Nitrification

After a utility has determined that nitrification is occurring, the utility must respond in a timely and effective manner. However, to prevent or at the very least minimize recurring nitrification, it is also important to identify and address the cause of the nitrification. Chapter 4 presents an overview of the causes of nitrification in drinking water distribution systems.

Determining the cause of nitrification requires utilities to examine existing data and create new data, where necessary, to assess the impacts of the following on nitrification:

- Treatment plant operation (quality of water entering the system)
- Finished water quality in the system
- Distribution system operation
- Storage tank and reservoir management
- Infrastructure materials and condition (distribution piping)
- Blending of waters from different sources and/or treatment facilities
- Chemical boosting, either chlorine, ammonia or both

DEVELOPING A NITRIFICATION RESPONSE PLAN

When there is indication that nitrification is occurring or that conditions are right for nitrification to occur, it is important to respond in a timely manner. Delaying response can result in further deterioration of water quality. However, it is also important that a response be well planned and coordinated to prevent spreading the problem to other parts of the distribution system. It is equally important for the response to be effective. In instances when the first response is ineffective, it is critical to understand the causes of nitrification and the possible reason(s) the response may not have been effective. Effective long-term solutions to nitrification may involve a combination of operational and capital improvements. Ideally, the underlying system conditions (e.g., dead ends, poorly mixed reservoirs, hydraulic conditions that limit tank turnover, etc.) that impact nitrification would be addressed to minimize the need for a nitrification response. However, that may not always be practical or may take several years to implement. It is recommended that, where practical, utilities develop a plan to address these underlying system conditions. This section discusses development of a nitrification response plan that identifies effective and appropriate levels of short-term response based on system-specific conditions.

Prior to development of a nitrification response plan, it is critical to develop a nitrification monitoring program and establish system-specific and possibly area-specific baseline concentrations for nitrification monitoring parameters (e.g., total chlorine, nitrite, free ammonia, total ammonia, nitrate, and HPC), alert levels, and action levels. The nitrification monitoring plan will identify system- or area-specific alert and action levels, which can be used to determine the appropriate level of nitrification response. Table 8-3 provides example baseline concentrations, alert levels, and action levels. It is intended to be used only as a guide, and it is recommended that utilities determine the combination of system-specific alert levels and action levels and develop their own nitrification response plans accordingly.

After the baseline concentrations are determined, alert levels and action levels can be established to develop a nitrification response plan. Table 8-3 provides example alert and action levels provided by one utility. Note that alert and action levels are system specific and should be based on baseline water quality and operational conditions. The level of response should be commensurate with the alert level or action level. That is, an alert-level response will be less severe than an action-level response. Consider the

Table 8-3 An example of a nitrification alert and action levels

Parameter	Target at Reservoir Inlets	Target at TCR Sampling Stations	Alert Level	Action Level 1 (Operational Responses Required)	Action Level 2 (Regulatory Violation Possible)
Total chlorine (mg/L)	>1.7	>1.5	1.0–1.5*	<1.0	<0.2
Nitrite (mg/L N)	<0.01	<0.01	>0.02	>0.03	>0.05
HPC-R2A (cfu/mL)	<500	<500	>500	>1000	>1000
Free ammonia (mg/L N)	<0.1	<0.15	>0.15	>0.2	NA†

Source: Based on information provided by Charlotte D. Smith & Associates, 2003.
cfu, colony-forming units; HPC, heterotrophic plate count; NA, not applicable; TCR, Total Coliform Rule.
* Use 1.0 for TCR sampling stations representative pipes with relatively of low flow
† NA: Assumes free ammonia-N has been converted to nitrite-N.

example for a water system that uses a single alert level and multiple action levels (action level 1 and action level 2). If the nitrite concentration in a storage tank increases above the alert level, the appropriate response may be increased monitoring. However, if the nitrite concentration continues to increase and exceeds action level 1, the response may be to initiate deep cycling (i.e., increase drawdown) to ensure maximum volume turnover. Finally, if the nitrite concentration exceeds action level 2, it may be necessary to initiate breakpoint chlorination of the storage tank. Table 8-4 presents an example of alert-level and action-level responses presented as programmatic and operational activities.

A decision tree can be a useful tool in developing a response plan, as discussed in the following section. Figure 8-4 presents an example nitrification response decision matrix. The decision tree is based on system-specific action levels and is provided only as an example. Utilities using this manual to develop a nitrification response plan should determine their own combination of system-specific alert-level and action-level responses.

If nitrification is suspected in an area of the distribution system or in a storage tank, there are several options for remedying nitrification, and the level of response should be proportional to the extent of the nitrification event. These responses, in order of increasing operational effort, include

- Increased monitoring (suspected or possible nitrification)
- Check and optimize ammonia feed rates at treatment plants and/or booster chloramination stations
- Check chloramine demand/decay rates; compare decay rates in distribution system to those in plant effluent
- Increase turnover in storage tanks (e.g., lower operational levels)
- Flush the affected area
- Breakpoint chlorination of targeted areas within the distribution system
- Drain and disinfect storage tanks
- Switch to a more stable water source (if available)
- Take excess storage out of service (where possible, seasonally or long-term)
- Breakpoint chlorination of the entire distribution system
- Engineering improvements to reduce water age, improve mixing, and boost monochloramine

Many of these responses are discussed in greater detail in the following sections. Figure 8-5 presents the results of a utility survey regarding the effectiveness of various nitrification responses (Kirmeyer et al. 2004). The figure is based on the responses of 29 utilities and reflects the judgment of the effectiveness of various responses based on the

Table 8-4 Examples of nitrification responses

Action Level	Programmatic Activities	Operational Activities
Action Level 1 (Alert)	• Increase water quality monitoring and operational vigilance • Initiate internal communications • Review historical monitoring trends • Review recent operating conditions • Prepare for situation escalation • Identify possible operational responses if action level 2 is attained.	• Increase monitoring (locations and/or frequency) • Increase reservoir cycling/turnover to decrease water age • Perform distribution system conventional flushing (to decrease water age)
Action Level 2 (Operational Responses)	• Evaluate which operational alternatives are appropriate for the situation • Initiate action level 2 operational response(s): – Continue increased operational and water quality monitoring. – Initiate increased communication with utility staff and, if appropriate, public notification (i.e., for increased flushing and potentially dirty water, loss of pressure, or other causes of complaints). – After action is initiated, document water quality changes for the evaluation of the success of response in mitigating nitrification. • Evaluate and document effectiveness of response • Evaluate and document any conditions that influenced the success of the response • Evaluate and implement any operating strategies for nitrification prevention that may help prevent recurrence of nitrification in this location	• Increase reservoir cycling/turnover to decrease water age • Enhance reservoir mixing through operations (i.e., filling the tank in a single cycle rather than several fill cycles over a period of time) • Operate finished water storage reservoirs with reduced volume or basin out of service • Enhance reservoir mixing with mechanical mixers • Perform localized, specific distribution system unidirectional flushing (to reduce chlorine demand from loose sediment or corrosion products on pipe walls). This means conducting a small-scale unidirectional flush of an area not scheduled in the routine system-wide unidirectional flushing program.
Action Level 3 (Regulatory Violation Possible)	• Determine whether regulatory violation is imminent • Determine whether emergency communications and actions are necessary • If regulatory violation occurs, follow appropriate customer notification requirements in accordance with state or provincial requirements	• Breakpoint-chlorinate a tank or reservoir • Inspect interior of reservoir using an ROV or diver to determine if there is a need for cleaning and/or any conditions requiring correction. • Completely drain and refill a tank or reservoir; clean and disinfect if needed. • Breakpoint-chlorinate one or more pressure zones • Switch water supplies, if possible

Source: Smith et al. 2003

number of utilities that use that response. For example, slightly more than 40 percent of the utilities surveyed practice "breakpoint chlorination of reservoirs/tanks and problem areas." Of those utilities, approximately 20 percent determined breakpoint chlorination was essential to control nitrification. Approximately 40 percent rated it very effective, and the remaining 40 percent rated it effective.

Following any response activity (e.g., targeted breakpoint chlorination of the affected area followed by flushing), total chlorine residual and other key indicators of nitrification should be resampled after the activity. The length of time following a response before remonitoring is system specific and activity specific; however, a maximum of 2 days following the response is recommended. If it does not appear that the nitrification problem

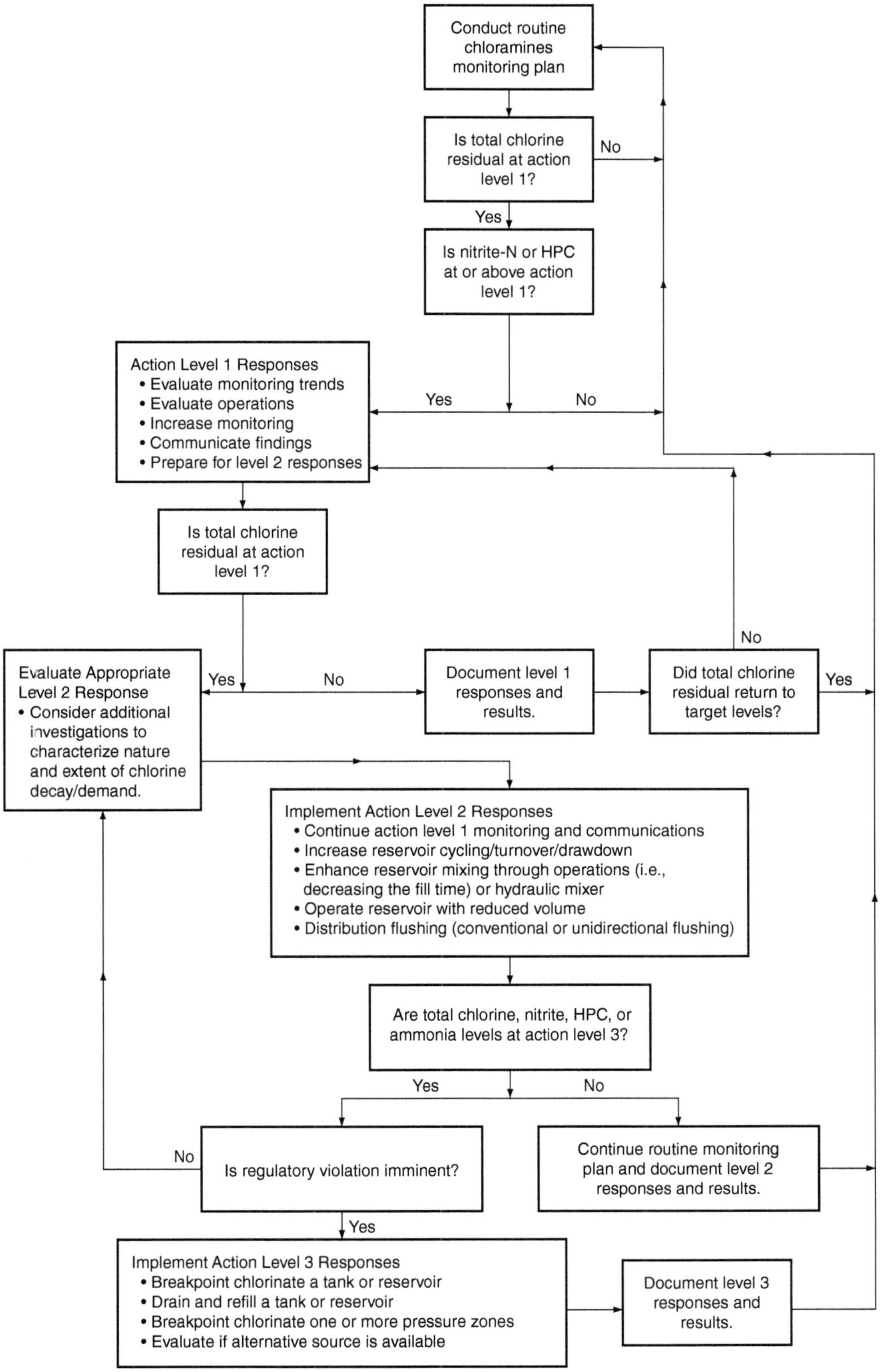

Source: Smith et al. 2003

Figure 8-4 Example of nitrification response decision tree

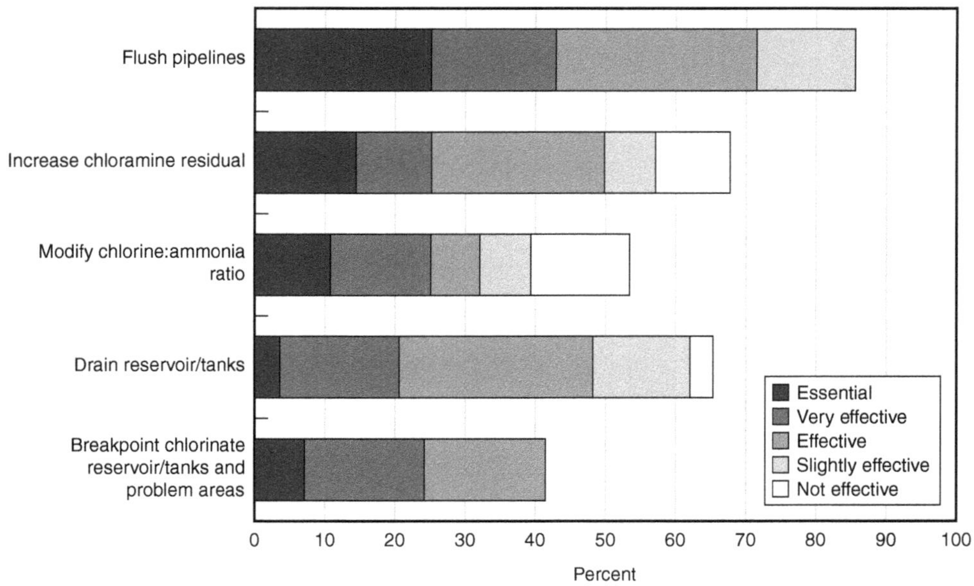

Source: Kirmeyer et al. 2004

Figure 8-5 Utility survey of effectiveness of various nitrification responses

has been solved, additional remedial activities may be required. For example, following breakpoint chlorination of a storage tank, Irvine Ranch Water District of California resamples within 12 to 30 h. If water quality parameters still exceed the action level, the tank is rechlorinated (Irvine Ranch Water District 2003). After water quality returns to baseline conditions, the increased monitoring conducted prior to and following the response can be reduced, and the system can return to its normal nitrification monitoring program.

RESPONSES TO DISTRIBUTION SYSTEM NITRIFICATION EPISODES

Once nitrification has begun to occur, it is critical to respond in a timely and effective manner. This section describes responses to nitrification in the distribution system, including breakpoint chlorination and flushing. It is a guide intended to present responses to nitrification that can be implemented in a timely manner to eliminate nitrification and minimize water quality degradation. Operational strategies and engineering improvements to prevent nitrification are discussed in Chapters 7 and 9, respectively.

Breakpoint Chlorination

Breakpoint chlorination may be the most effective short-term nitrification response but may not prevent nitrification from reoccurring (Rosenfeldt et al. 2009; Carrico et al. 2008; Odell et al. 1996). Breakpoint chlorination refers to the addition of chlorine to water at a concentration whereby all the ammonia has been oxidized to nitrogen, resulting in a free chlorine residual. Recall that nitrifying bacteria are ubiquitous in the environment and that suppression of their growth is something that must be done constantly.

Figure 8-6 presents an ideal breakpoint curve in laboratory or ultrapure water without chlorine-consuming substances such as natural organic matter (NOM) or nitrite and after equilibrium has been established. In field applications, nitrite, which is typically present in nitrifying waters, will also need to be oxidized. Nitrite exerts a free chlorine demand of 5:1 (i.e., 0.2 mg/L nitrite as N will exert 1.0 mg/L free chlorine demand as Cl_2).

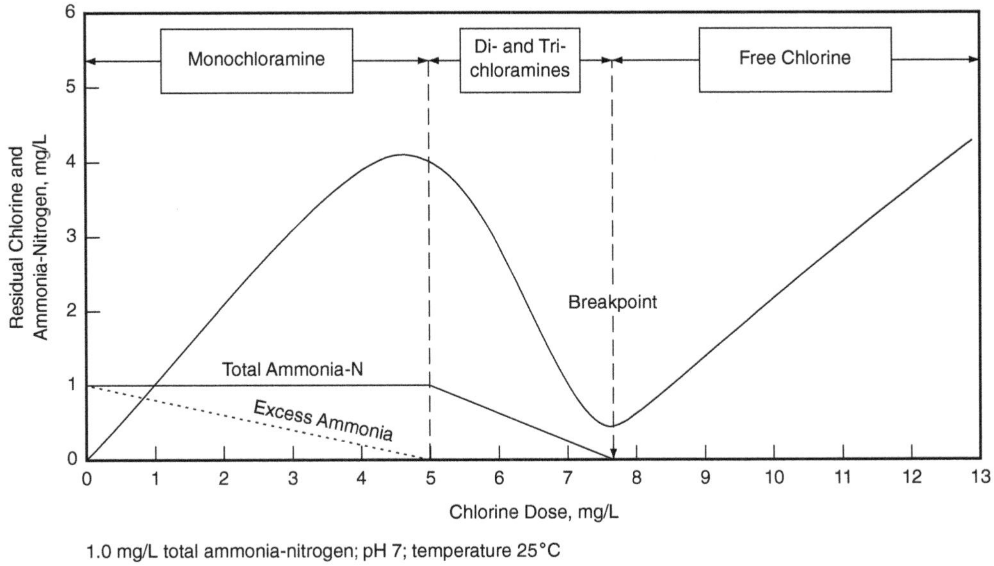

Adapted from Wolfe et al. 1984

Figure 8-6 Ideal breakpoint curve

The y-axis in the breakpoint curve represents the measured total chlorine residual, regardless of the type of chlorine (e.g., free chlorine, monochloramine, di- or trichloramine). As Figure 8-5 demonstrates, at weight ratios of less than about 5:1 (chlorine dose on the x-axis in Figure 8-5 is equivalent to the chlorine-to-ammonia-N [Cl_2:NH_3-N] weight ratio as the initial concentration of total ammonia in this example is 1 mg/L), there is an excess of free ammonia (a critical component in the onset of nitrification), and monochloramine is the predominant combined chlorine species formed. It is preferred that the Cl_2:NH_3-N weight ratio be maintained between approximately 4.5:1 and 5:1, although this can vary depending on the system. This keeps the amount of free ammonia entering the system to a minimum and prevents formation of dichloramine and trichloramine, which are formed at ratios greater than about 5:1. Dichloramine and trichloramine can cause taste and odor issues in the system, they are unstable, and they provide no disinfection benefit.

During breakpoint chlorination, the ammonia molecule becomes progressively more chlorinated, i.e., $NH_3 \rightarrow NH_2Cl \rightarrow NHCl_2 \rightarrow NCl_3$, as the Cl_2:NH_3-N weight ratio increases and the amount and type of chlorine present changes according to Figure 8-5 (Kirmeyer et al. 2004). The following equation presents the overall breakpoint reaction (Woolschlager et al. 2001):

$$2NH_3 + 3HOCl \rightarrow N_2 + 3HCl^- + 3H_2O \qquad \text{(Eq. 8-1)}$$

When chlorine is present at ratios in excess of 5:1, monochloramine is formed but quickly reacts with excess free chlorine (hypochlorous acid) to form dichloramine that rapidly decomposes (Kirmeyer et al. 2004; White 1999). Monochloramine in the presence of excess hypochlorous acid forms dichloramine according to the following reaction:

$$NH_2Cl + HOCl \rightarrow NHCl_2 + H_2O \qquad \text{(Eq. 8-2)}$$

Dichloramine will undergo a series of complex decomposition and oxidation reactions in the presence of excess free chlorine to form nitrogen-containing products, including nitrogen gas (N_2), nitrate (NO_{3-}), nitrous oxide (N_2O), nitric oxide (NO), and others (Kirmeyer et al. 2004). Dichloramine may also react with excess free chlorine to form trichloramine, as follows (White 1999):

$$NHCl_2 + HOCl \rightarrow NCl_3 + H_2O \qquad \text{(Eq. 8-3)}$$

At a Cl_2:NH_3-N weight ratio of 7.6:1 (i.e., the dip in the breakpoint curve), all of the available ammonia is oxidized to N_2 and other oxidized nitrogen products and chlorine residuals are greatly reduced or eliminated. At this point, any remaining chlorine residual occurs as dichloramine with some trace amounts of monochloramine and free chlorine (Kirmeyer et al. 2004). This is known as the breakpoint because above this weight ratio, free chlorine is the predominant chlorine residual present, and nitrogen species have been oxidized to form nitrogen gas. Beyond that point, any additional chlorine added results in an equal and proportional increase in the free chlorine residual (Wolfe et al. 1984).

To achieve breakpoint chlorination in ultrapure waters, Cl_2:NH_3-N ratios of greater than about 7.6:1 are needed. However, in drinking water distribution systems, the breakpoint Cl_2:NH_3-N ratio may be much higher than 7.6:1, depending on pH, temperature, and the presence of other substances like nitrite and organic nitrogen that can react with free chlorine. In nitrifying distribution systems, nitrite may exhibit a significant chlorine demand and significantly impact the required breakpoint chlorination dose. Nitrite is oxidized by free chlorine according to the following reaction (White 1999):

$$NO_{2^-} + HOCl \rightarrow NO_{3^-} + H^+ + Cl^- \qquad \text{(Eq. 8-4)}$$

Kirmeyer et al. (2004) reported breakpoint Cl_2:NH_3-N ratios ranging from 7:1 to 16:1 in a survey of 68 chloraminated systems. A 10:1 ratio is often used for convenience of calculations and generally matches actual chlorine requirements quite well. Additional chlorine may also be added at a ratio of 5:1 (Cl_2:NO_2-N) to account for the chlorine demand exerted by nitrite. Site-specific breakpoint ratios should first be determined experimentally for each water supply; once determined, they can be used for all breakpoint operations in the system. It is also important to note that breakpoint reactions do not occur instantaneously because their kinetics are highly dependent on pH. At a relatively high pH of 8.5 or above, typically recommended for corrosion control and chloramine stability, it takes several hours for the breakpoint equilibrium demonstrated in Figure 8-5 to develop. Therefore, monitoring of resulting free chlorine levels should be done for several hours or overnight after chlorine dosing has been completed to allow time for mixing and all reactions to run to completion.

The goal of breakpoint chlorination relative to nitrification response is threefold (Dennis et al. 1991):

- First, breakpoint chlorination oxidizes any free ammonia in the distribution system, depriving nitrifying bacteria of a nutrient source.
- Second, upon oxidation of the free ammonia and conversion of dichloramine and trichloramine to free chlorine and nitrogen, the free chlorine acts as a disinfectant and inactivates the nitrifying bacteria in the system.
- Third, nitrite is oxidized to satisfy any nitrite demand for either free chlorine or chloramine. The length of time and free chlorine residual at which breakpoint chlorination should be maintained depends on water quality and system needs and goals.

If factors such as poor tank mixing, excessive water age, or inadequate control of the Cl_2:NH_3-N ratio are the root cause of the nitrification event, breakpoint chlorination will not be an effective long-term solution. It may, however, be an effective short-term solution. In these cases, utilities should consider more permanent control strategies, such as changes in operation or engineering improvements.

Breakpoint chlorination of the entire system. Breakpoint chlorination of the entire distribution system involves discontinuation of ammonia addition and adjusting the chlorine dose at the treatment plant (and/or booster stations) to achieve a free chlorine residual throughout the distribution system.

Many utilities throughout the United States practice routine, annual, or semiannual system-wide breakpoint chlorination as a preventive measure (Kirmeyer et al. 2004). More than 40 percent (31 of 70) of chloraminated utilities responding to the 2017 AWWA Disinfection Survey reported the practice of a temporary switch to free chlorine for nitrification control. Durations of the free chlorination intervals reported ranged from 1 week to 5 months (AWWA Water Quality and Technology Division Disinfection Systems Committee 2017). Chapter 4 provides additional detail on this committee survey.

Breakpoint chlorination of the entire distribution system is an appropriate response to a widespread nitrification episode in the distribution system but may not be effective as a nitrification prevention measure. Carrico et al. (2008) observed that nitrification returned within 3 months following a 30-day conversion to free chlorine in one system and within 1 month following 4-month conversion to free chlorine in another. Similarly, Rosenfeldt et al. (2009) found that bacterial activity returned shortly after a return to chloramines following a 2- to 3-week free chlorine conversion period.

Target free chlorine concentrations entering the distribution system and at the end of the distribution system will vary from system to system, and preferences are largely regional. However, a sufficient free chlorine residual in finished water to achieve a residual of greater than 0.5 mg/L at the ends of the system is a typical goal.

The extent to which systems can achieve these targets is dependent on several system factors. The US Environmental Protection Agency (USEPA) has established a maximum residual disinfectant level of 4 mg/L for chlorine based on a running annual average calculated across all sample sites. This may limit the free chlorine concentration entering the system and make it difficult to achieve a free chlorine residual of 0.5 mg/L at the ends of the system. In these instances, a longer breakpoint period (maybe 1–3 months as opposed to 1 or 2 weeks) or use of booster chlorination may be necessary. The length of time over which utilities practice system-wide breakpoint chlorination depends on many system-specific factors, including system size, water temperature, pH, extent of the nitrification problem, and individual state requirements.

The procedure by which a utility breakpoint-chlorinates the entire distribution system will be largely system specific. Figure 8-7 presents an example system-wide breakpoint chlorination protocol. When using this example to develop a system-specific procedure, it is important to realize that shutting off the ammonia feed system will result in Cl_2:NH_3-N ratios in excess of 5:1 at the chlorinated/chloraminated water interface. To minimize the potential for water quality degradation, including taste and odor, it is recommended that this be accompanied by flushing and efficient storage cycling to more quickly displace chloraminated water in the distribution system and particularly in the nitrifying areas of the system.

After resuming ammonia feed, the same strategies (e.g., flushing, lowering water levels in tanks, etc.) that were used to accelerate the movement of chlorinated water should be used to increase movement of chloraminated water throughout the distribution system and particularly in those areas with high water age.

In the early 1990s, Ann Arbor, Mich., experienced isolated nitrification events within the distribution system (Skadsen 1993). The events were characterized by decreases in chloramine residual (from an average of 3.2 mg/L to an average of 1.4 mg/L) and increases in HPC (from an average of 68 cfu/mL to an average of 2,000 cfu/mL) and nitrite concentrations (from an average of 16.6 µg/L to an average of 86.6 µg/L). The city attempted several solutions to remedy the nitrification, including adjusting the Cl_2:NH_3-N ratio, increasing the chloramine residual (to 8.0 mg/L), low-velocity flushing of affected areas,

Step 1.
Conduct background monitoring (chlorine, nitrite, HPC R2A).

Step 2.
Determine chlorine dose required to achieve the target free chlorine residual in the problem areas of the distribution system.

Step 3.
Conduct any necessary customer or state agency notifications.

Step 4.
Reduce tanks levels to lowest level possible or remove tanks from service.

Step 5.
Have flushing crews ready.

Step 6.
Discontinue the ammonia feed at the water treatment plant and any booster disinfection stations and adjust chlorine to the dose determined by step 2.

Step 7.
Breakpoint storage tanks.

Step 8.
Measure free chlorine and HPC in the known nitrifying areas of the distribution system.

Step 9.
Continue to operate under free chlorine conditions until desired free chlorine residual and baseline HPC levels are achieved in the nitrifying areas of the distribution system.

Step 10.
Resume ammonia feed.

Source: Golden State Water Company

Figure 8-7 An example of a system-wide breakpoint chlorination protocol

and breakpoint chlorination of the entire system. Breakpoint chlorination was shown to be the only effective strategy for controlling nitrification in the Ann Arbor distribution system. The breakpoint period extended approximately 10 weeks. Almost immediately, the city saw a reduction in HPC and nitrite concentrations in the nitrifying areas and was able to maintain a free chlorine residual throughout the distribution system. It is interesting to note, however, that the city experienced increases in HPC counts in nonnitrifying areas of the distribution system during the breakpoint period. It was theorized that this increase was the result of granular activated carbon (GAC) carryover in the finished water. Fine particles, such as GAC, are capable of harboring bacteria. Monochloramine is a superior disinfectant compared to free chlorine for such bacteria, and as such, elevated HPC counts were observed (Stewart et al. 1990; LeChevallier et al. 1984, 1987, 1988).

Breakpoint chlorination of the affected area of the distribution system. Where nitrification is localized in an isolated part of the distribution system, targeted breakpoint chlorination can be as effective as system-wide chlorination and can minimize disruptions to treatment plant operations and negative impacts on distribution system water quality.

When conducting breakpoint chlorination of a targeted area, it is important to hydraulically isolate the affected area, to the extent possible, to minimize the mixing of chlorinated and chloraminated water. The primary reason for this is that the Cl_2:NH_3-N weight ratio at the chlorinated water and chloraminated water interface will often exceed 5:1 and can result in dichloramine and trichloramine formation and the corresponding destruction of any chlorine residual, free or combined (Mahmood et al. 1999; Muylwyk et al. 1999; Barrett et al. 1985).

Breakpoint chlorination of a hydraulically isolated area of the distribution system requires that there is a location available to inject the chlorine, such as an injection vault, booster chlorination station, or storage tank or reservoir. It also requires that the affected area be small enough that it can be reasonably contained by manipulation of distribution system valves. This response method typically involves addition of chlorine at a dose sufficient to achieve a 1- to 2-mg/L free chlorine residual. Many years ago, the Metropolitan Water District of Southern California (MWDSC) applied a dose sufficient to achieve a free chlorine residual of approximately 0.5 mg/L in the distribution system. However, MWDSC no longer practices breakpoint chlorination because they changed the Cl_2:NH_3-N ratio from 3:1 to 5:1 and increased the total chlorine residual at the plant effluent from 1.5 to 2.5 mg/L (Krasner 2012).

Case Study—Breakpoint chlorination in distribution system for nitrification control

A study conducted by Carrico et al. (2008) evaluated the effectiveness of switching to free chlorine for 1 month to control and prevent nitrification in isolated sections of cast-iron pipe. The study investigated the duration and degree of the resulting temporary chlorine residual decrease after breakpoint chlorination and after switching back to free chlorine, and the total trihalomethanes levels during the periods that the disinfectants are switched.

The effectiveness of breakpoint chlorination was evaluated in three distribution systems that switch from chloramine to free chlorine during the month of March each year. Breakpoint chlorination is also accompanied by a unidirectional flushing program in all three systems. Two of the distribution systems are in North Carolina, and one is in South Carolina.

This study revealed that breakpoint chlorination in the distribution system is only effective as a short-term response. Carrico et al. (2008) observed that nitrification returned within several months of the breakpoint period. This indicates that free chlorine is effective for stopping nitrification but that it is likely to return after the return to chloramine again. Ammonia-oxidizing microorganisms (AOM) were found attached to pipe-wall specimens that were removed from the distribution system both during and at the end of the free chlorination period, suggesting that the AOM survived exposure to free chlorine and the presence of biofilms may have offered protection to the AOM from disinfectant residual exposure or that AOM were reintroduced into the system through some other mechanism. Therefore, flushing before or during the free chlorination period might remove the biofilm from the pipe surface and thus expose the pipe surface to free chlorine.

This study also found that long residence time or high water age in the distribution system is not the only contributing factor to nitrification. A cause could be the cast-iron piping. Westbrook et al. (2007) mention that cast-iron piping can be the cause when chloramine reacts with corrosion products from the corroded pipe material and releases ammonia.

Nitrification was still observed at one monitoring site within one of the distribution systems even 1 week after the free chlorination began. This is an indication that residence time plays an important role too.

The temporary breakpoint (free) chlorination in this study also caused increased DBP levels and periods of low disinfectant residuals in the distribution system.

The significant chlorine dose required to achieve breakpoint will likely result in taste and odor issues, as well as high DBP concentrations, and may result in increases in bacterial and/or coliform counts. If at all feasible, it is recommended that the area containing nitrified water be flushed from the system during the introduction of breakpoint chlorination. This serves three purposes. First, it will prevent customers from receiving drinking water of poor quality. Second, flushing will help to remove the water of poor quality and may remove excess system sediments and biofilms. Finally, flushing will aid in drawing the chlorinated water into the affected areas of the system.

Superchlorination. Superchlorination involves dosing chlorine at concentrations sufficient to achieve a free chlorine residual of 50 mg/L or more for up to 24 h; however, a much shorter duration is typically utilized. The typical hold time is approximately 2 h. Even more so than other targeted burnout activities, it is critical that the response area be relatively small and easily isolated from the remainder of the distribution system. This approach is sometimes used as a last resort response to nitrification episodes that cannot be remedied by other methods discussed in this chapter, such as old areas of the distribution system with severe tuberculation. Even in these cases, superchlorination may only temporarily relieve nitrification.

When superchlorination is practiced, customers in the affected area should be notified of the superchlorination activities and advised that water service will not be available until such activities are complete. When applicable, utilities should notify customers 3–4 days in advance that they will be without water during the superchlorination event. Prior to addition of chlorine, utility personnel should shut the valves at the meter box into each home in the affected area. The line is purged during chlorine addition to improve the effectiveness of the response. After the targeted hold time has been achieved, the distribution main is flushed and dechlorinated (for more on dechlorination, see the Flushing section of this chapter). Before valves are opened to restore service to individual customers, hose bibs at each customer residence or business are opened to flush any remaining superchlorinated water from the service line.

Impact on water quality and Disinfectant/Disinfection Byproducts Rule compliance. Introduction of free chlorine into a chloraminated distribution system may result in an increase in DBP concentrations, specifically trihalomethanes (THMs) and haloacetic acids (HAAs). For example, total THM concentrations increased by 50 to 100 percent following a temporary month-long conversion to free chlorine in one system (Carrico et al. 2008; Vikesland et al. 2007). In another example, chloroform levels increased from approximately 30 µg/L to 110 µg/L at one location in a distribution system during a 1-month breakpoint period (Rosenfeldt et al. 2009). Because of concerns over short-term exposure to DBPs, some utilities have ceased to practice scheduled breakpoint chlorination of the distribution system (Mann et al. 1998).

At the chlorine–chloramine water interface, the Cl_2:NH_3-N weight ratio is likely to be in excess of 5:1; as a result, systems may experience short-term taste and odor and disinfectant residual problems. Carrico et al. (2008) observed total chlorine concentrations <0.5 mg/L for a period of up to 12 h at an individual monitoring location following conversion to free chlorine and residuals less than 0.5 mg/L for a period of approximately 8 days in a distribution system storage tank.

Introduction of free chlorine into the distribution system may also result in sloughing of biofilms, causing short-term increases in HPC or coliform concentrations (Odell et al. 1996). Skadsen (1993) also noted an increase in HPC concentrations in nonnitrifying areas during breakpoint chlorination, resulting from the decreased effectiveness of free chlorine for some bacteria. To minimize these impacts, some systems flush free chlorinated water from the distribution system (see the Flushing discussion presented in this chapter). Rosenfeldt et al. (2009) observed a decrease in bacterial counts during breakpoint chlorination; however, bacterial counts returned to previous levels or higher within approximately 1 month of returning to chloramine.

If a system routinely, e.g., every August, switches from chloramines to free chlorine for a scheduled breakpoint period and is required to collect Stage 1 or Stage 2 Disinfectant/Disinfection Byproducts (D/DBP) Rule compliance samples during that period, those analytical results are to be included in the compliance determination (63 FR 69390). In this instance, an increase in DBP concentrations could place a system in jeopardy of violating the Stage 1 or Stage 2 D/DBP Rule maximum contaminant levels. However, if a switch to free chlorine is done as an immediate or emergency response to a nitrification event, samples collected during that period may not be required to be included in the compliance determination.

The Stage 1 and Stage 2 D/DBP rules require monitoring to be conducted "during normal operating conditions." Thus, a system that routinely switches to free chlorine for a breakpoint period is under normal operating conditions. On the other hand, consider a chloraminated system that is experiencing nitrification in localized areas of the distribution system and opts to convert to free chlorine for 2 weeks to correct the problem as part of a response activity rather than typical operating conditions. Under this scenario, if the utility is required to collect samples during the period when operating with free chlorine, the state may allow the sampling to be postponed until normal operating conditions return or may decide that those samples would not be used for determining compliance (USEPA 2001).

Breakpoint chlorination is more appropriate as an emergency response to nitrification rather than a long-term solution. As such, it is recommended that utilities explore operational and engineering strategies for long-term nitrification control. However, it is recommended that utilities planning to use breakpoint chlorination as a mitigation response consult with their state agency to determine whether monitoring conducted during a breakpoint chlorination period will be used for compliance determination.

Long-term effectiveness of breakpoint chlorination. Breakpoint chlorination has been demonstrated in many instances to be effective for the control of nitrification. However, there are several studies that indicate it is only effective as a short-term response and is not likely to prevent nitrification from reoccurring (Rosenfeldt et al. 2009; Carrico et al. 2008; Odell et al. 1996).

Carrico et al. (2008) observed that nitrification returned within approximately 3 months following breakpoint chlorination. Others have observed similar results. Rosenfeldt et al. (2009) noted that bacterial counts returned to prebreakpoint levels within approximately 1 month of returning to chloramines.

There are reasons for the limited effectiveness of a brief free-chlorine conversion relative to other long-term nitrification control strategies. It has been observed that free chlorine application after full monochloramine biofilm formation (as would be expected in a drinking water distribution system) results in minimal free chlorine biofilm penetration (Lee et al. 2011). Put more simply, breakpoint chlorination may starve the nitrification process of a critical nutrient (free ammonia), but it does little to inactivate nitrifying bacteria, specifically AOM, embedded in distribution biofilms. Therefore, there may be short-term benefits associated with nutrient deprivation and AOM inactivation at the biofilm surface, but AOM within the biofilm remain viable, and nitrification is likely to return within a short period of time following a return to chloramines.

Coupled with short-term increases in distribution system DBP concentrations and reductions in disinfectant residual along the chlorine-chloramine boundary, breakpoint chlorination might be considered a last resort in many cases of nitrification. Further, where a free chlorine period is required by statute, it is possible that water systems are conducting the free chlorination during periods when DBPs are not as likely to be a concern (i.e., cooler water temperature months), which may be counterproductive to controlling nitrification.

Monitoring. Utilities should monitor free chlorine as well as total chlorine and continue to monitor for key nitrification parameters (e.g., nitrite, monochloramine, and HPC R2A) during breakpoint chlorination periods. In addition, utilities may wish to increase monitoring frequency in known problem areas in the distribution system as well as storage facilities. Following breakpoint activities, increased monitoring in known nitrifying areas should continue until it is evident the response was effective. Chapter 6 identifies key nitrification monitoring parameters and discusses the relevance of each during breakpoint chlorination periods.

Necessary approvals and customer notification. Prior to conducting breakpoint chlorination of the entire distribution system, whether scheduled or as an emergency response to a nitrification episode, the appropriate regulatory agency should be notified, and approval requested. Breakpoint chlorination of a targeted distribution system area, a storage tank, or reservoir may be considered maintenance activities and not require notification in some states. Contact the appropriate regulatory agency to determine when it is necessary to seek approval for breakpoint chlorination activities.

Customers may need to be notified when a temporary change from chloramines to free chlorine is planned due to noticeable changes in water taste and odor. Customers should be informed that they may notice changes in taste and odor or color. When superchlorination is practiced, customers should be advised several days in advance if they will not have water service during the superchlorination period.

Flushing

Flushing can be an effective practice for preventing nitrification. However, flushing can also be an effective emergency response to both nitrification and indicators of nitrification, such as declining chloramine residuals, increases in free ammonia, nitrite, and HPC concentrations. Properly designed flushing with a velocity selected for the purpose, pipe material, and pipe condition can remove distribution system biofilms (including AOM and NOB) and sediments, reduce disinfectant demand, and bring fresher water into areas of the distribution system with low flow, which results in an increase of residuals. Additional guidance on flushing program design can be found in AWWA Manual M68, *Water Quality in Distribution Systems*.

Flushing is usually performed based on water quality parameters such as total chlorine residual, nitrite, HPC, odor, etc., and is also based on customer water quality complaints. A well-planned routine flushing program that is implemented on a regular basis can help decrease the possibility of water quality deterioration and consequently prevent nitrification.

There are basically two types of system-wide flushing: conventional and unidirectional flushing. These two types and the differences in each approach are discussed in Chapter 7. A third alternative is spot flushing. Spot flushing, as opposed to system-wide flushing, may be a more appropriate response to nitrification. Spot flushing can be either conventional or unidirectional flushing or a combination of the two methods. Spot flushing (or emergency flushing) is often performed in response to customer complaints for color, taste, or odor problems and in response to other water quality problems, such as insufficient disinfectant residual, evidence of nitrification, or positive coliform results. Spot flushing, including the use of automatic flushers, is mainly used to respond to acute water quality problems. This type of flushing is not necessarily used for pipe cleaning but rather for replacing a large volume of water from low-demand areas (e.g., dead ends) and areas that show signs of nitrification or other water quality problems.

Flushing the affected area. Where nitrification has occurred in the distribution system, flushing can be an appropriate response to remedy the situation. Care must be taken to ensure the affected area is flushed completely and the conditions that led to the nitrification incident are mitigated. Although both conventional and unidirectional flushing can achieve these objectives, unidirectional flushing offers the best chance of mitigation and preventing (or at least minimizing) the potential for future nitrification incidents. Unidirectional flushing, as previously discussed, results in higher pipe scouring velocities and will likely remove more of the system biofilms, sediments, and corrosion byproducts that may have caused the nitrification episode. A well-planned unidirectional flushing response plan can also help to prevent movement of water that has begun to undergo nitrification to other previously unaffected areas of the system.

Where excessive water age is the likely cause of nitrification, flushing offers a short-term solution. However, unless a maintenance flushing program is initiated, flushing will not offer a long-term solution. In these cases, more permanent engineering solutions may be most effective for preventing future nitrification episodes.

Flushing after breakpoint chlorination. As discussed previously in this chapter, breakpoint chlorination can also be an appropriate response to nitrification. Following breakpoint chlorination, excess free chlorine will remain in the distribution system. When operations return to chloramination, the free chlorine-chloramine boundaries in the distribution system will have Cl_2:NH_3-N ratios in excess of the recommended 4.5:1–5:1. In these regions of the distribution system and wherever chlorinated and chloraminated water mix, the potential exists to form the less effective and more problematic (from a taste and odor perspective) dichloramine and trichloramine. One way in which utilities can minimize this impact is to utilize flushing to move the chlorine-chloramine interface before and after breakpoint chlorination. This can be effective for distribution system piping; however, other strategies, such as breakpoint chlorination of a storage tank or reservoir prior to conversion of the system, may be more effective.

When the entire system is breakpoint chlorinated, a combination of flushing and operational changes may be the most effective approach. This can help to reduce water losses to flushing and still ensure that the switch to chloramines is done with as little impact as possible on distribution system water quality. When only the affected area (area in which nitrification has occurred or potential for nitrification has been identified) is breakpoint chlorinated, it may be beneficial to flush as much of the chlorinated water from the system as is possible.

Impacts on water quality, regulatory compliance, and customer satisfaction. Generally, flushing can improve distribution system water quality, help to achieve regulatory compliance, and result in increased customer satisfaction. Flushing can remedy water quality problems such as nitrification, color, taste, and odor. A routine, properly designed system-wide flushing program can result in sustainable improvements in distribution system water quality.

A poorly implemented flushing program, on the other hand, can contribute to a deterioration in distribution system water quality. Improperly sequenced flushing activities can move degraded water within the distribution system, resulting in a spread of the problem rather than a solution. For this reason, flushing activities should be carefully planned and sequenced to avoid potential exacerbation of water quality problems.

Flushing can aid utilities in achieving regulatory compliance. Routine flushing or flushing as a response to nitrification or the indicators of potential nitrification can prevent loss of disinfectant residual. Flushing can help to reduce water age in system dead ends or areas with low water demand, resulting in lower DBP concentrations. (Spot flushing in the vicinity of a sampling location solely for the purpose of improving sampling results is not recommended; routine maintenance flushing or flushing of areas to prevent water quality degradation is recommended.) Flushing can improve microbiological water quality by periodically removing system biofilms, sediments, and corrosion byproducts that can serve as a nutrient source for microbes in the distribution system. Flushing following a line break can minimize the potential for microbiological contamination of the distribution system.

As previously mentioned, flushing generally increases customer satisfaction because of enhanced water quality. However, this may not always be the case. A poorly implemented flushing program can result in water of poor quality being drawn into larger regions of the system and result in customer complaints. Further, even a well-planned and sequenced flushing program can disturb sediments in the system and result in short-term deterioration in water quality at the tap. Finally, depending on the location of hydrants where flushing occurs, streets may temporarily be closed, and large volumes of water may drain onto consumers' property. The inconvenience that results may be a source of customer complaints.

Steps can be taken to minimize the negative impacts associated with flushing. Where flushing is required in high-traffic areas, it is recommended that flushing occur during nonpeak hours. AWWA offers several resources on flushing, including DVDs (*WSO Flushing and Cleaning*, AWWA 2016; *Unidirectional Flushing*, AWWA 2002) and print materials (Kirmeyer et al. 2000).

Monitoring. Chapter 6 identifies the key nitrification monitoring parameters that a water utility should monitor for in the distribution system: total chlorine or monochloramine, nitrite-N, free ammonia-N, temperature, and pH (see Table 6-2). These are the parameters that a utility should monitor for during flushing activities in the distribution system. Utilities should make it a common practice to collect samples for the appropriate parameters before and after flushing activities, as well as flow and pressure measurements to determine the flushing velocity and volume of water flushed. Hence, the water utility can determine the effectiveness of flushing as a response to nitrification.

Dechlorination. Dechlorination must occur where large volumes of chloraminated or chlorinated water are discharged into a storm drain or directly to the environment, which is basically the case when water is flushed out of a hydrant or a flush-out. Local regulatory agencies must be contacted to obtain discharge requirements and information regarding discharge permits that the utility must obtain before a discharge to the storm drain is allowed. Local regulatory agencies may require that only dechlorinated water with no detectable chlorine residual enter the storm drain.

Dechlorination is best performed by adding a controlled amount of dechlorination chemical to the flow of water that will be discharged to the storm drain. Dechlorination can be accomplished with several different dechlorination chemicals. Water utilities most frequently use sodium thiosulfate, sodium sulfite, and sodium bisulfite. Other dechlorination alternatives include sulfur dioxide, sodium metabisulfite, calcium thiosulfate, ascorbic acid (vitamin C), and sodium ascorbate (buffered form of vitamin C).

AWWA offers several resources on dechlorination. The Water Research Foundation guidance manual (Tikannen et al. 2001) discusses advantages and disadvantages of all dechlorination chemicals previously mentioned.

The series of pictures in Figure 8-8 shows an example of how dechlorination is performed.

Increase Chloramine Residual

Increasing chloramine residual is primarily used as a nitrification prevention measure. A loss of chloramine residual is typically caused by autodecomposition or consumption of chloramine by NOM or other substances in the water. A decrease in chloramine residual can also indicate the beginning of nitrification. During a nitrification episode, an abnormal decrease in chloramine residual should be investigated before any adjustments are made

Put on a safety vest and hardhat. Place "Flooded" signs and cones around work area.

Install the fire hose to the diffuser and then to the fire hydrant.

Prepare the dechlorination solution.

Dry dechlorination chemical is added to the container and then water is added to dissolve the chemical.

Dechlorination container is placed between hydrant and nearest storm drain. Open valve on container to allow solution to drip out.

Dechlorination container is placed on the curb in a position such that the dechlorination solution will drip into the water stream.

Figure 8-8 Steps in the dechlorination procedure

Turn on the fire hydrant.

Check chlorine residual of hydrant water (at start up).

After adding the sodium thiosulfate at the appropriate application rate, the water should be tested for chlorine residual at approximately 10 ft downstream to allow for sufficient contact with sodium thiosulfate, for chlorine residual. The result of the chlorine residual test should be zero (water remains clear after adding the reagent). If there is a detectable residual, the solution application rate should be increased until no residual is detected in the water 10 ft downstream of the point of application.

Check chlorine residual of hydrant water (during and at end of flushing). Record all chlorine residual results.

Source: Golden State Water Company

Figure 8-8 Steps in the dechlorination procedure

to the dose used at the water treatment plant because an increase in chloramine residual has been demonstrated not to be effective to correct a nitrification episode (Skadsen 2002).

A reduction in chloramine residual may also originate from operational problems, e.g., the chloramine residual entering the system is too low or the Cl_2:NH_3-N weight ratio is too low. Consequently, the operator should review the following: chlorine and ammonia feed rates, free and total chlorine residual data (both dose set point and residual entering the system), free ammonia dose and concentration at the treatment plant and entering the distribution system, calibration of the chemical feed pumps, and chlorine and ammonia analyzers. If the operator determines that the chloramine residuals, ammonia measurements, and Cl_2:NH_3-N ratios appear to be within the targeted range, the chloramine residual and the Cl_2:NH_3-N ratio may not be a contributing factor for the nitrification event and therefore adjustments to the chloramine residual targeted in finished water and across the distribution system will be ineffective for nitrification control.

RESPONSES TO NITRIFICATION EPISODES IN DISTRIBUTION SYSTEM STORAGE FACILITIES

When nitrification occurs in a storage facility or reservoir, the response may be significantly different than when nitrification occurs in the distribution system. First, early detection of nitrification in a storage tank can prevent spread of nitrification to other areas of the distribution system. Second, when detected in a timely manner, the nitrification can be kept isolated in the storage facility, making the response more efficient. For these reasons, it is critical that distribution system storage facilities be included when developing a nitrification monitoring plan. This section discusses responses to nitrification in distribution storage facilities.

Breakpoint Chlorination of Storage Tanks and Reservoirs

Breakpoint chlorination refers to the addition of chlorine to water at a concentration whereby all the ammonia has been oxidized to nitrogen, resulting in a free chlorine residual. When nitrification occurs in a storage tank or reservoir, breakpoint chlorination can be an effective response measure. Determining the appropriate chlorine dose to achieve breakpoint chlorination in a storage tank or reservoir requires utilities to

- determine the starting volume of water in the tank/reservoir,
- determine the added volume in the tank/reservoir (new water added during injection at the inlet),
- measure the total chlorine concentration in the reservoir and in the tank inflow,
- measure the total ammonia in the reservoir and in the tank inflow,
- measure the nitrite in the reservoir and in the tank inflow, and
- identify target free chlorine residual in the tank/reservoir after breakpoint.

In poorly mixed tanks, it is recommended that the total chlorine, total ammonia, and nitrite concentrations be measured at multiple depths in the storage tank and the analytical results averaged. In such a case, mixing or recirculation equipment may be required to effectively breakpoint-chlorinate the tank.

A free chlorine residual of at least 1.0 mg/L is recommended for most storage tanks. The duration of the breakpoint period can vary depending on the ability of a system to isolate and remove a storage facility from service given pressure and emergency requirements. Where it is possible to remove a tank from service, a period of 12–24 h is recommended. The Irvine Ranch Water District (2003) targets a free chlorine residual of 2.0 mg/L in nitrifying storage tanks and returns the tank to service within 12–30 h if monitoring indicates breakpoint chlorination has occurred and nitrification has ceased. The East Bay Municipal Utility District (EBMUD) adds sodium hypochlorite at the tank inlet to achieve an initial target free chlorine residual of 2.0–2.5 mg/L free chlorine once the tank is full and targets a residual of greater than 1 mg/L after a period of approximately 24 h (Hill and Teefy 2012). Lower target free chlorine residuals (e.g., 0.5 mg/L) have been used for large water storage facilities. These examples are merely guidelines, and the required duration is a function of free chlorine dose and time. A discussion of AOM inactivation by free chlorine is included in Chapter 5 to inform decision-making on how long to breakpoint chlorinate.

The following shows how to calculate the amount of sodium hypochlorite needed to breakpoint a reservoir and achieve the target free chlorine concentration at the end of the procedure. Sodium hypochlorite is added to the inlet line as the tank is filling. This helps to mix the hypochlorite with the contents of the reservoir. In general, for each 1 mg/L of ammonia, 10 mg/L of chlorine is needed; for each 1 mg/L nitrite, 5 mg/L chlorine is needed. This will theoretically result in water with zero chlorine residual; additional hypochlorite is needed to achieve the target free chlorine residual. The two volumes of water (1, water inside the tank at the start and 2, water used for filling the tank) are considered separately because they have different starting concentrations of each parameter. It is assumed that the sodium hypochlorite has a strength of 12.5 percent and a specific gravity of 1.15. Other values can be substituted as needed in the following equations.

First, determine the amount of hypochlorite needed for the water inside the tank:

$$\text{Hypo}_{\text{-tank}} = \left(100/(12.5\times1.15)\right) \times V_{\text{tank}} \times \left[\left(\text{Cl}_{2\text{-target}} - \text{Cl}_{2\text{-tank}}\right) + \left(\text{NH}_{3\text{-tank}} \times 10\right) + \left(\text{NO}_{2\text{-tank}} \times 5\right)\right] \quad \text{(Eq. 8-5)}$$

Where:

$\text{Hypo}_{\text{-tank}}$ = hypochlorite needed to dose water in tank, gal
V_{tank} = starting volume in tank, million gal (MG)
$\text{Cl}_{2\text{-target}}$ = target free chlorine residual when finished, mg/L
$\text{Cl}_{2\text{-tank}}$ = measured chlorine residual inside tank, mg/L
$\text{NH}_{3\text{-tank}}$ = measured total ammonia inside tank, mg/L
$\text{NO}_{2\text{-tank}}$ = measured nitrite inside tank, mg/L
12.5 = strength of sodium hypochlorite solution (% W/V)
1.15 = specific gravity of sodium hypochlorite solution (kg/L)

Similarly determine the amount of hypochlorite needed for the influent water:

$$\text{Hypo}_{\text{-inf}} = \left(100/(12.5\times1.15)\right) \times V_{\text{inf}} \times \left[\left(\text{Cl}_{2\text{-target}} - \text{Cl}_{2\text{-inf}}\right) + \left(\text{NH}_{3\text{-inf}} \times 10\right) + \left(\text{NO}_{2\text{-inf}} \times 5\right)\right] \quad \text{(Eq. 8-6)}$$

$\text{Hypo}_{\text{-inf}}$ = hypochlorite needed to dose the influent water, gallons
V_{inf} = volume of influent water, MG
$\text{Cl}_{2\text{-target}}$ = target free chlorine residual when finished, mg/L
$\text{Cl}_{2\text{-in}}\text{f}$ = measured chlorine residual in the influent water, mg/L
$\text{NH}_{3\text{-inf}}$ = measured total ammonia in the influent water, mg/L
$\text{NO}_{2\text{-inf}}$ = measured nitrite in the influent water, mg/L

Finally, the amount of hypochlorite needed to achieve the target free chlorine residual is the sum of the two:

$$\text{Hypo}_{\text{-total}} = \text{Hypo}_{\text{-tank}} + \text{Hypo}_{\text{-inf}} \quad \text{(Eq. 8-7)}$$

Exhibit 8-1 provides an example breakpoint dose calculation. There are two methods by which breakpoint chlorination can be achieved in a storage tank or reservoir: injection at the inlet while the tank is filling or addition through access hatches or recirculation/ mixing system while the tank is isolated and mechanically mixed.

Injection of chlorine at the inlet while the tank is filling. Injection of chlorine into the tank inlet may be difficult if no facilities, such as an injection vault and/or inline mixer, are available. Injection at the inlet generally requires a portable pump with a hose adapted to the pump discharge and injection port and a small generator for power. The pump suction can draw from buckets or carboys. Some utilities have vehicles equipped with portable hypochlorite feed systems. Guistino (2004b) identified this as one of the easiest and most efficient means by which to breakpoint chlorinate a storage tank or reservoir.

Exhibit 8-1 Example Breakpoint Dose Calculation

Big City operates at 1-MG storage tank and suspects nitrification is occurring in the tank. Nitrite is 0.25 mg/L, well above the baseline concentration of 0.01 mg/L. Total ammonia is 0.04 g/L, which is well below the baseline concentration of 0.15 mg/L. The total chlorine residual in the tank is 0.15 mg/L, which is well below the baseline concentration of 0.5–1.0 mg/L for this tank. The inflow has a total nitrite concentration of 0.08 mg/L, a total ammonia concentration of 0.4 mg/L, and a total chlorine concentration of 1.5 mg/L. Calculate the required breakpoint dose to achieve a 1.0 mg/L free chlorine residual in the tank. Assume the tank contains 0.5 million gal (MG) at the start of the breakpoint operation, 0.5 MG will be pumped into the tank, and use of a 12 percent (trace percent) sodium hypochlorite solution (SG = 1. 15).

The effectiveness of this method can be compromised if the mixing characteristics of the tank are such that the chlorine does not mix into all layers of the tank. Chapter 9 provides additional information on tank mixing and mixing systems. In storage tanks in which nitrification is occurring because of stratification in the storage tank (i.e., upper regions of the tank consist of much older water), sufficient tank volume must be drained to ensure complete mixing of the storage tank while filling. Therefore, the tank level must be lowered to allow for sufficient fill time to completely mix the storage tank. However, it is also important to minimize or eliminate the possibility that water of poor quality is drained into the distribution system; this could result in nitrification in other parts of the system or water of poor quality reaching customers. Guistino (2004b) recommended a minimum available volume of 50 percent to ensure adequate mixing. However, this level is tank specific and is dependent on the mixing characteristics of the storage facility. A lower tank level, such as 25 percent, is recommended, if possible. Figure 8-9 presents an example procedure to breakpoint chlorinate a storage tank or reservoir by addition of chlorine at the inlet while the tank is filling.

After breakpoint chlorination, water drawn from the tank will contain a free chlorine residual. This may result in Cl_2:NH_3-N weight ratio greater than 5:1 at the interface of water from the tank and water in the distribution system. As a result, flushing is recommended, where practical, to simulate increased demand and quickly remove free chlorine-containing water from the tank and distribution system. In certain instances, distributing water to customers with free chlorine may be acceptable, provided the residual is not too high (greater than approximately 1.0 mg/L Cl_2).

Addition of chlorine through hatches or recirculation system. Chlorine can be added through access hatches or through recirculation equipment. Recirculation or mixing equipment may be necessary if the shape of the reservoir is such that it is not possible to disperse the chlorine over the entire surface area.

There are several methods by which chlorine can be added through access hatches, each with its own advantages and disadvantages as described in Table 8-5. The drag-the-bag and broadcast spray methods are generally not recommended unless the storage

Step 1.
Lower reservoir to 50% of total volume (or as low as possible).

Step 2.
Measure water quality inside tank/reservoir. Calculate required breakpoint chlorination dose.

Step 3.
Begin filling the tank as quickly as possible and add hypochlorite as quickly as possible during the fill cycle.

Step 4.
Isolate the tank or reservoir and let sit for 24 hours (if possible).

Step 5.
Measure chlorine residual to ensure the target free chlorine residual was achieved prior to returning to service.

Step 6.
As long as free chlorine residual is not too high (> ~1.0 mg/L), return tank/reservoir to service.

Figure 8-9 An example of a storage tank breakpoint chlorination procedure

facility configuration is such that the pour-in-the-hatch method is not likely to be effective, i.e., in very large reservoirs. With the pour-in-the-hatch method, it is critical that the sodium hypochlorite be well mixed within the tank. Sodium hypochlorite is denser than water, and it has been reported that slug doses will settle to the bottom of the tank or reservoir (San Francisco Public Utilities Commission 2004). Carlomagno et al. (2005) describe a similar occurrence and the need for a mixer to disperse the sodium hypochlorite throughout the storage tank. Personal protective equipment (PPE) is required with all the identified methods; in particular, the broadcast spray and drag-the-bag methods present an increased risk of chlorinous vapor production and require respiratory protection. Of course, other relevant safety procedures must be followed (such as fall protection) when working at the top of a tank.

As listed in Table 8-5, it is not necessary to remove a tank or reservoir from service when the broadcast spray method is used. However, in those instances where the reservoir or tank level is not lowered with this method, the tank should be at least temporarily removed from service to allow for adequate settling of the sodium hypochlorite solution through the water column.

In 1992, the Fort Worth (Texas) Water Department (FWWD) experienced severe drops in chloramine residual at two of its finished water reservoirs due to nitrification. The Westland (5 MG) and Caylor (5 MG) reservoirs had been sized based on future demands, resulting in excessive detention times within the reservoirs. FWWD opted to install rechlorination and recirculation equipment at the reservoirs. In this case, chlorine was introduced in the bottom of the tank and rechlorinated water was drawn off the top of the tank for distribution. The recirculation equipment provided a small, continual vortex in the tank, allowing a chlorine residual to be maintained (Kirmeyer et al. 1995). It was reported that this was an effective means of controlling nitrification.

Depending on the degree of nitrification and the chlorine dose used, it is recommended to isolate the tank from the system during breakpoint chlorination. If rechlorination is practiced, as in the FWWD example previously discussed, it may also be necessary to add ammonia to the tank discharge prior to the distribution system to maintain the Cl_2:NH_3-N ratio and prevent dichloramine and trichloramine formation.

Table 8-5 Comparison of breakpoint chlorination methods for storage tanks and reservoirs by addition through access hatches

Method	Procedure	Advantages	Disadvantages
Pour-in-the-hatch while the tank is filling	1. Measure water quality, and calculate required chlorine dose. 2. Drop reservoir to low/target level. 3. Confirm positive/inflow into reservoir. 4. Pour sodium hypochlorite solution into hatch. 5. Wait until high/target level is achieved. 6. Take reservoir out of service and let sit for 24 h (if possible). 7. Measure free chlorine residual. 8. Return reservoir to service if OK.	• Quick and easy • Good for small tanks	• Requires good tank mixing • Only works when hatch is directly over inlet • May be difficult if the hatch is not easily accessible • Respiratory protection needed
Broadcast spray	1. Measure water quality, and calculate required chlorine dose. 2. Drop reservoir to low/target level (if applicable). 3. Isolate reservoir (if practicable). 4. Broadcast sodium hypochlorite over surface of reservoir. 5. Fill reservoir (if applicable). 6. Measure free chlorine residual. 7. Return to service if OK.	• Good for medium-sized tanks and below-grade reservoirs • Broadcasts high chlorine dose over entire surface of reservoir or tank • Good mixing as higher density water (containing free chlorine) drops through water column • No need to take reservoir out of service (but recommended) • Use of "spray sticks" can minimize exposure to hypochlorite. • Venturi nozzles can eliminate need for pumps.	• Need nozzle with sufficient pressure to spray over entire surface of reservoir or use a boat if reservoir allows access. May require chemical- resistant pump that can pump into high-pressure supply • Requires respiratory protection • Not recommended in most cases
Drag-the-bag	1. Measure tank/reservoir water quality, and calculate required chlorine dose. 2. Take reservoir out of service (if possible). 3. Use 65–75 percent calcium hypochlorite tablets. 4. Use mesh bag (attached to boat for large reservoirs). 5. Drag bag for several hours. 6. Measure free chlorine residual. 7. Repeat as necessary to achieve desired free chlorine residual. 8. Return reservoir to service	• Low chance of overdosing • Use of granular calcium hypochlorite eliminates chlorine fumes	• Awkward • Not very efficient • Time consuming • Need to disinfect boat • Not recommended in most cases

Source: Guistino 2004b

Deep Cycling of Storage Tanks and Reservoirs

Many utilities implement deep cycling of storage tanks and reservoirs to increase turnover and prevent or respond to nitrification. However, the ability of a utility to implement deep cycling of storage facilities is dependent on system pressure and emergency storage requirements. In 1987, the Philadelphia Water Department lowered tank levels to improve turnover and prevent nitrification. This strategy was effective for several years, until system pressure requirements necessitated higher operating levels (Odell et al. 1996). The Irvine Ranch Water District (2003) and the San Francisco Public Utilities Commission (Smith 2003) both include deep cycling of storage tanks as a level 2 response in their nitrification action plans. The effectiveness of deep cycling also depends on storage tank configuration. Kirmeyer et al. (2004) reported that deep cycling is only effective in well-mixed storage tanks.

Draining and Disinfection of Storage Facilities

When other responses to nitrification are ineffective, draining and disinfection may be an appropriate response to nitrification in small storage facilities (i.e., less than 1 MG). Much like flushing, draining a storage facility to waste simply removes the nitrified water from the system. The ability to use this approach is dependent on the ability of the utility to maintain adequate system pressures and meet emergency flow requirements with the affected tank offline. It is also avoided in water-scarce regions. Draining the tank without disinfection usually will not be effective for the control of nitrification. In fact, some states require that drained storage facilities be disinfected prior to returning them to service. ANSI/AWWA Standard C652, Disinfection of Water Storage Facilities, provides guidelines for disinfection of all equipment used to clean storage facilities. Chapter 7 discusses routine cleaning and tank maintenance as a nitrification prevention measure in greater detail.

Impact on Water Quality, Regulatory Compliance, and Customer Notification

Introduction of free chlorine into a chloraminated distribution system may result in an increase in chlorinated DBP concentrations, specifically THMs and HAAs. At the chlorine-chloramine water interface, the Cl_2:NH_3-N ratio is likely to be in excess of 5:1. As a result, systems may experience short-term reductions in disinfectant residual and taste and odor problems (Rosenfeldt et al. 2009; Carrico et al. 2008).

Breakpoint chlorination of a targeted area of a distribution system or storage tank or reservoir may be considered a maintenance activity and not require notification in some states. Contact the appropriate regulatory agency to determine when it is necessary to seek approval for breakpoint chlorination activities.

Monitoring

Utilities should monitor free chlorine during breakpoint chlorination of storage tanks to assure target residual levels are maintained for the duration of the breakpoint activity. Upon returning the tank to service, the tank should be resampled within 24 to 48 h for key nitrification parameters to determine if the response was effective. Chapter 6 identifies key nitrification monitoring parameters and discusses the relevance of each during breakpoint chlorination periods.

KEY TAKEAWAYS

Nitrification is most likely to initiate in distribution system dead ends, storage facilities, and other low-flow areas. Indicators that nitrification may be occurring include low total chlorine, high nitrite, and/or low free ammonia concentrations. Other observations, such as high nitrate concentrations, high HPC R2A counts, and depressed pH, can also be indicative of nitrification.

After it has been determined that nitrification is occurring, it is important to determine the cause of the nitrification episode. Determining the cause of nitrification and preventing recurring nitrification require an assessment of the impacts of distribution monitoring, finished water quality, treatment plant operation, distribution system and storage tank operation, and distribution piping on nitrification. Prevention of nitrification, either by operational or physical system improvements, is the preferred method of dealing with nitrification. However, when nitrification is occurring in a distribution system or storage facility, it is critical that a timely and effective response be initiated. Developing a nitrification response plan is critical to achieving this objective. A well-coordinated response plan should include a combination of alert levels and action levels requiring varying degrees of response. Alert levels and action levels should be developed based on system-specific baseline concentrations of critical nitrification monitoring parameters (e.g., total chlorine, free ammonia, and nitrite).

The levels of operational response to nitrification vary depending on the alert level or action level and the degree to which nitrification is occurring in the distribution system. The most basic response includes increased monitoring and verification of chlorine and ammonia feed rates at the treatment plant or booster disinfection stations. The next level of response typically includes targeted flushing, breakpoint chlorination of the affected tank or area, increased tank cycling (lower tank operating levels), and cleaning and disinfecting storage tanks. For widespread nitrification, breakpoint chlorination of the entire distribution system may be necessary.

REFERENCES AND BIBLIOGRAPHY

American National Standards Institute/American Water Works Association. Disinfection of Water Storage Facilities. ANSI/AWWA C652. Denver: AWWA.

AWWA (American Water Works Association). 2002. *Unidirectional Flushing DVD*. Denver: AWWA.

AWWA Disinfection Systems Committee. 2008. "Committee Report: Disinfection Survey, Part 2—Alternatives, Experiences, and Future Plans." *Journal AWWA*, 100:011:110–124.

AWWA. 2016. *WSO Flushing and Cleaning DVD*. Denver: AWWA.

AWWA. 2017. Water Utility Disinfection Survey Report. Report prepared for AWWA by Cornwell Engineering Group, 2018

AWWA. 2017. Manual M68, *Water Quality in Distribution Systems*. Denver: AWWA.

Barrett, S.E., M.K. Davis, and M.J. McGuire. 1985. "Blending Chloraminated and Chlorinated Waters." *Journal AWWA*, 77:1:50–61.

Carlomagno, B., A. Wong, M. Quady, C. Smith, and W. Grayman. 2005. "Evaluation of Mixing Improvements in Distribution System Reservoirs for Nitrification Control and Prevention." In *Proc. of the AWWA Annual Conference and Exposition*. Denver: AWWA.

Carrico, B.A., F.A. DiGiano, N.G. Love, P. Vikesland, K. Chandran, M. Fiss, and A. Zaklikowski. 2008. "Effectiveness of Switching Disinfectants for Nitrification Control." *Journal AWWA*, 100:10:104–115.

Dennis, J.P., D.C. Rauscher, and D.A. Foust. 1991. "Practical Aspects of Implementing Chloramines." In *Proc. of the AWWA Annual Conference and Exposition*. Denver: AWWA.

Guistino, J. 2004a. Breakpoint Chlorination Spreadsheet Tool. Personal communication. Contra Costa Water District, Concord, Calif.

Guistino, J. 2004b. Breakpoint Chlorination of Reservoirs—A Nitrification Band-Aid. Presented at the 2004 AWWA Cal-Nevada Section Spring Conference.

Harrington, G.W., D.R. Noguera, C.C. Bone, A.I. Kandou, P.S. Oldenburg, J.M. Regan, and D. Van Hoven. 2003. *Ammonia from Chloramine Decay: Effects on Distribution System Nitrification*. Denver, Colo.: AWWA Research Foundation.

Hill, C.H., and S. Teefy, personal communication, 2012.

Irvine Ranch Water District. 2003. *Nitrification Action Plan*. July 7, 2003.

Kirmeyer, G.J., M. Friedman, J. Clement, A. Sandvig, P.F. Noran, K.D. Martel, D. Smith, M. LeChevallier, C. Volk, E. Antoun, D. Hiltebrand, J. Dyksen, and R. Cushing. 2000. *Guidance Manual for Maintaining Distribution System Water Quality*. Denver, Colo.: AWWA Research Foundation.

Kirmeyer, G., K. Martel, G. Thompson, L. Radder, W. Klement, M. LeChevallier, H. Baribeau, and A. Flores. 2004. *Optimizing Chloramine Treatment*, 2nd ed. Denver, Colo.: AWWA Research Foundation.

Kirmeyer, G.J., L.H. Odell, J.G. Jacangelo, A. Wilczak, and R.L. Wolfe. 1995. *Nitrification Occurrence and Control in Chloraminated Water Systems*. Denver, Colo.: AWWA Research Foundation and American Water Works Association.

Krasner, personal communication, 2012.

LeChevallier, M.W., T.M. Babcock, and R.G. Lee. 1987. "Examination and Characterization of Distribution System Biofilms." *Applied and Environmental Microbiology*, 53:12: 2714–2724.

LeChevallier, M.W., C.D. Cawthon, and R.G. Lee. 1988. "Inactivation of Biofilm Bacteria." *Applied and Environmental Microbiology*, 54:10: 2492–2499.

LeChevallier, M.W., T.S. Hassenauer, A.K. Camper, and G.A. McFeters. 1984. "Disinfection of Bacteria Attached to Granular Activated Carbon." *Applied and Environmental Microbiology*, 48:5: 918–923.

Lee, W., D.G. Wahman, P.L. Bishop, and J.G. Pressman. 2011. "Free Chlorine and Monochloramine Application to Nitrifying Biofilm: Comparison of Biofilm Penetration, Activity, and Viability." *Environmental Science & Technology*. 45:4:1412–1419.

Mahmood, F., J. Pimblett, N. Grace, and B. Utne. 1999. "Combining Multiple Water Sources and Disinfectants: Options for Water Quality Compatibility in Distribution Systems." In *Proc. of the AWWA Water Quality Technology Conference.* Denver: AWWA.

Mann, R.W., R.L. Wolfe, S.E. Barrett, and N.I. Lieu. 1998. "Optimizing Chloramine Residuals to Prevent Nitrification in Southern California Water Utilities."

Muylwyk, Q., A.L. Smith, and J.A. MacDonald. 1999. "Implications on Disinfection Regime When Joining Water Systems: A Case Study of Blending Chlorinated and Chloraminated Water." In *Proc. of the AWWA Water Quality Technology Conference.* Denver: AWWA.

Odell, L.H., G.J. Kirmeyer, A. Wilczak, J.G. Jacangelo, J.P. Marcinko, and R.L. Wolfe. 1996. "Controlling Nitrification in Chloraminated Systems." *Journal AWWA*, 88:7:86–98.

Rosenfeldt, E.J., C. Baeza, and D.R.U. Knappe. 2009. "Effect of Free Chlorine Application on Microbial Quality of Drinking Water in Chloraminated Distribution Systems." *Journal AWWA*, 101:10:60–70.

San Francisco Public Utilities Commission, personal communication, 2004.

Skadsen, J. 2002. "Effectiveness of High pH for Control of Nitrification." *Journal AWWA*, 94:7: 73–83.

Skadsen, J. 1993. "Nitrification in a Distribution System." *Journal AWWA*, 85:7:95–103.

Smith, C.D. 2003. *Nitrification Response Plan, Draft Final.* Prepared for the San Francisco Public Utilities Commission. Charlotte D. Smith and Associates, Inc.

Stewart, M.H., and N.I. Lieu. 1997. "Nitrification in Chloraminated Drinking Water and Its Association With Biofilms." In *Proc. of the AWWA Water Quality Technology Conference.* Denver: AWWA.

Stewart, M.H., R.L. Wolfe, and E.G. Means. 1990. "Assessment of the Bacteriological Activity Associated with GAC Treatment of Drinking Water." *Applied and Environmental Microbiology*, 56:12:3862–3869.

Tikannen, M.T., J.H. Schroeter, L.Y.C. Leong, and R. Ganesh. 2001. *Guidance Manual for Disposal of Chlorinated Water.* Denver: AWWA Research Foundation.

USEPA (United States Environmental Protection Agency). 2001. *Stage 1 Disinfectants/ Disinfection Byproducts Rule: Frequently Asked Questions.* Washington: USEPA Department of Water.

Vikesland, P., N. Love, M. Fiss, A. Zaklikowski, K. Chandran, F. DiGiano, and B. Carrico. 2007. "Water Quality Impacts of Disinfectant Switching Practices at Pilot and Full-Scale." In *Proc. of the AWWA Annual Conference and Exposition*. Denver: AWWA.

Westbrook, A., F.A. DiGiano, and W. Zhang. 2007. "Chloramine Decay Rates at Pipe Surfaces and in Bulk Water in a Bench-Scale Pipe Section Reactor." In *Proc. of AWWA Distribution System Research Symposium*, Reno, Nev. Denver: AWWA.

White, G.C. 1999. *Handbook of Chlorination and Alternative Disinfectants*, 4th ed. New York: John Wiley & Sons.

Wilczak, A., J.G. Jacangelo, J.P. Marcinko, L.H. Odell, G.J. Kirmeyer, and R.L. Wolfe. 1996. "Occurrence of Nitrification in Chloraminated Distribution Systems." *Journal AWWA*, 88:7:74–85.

Wolfe, R.L., N.I. Lieu, G. Izaguirre, and E.G. Means III. 1990. "Ammonia-Oxidizing Bacteria in a Chloraminated Distribution System: Seasonal Occurrence, Distribution, and Disinfection Resistance." *Applied Environmental Microbiology*, 56:2:451–462.

Wolfe, R.L., E.G. Means III, M.K. Davis, and S.E. Barrett. 1988. "Biological Nitrification in Covered Reservoirs Containing Chloraminated Water." *Journal AWWA*, 80:9:109–114.

Wolfe, R.L., N.R. Ward, and B.H. Olsen. 1984. "Inorganic Chloramines as Drinking Water Disinfectants: A Review." *Journal AWWA*, 76:5: 74–88.

Woolschlager, J.E., B.E. Rittmann, P. Piriou, L. Kiene, and B. Schwartz. 2001. "Using a Comprehensive Model to Identify the Major Mechanisms of Chloramine Decay in Distribution Systems." *Water Science and Technology: Water Supply*, 1:4:103–110.

Chapter 9

Engineering Changes and Capital Improvements for Nitrification Prevention

Tarrah Henrie, Sarah Richardson, Meric Selbes

INTRODUCTION

Chapters 6–9 are intended to walk the reader through the order of actions that should be taken to resolve nitrification issues identified during a routine monitoring program, as detailed in Figure 9-1. Before making any operational changes or capital improvements, a utility may want to increase monitoring frequency as outlined in Chapters 6 and 8. The results of monitoring are critical to diagnosing the locations and causes of nitrification. Once data have been collected, the utility should experiment with operational and treatment changes that are explained in Chapters 7 and 8. It is ideal to make one change at a time and monitor the impact of the individual change to isolate its impact on nitrification as opposed to making multiple changes at once that make it difficult to determine what is effective. When operational practices like increasing tank turnover are not sufficient for nitrification prevention, utilities should consider implementing the capital improvements that are the topic of this chapter. Please refer to Table 9-1 for the list of key points discussed in this chapter.

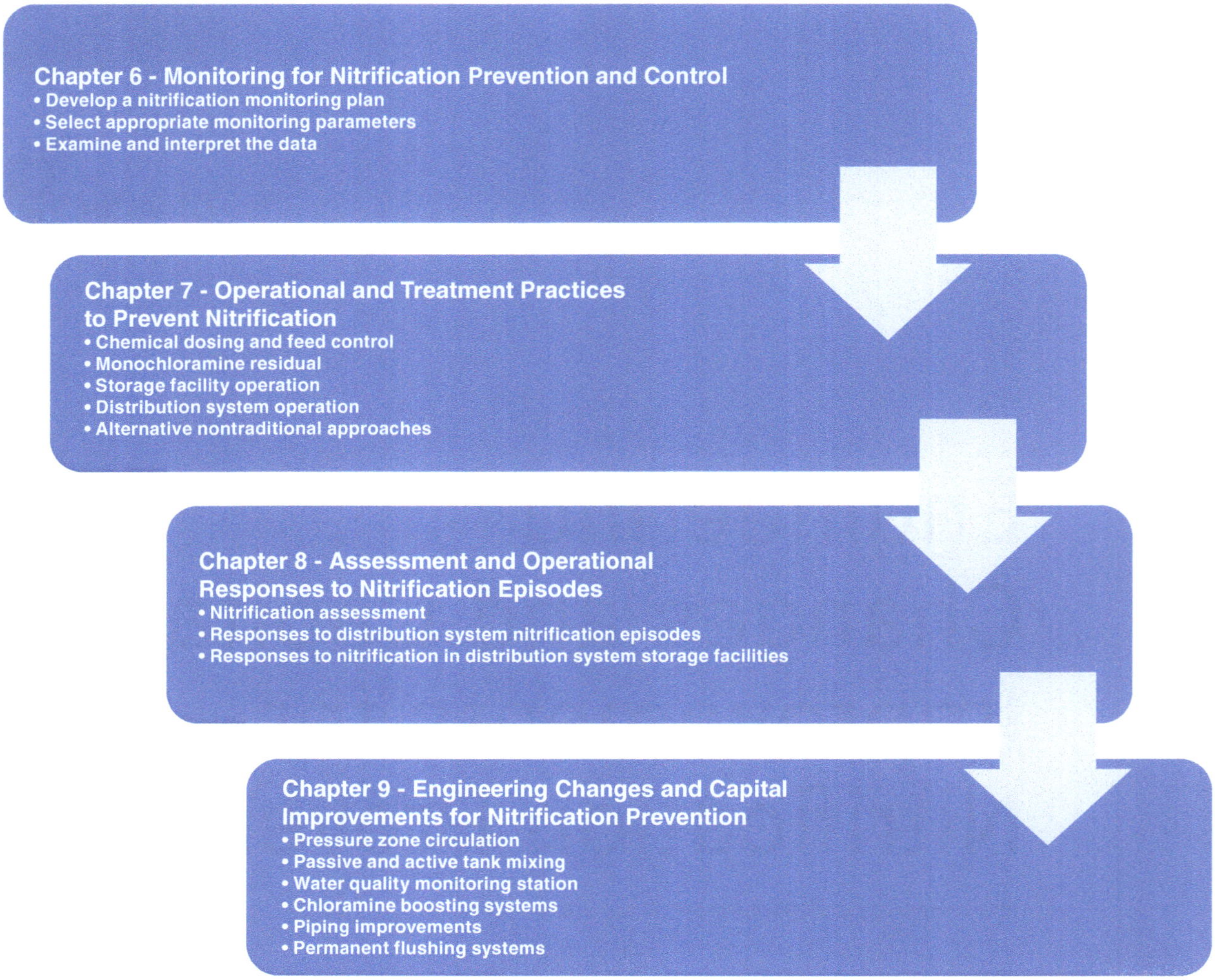

Source: Tarrah Henrie

Figure 9-1 Chapter flow chart

As presented in Table 9-1, there are several alternative strategies that can be implemented in the distribution system to prevent nitrification. While some of the strategies target reducing water age, other strategies maintain chloramine residual to prevent nitrification. On the other hand, water quality monitoring stations can provide real-time water quality data to guide operators as to when and where to take an action as soon as signs of nitrification are observed. The alternatives for mitigating nitrification are discussed further in the sections below.

It is important to note that there is no "silver bullet" strategy that would prevent nitrification in every distribution system. Each distribution system is different; the impacts of each strategy on nitrification must be assessed thoroughly. Depending on the location and frequency of nitrification, the strategy to prevent nitrification will vary. It is beneficial to conduct a holistic distribution-system analysis and evaluate advantages and disadvantages of each strategy before implementation. In some cases, a combination of strategies may be necessary.

Table 9-1 Key points from Chapter 9

Pressure zone circulation	• In some cases, there are ways to operationally turn over tanks and prevent higher residence times. • If that is not possible due to pressure constraints or fire-flow restrictions, an operational zone valve could help. • Another option is to regulate pressure between zones and move water from a higher pressure zone to a lower pressure zone.
Tank mixing	• Inlets and outlets on existing tanks can be modified to achieve passive mixing, and new tanks can be designed with an inlet that is separate from the outlet, which can improve mixing conditions. • Active mixing can increase the homogenization regardless of where inlet and outlet piping is placed. • Active mixers are designed as needed and can be installed in an existing or new tank. • Active mixing alone may not be sufficient to address all nitrification issues. It maintains homogenous conditions in the tank, but it does not address the need for an adequate chlorine-to-ammonia ratio directly. It is appropriate for tanks that can be deep cycled regularly and have nitrification occurring in the tank but do not receive already nitrified water. • In rare instances when total trihalomethane (TTHM) concentrations are near or over the regulatory limit in chloraminated water, aeration systems can be used. Aeration simultaneously provides mixing and TTHM removal.
New tank design considerations	• Minimizing water age and improving mixing conditions should be considered. • One design option is to install multiple smaller tanks rather than one large tank. This allows for operational flexibility and seasonal variation. • When designing a new area of the distribution system, consider installing a booster system from a lower pressure zone to provide the required fire flow rather than installing additional storage.
Online water quality monitoring stations	• Water quality analyzers can be installed either individually or in combination to monitor the water quality in the distribution system where the system is prone to nitrification.
Chloramine boosting systems	• Chlorine can be added in the distribution system to form monochloramine by combining with the free ammonia that is already present. This can adjust the ratio of chlorine to ammonia back to a more optimal level, but the total chlorine residual would be limited by how much ammonia is present. • Recombining free ammonia may not be sufficient if the resulting chloramine residual is too low and a higher residual target is desired. In such instances, it may be necessary to add both chlorine and ammonia. • Manual and/or automated chlorine- and ammonia-boosting systems may be used.
Piping improvements	• Piping improvements in the distribution system can reduce nitrification. These improvements typically include relining or replacing corroded/old water mains, eliminating dead ends, and sizing/resizing the water mains.
Flushing systems	• To create an artificial water demand and reduce the water age in select locations of the distribution system, flushing valves can be installed. • Flushing trucks are available that are designed to minimize water waste by taking water from a hydrant, filtering it, and feeding the filtered water into another hydrant in the distribution system. Some of these also include chlorine and ammonia boosting.
Other technologies	• Some utilities are experimenting with ultraviolet light disinfection systems that can inactivate nitrifying bacteria.
Combinations of capital improvements	• It may be necessary to implement several of the options or use hybrid approaches to achieve target goals with respect to nitrification.

ENGINEERING CHANGES AND CAPITAL IMPROVEMENTS

Increased Circulation to Decrease Water Age

In this section, three methods for water circulation to reduce water age will be discussed. The first method is simply deep cycling storage facilities to prevent water from becoming stagnant in the storage tank. The second method is using an operational zone valve (OZV) to help with zones that have multiple tanks feeding into the same zone. The final method is circulating water from higher to lower zones to keep water from storage tanks moving.

Storage Tank Deep Cycling

Sometimes it can be as easy as changing the set points for the tank's fill cycle so that the tank can deep cycle. Tank cycling is discussed in more detail in Chapter 7. If deep cycling is not possible due to pressure constraints or fire flow restrictions, an OZV could help.

Operational Zone Valves

If multiple tanks feed into one pressure zone, having an OZV can allow water from slightly lower elevation tanks to drain and turn over more fully without pumping by isolating the area the higher elevation tank can serve. This allows for the lower elevation tank to feed a larger area and achieve a higher turnover. Another way to look at this situation is through water age. If there are areas of higher water age within a zone, there might be a way of optimizing the tank turnover to allow for a more consistent water age through installation of an OZV.

Pressure Zone Circulation

If the capital expense of an OZV is too high or there is not a practical location to install it, there is another option for turning over tanks with higher water age. Water can be regulated down to lower pressure zones to keep it from becoming stagnant in higher elevation tanks that may be maintained at a higher level due to fire flow or pressure requirements. In this approach, water is pumped to the higher zone to maintain necessary tank and fire-flow levels, then the water is regulated back down to the lower pressure zone to reduce the residence time within the higher elevation tank. While this option can require minimal capital investments, it can be expensive to operate due to additional pumping costs depending on the existing infrastructure (electricity costs, head difference, and amount of water to circulate). A life-cycle cost analysis may be useful to determine if more routine pressure zone circulation is more cost-effective than an OZV.

Considerations for Increased Circulation

Pressure zone circulation is a good option if existing infrastructure is in place. If not, there is a capital cost associated with the installation of the valves (zone valve or pressure-reducing valve) that depends on the diameter of the pipe. Furthermore, a vault or a housing may be necessary in some cases, which would contribute to the capital cost.

One of the consequences associated with pressure zone circulation through regulating water from higher zones is higher electrical costs. To convey the water from a lower pressure zone to a higher pressure zone, pumping is required. When the water is moved to the lower pressure zone again, the energy that was put in to increase pressure is lost. This results in increased energy consumption at the booster pump stations. Furthermore, water with signs of nitrification can be conveyed to an area where nitrification has not been observed previously. Depending on the extent of nitrification, there can be an increased number of complaints related to manganese mobilization and even hydrogen sulfide odor.

TANK MIXING

This section examines the options for tank improvements that can reduce water age, improve mixing and overall water quality, and decrease tank stratification. Tank stratification is a condition where water quality differs at different depths. In a stratified tank, the top of the tank often has higher temperatures, lower chlorine residual, more nitrification, and more disinfection byproducts compared to lower parts of the tank. It is important to keep in mind that adequate tank turnover must be achieved in conjunction with tank mixing improvements to efficiently control nitrification. Computational fluid dynamics models can be useful for understanding tank hydraulics and planning out mixing systems to lower hydraulic residence time and improve water quality.

Passive Tank Mixing

As discussed in Chapter 7, measurements of temperature and water quality gradients in tanks can be conducted to determine if an existing tank is stratified. With passive mixing systems, it is critical to have adequate tank turnover because mixing only occurs when the tank is filling. Tank inlets and outlets on existing tanks can be modified to improve passive mixing, and new tanks can be designed with an inlet that is separate from the outlet, which can improve mixing to some extent.

One example of a passive mixing system that can be used in new or existing tanks is duckbill style check valves. These valves open using distribution system pressure, so no additional electricity is required. As water is pushed through these valves, the water is mixed via jetting action. An example is shown in Figure 9-2. Some head loss is associated with these valves, but the design can be modified to lower head loss if necessary when less intense mixing is required.

Active Tank Mixing

Active tank mixing systems provide continuous mixing, as opposed to passive mixing systems that only provide mixing during tank cycling. They also can support in-tank chemical addition, which can be more challenging with passive mixing. Because of capital costs involved with active mixers, utilities need to identify which tank will benefit most from active mixing. Before deciding to install an active mixing system, it is critical to determine if mixing alone is likely to address the nitrification issue or if chloramine boosting along with active mixing is required. A study can be conducted by installing a temporary portable mixer and conducting depth measurements. Please refer to Figure 9-3 for an example of an active tank mixing and chloramine boosting decision flow chart.

Active tank mixing is beneficial for eliminating dead zones contributing to nitrification that arise in tanks, but on its own, it cannot improve the quality of water entering the tank or stop nitrification that began outside the tank. However, active mixing is a critical component of in-tank chloramine boosting.

Active tank mixing can make the water quality of the tank uniform throughout the day and independent of tank fill/draw cycles, as shown in Figure 9-4.

Several different styles and brands of active tank mixers that rely on different mixing mechanisms are available on the commercial market. For example, there are "sheet flow" mixers that consist of a pump installed at the bottom of the tank that pulls water from bottom few inches of the tank, directs it upward in a thin sheet, and induces flow indirectly through the tank to provide continuous mixing. Another style is an impeller style mixer that creates a vortex to continuously homogenize water. A compressed air–style mixer (shown in Figure 9-5) uses an air compressor to create large bubbles that pull water upward as they float to continuously blend tank water. Some states require certification for mixing devices installed inside tanks comply with NSF/ANSI 61: Drinking Water System Components—Health Effects.

Source: Red Valve

Figure 9-2 A duckbill check valve style passive mixing system

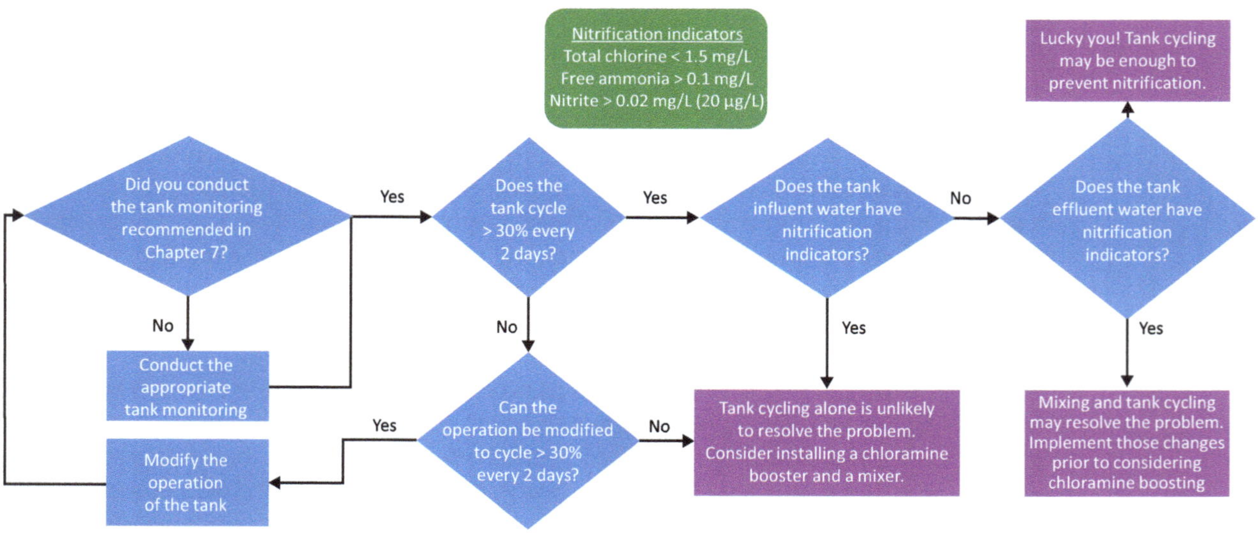

Source: Tarrah Henrie

Figure 9-3 An example of an active tank mixing and chloramine boosting decision flow chart: Note that the values associated with nitrification indicators and tank cycling can serve as starting point for many systems, but the most appropriate values can vary by system.

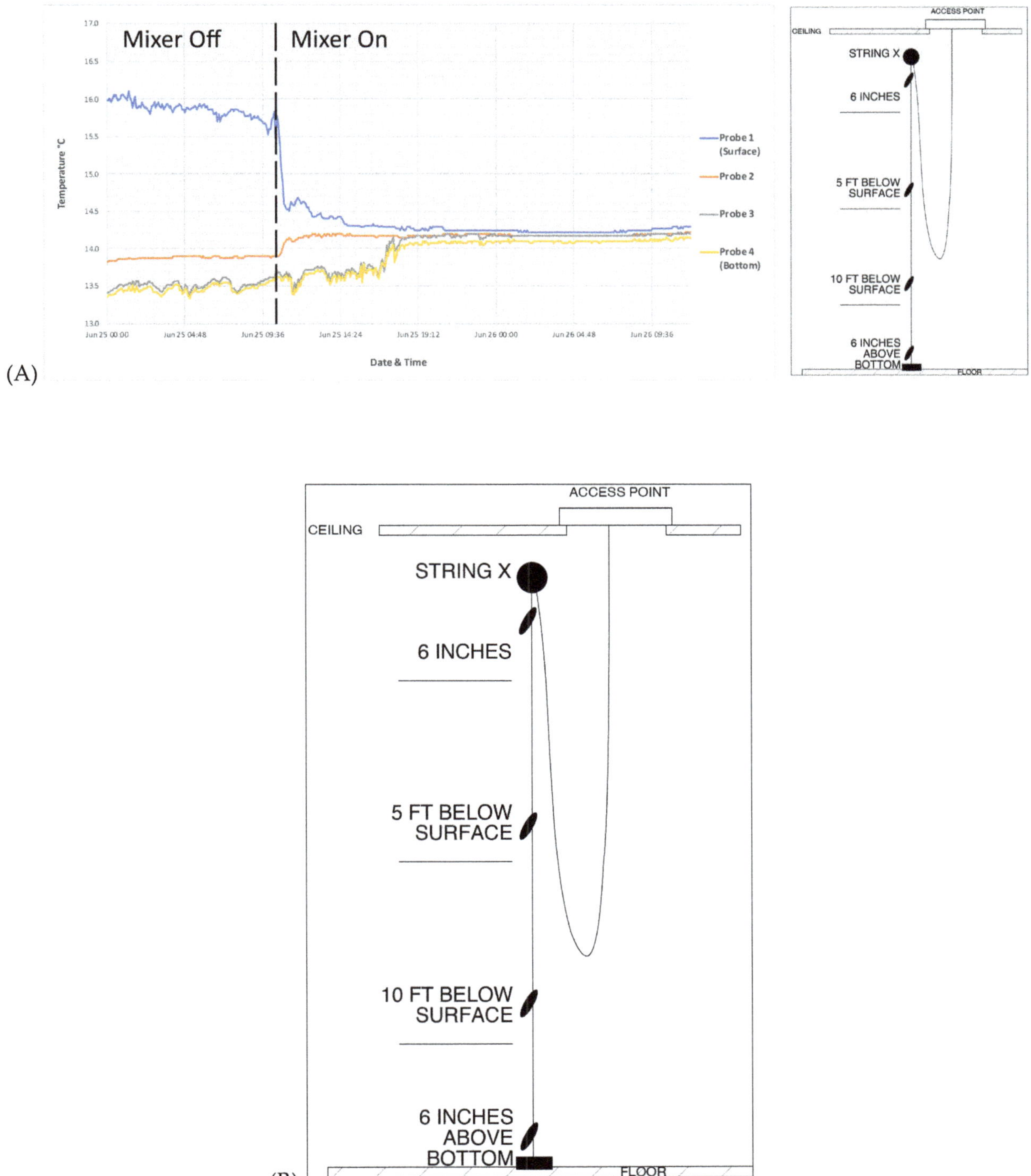

Source: IXOM Watercare

Figure 9-4 Temperature at different depths in a tank with and without active mixing (A) and the depth locations of the probes (B)

Source: IXOM Watercare

Figure 9-5 An example of a pump-style active mixer

Solar-powered installations can also be considered in locations with no electricity, as shown in Figure 9-6. They can also charge a battery that can power active tank mixers when the sun is not out.

Aeration Systems

Aeration is not typically used for mixing alone but can provide adequate mixing. In rare conditions where tanks in chloraminated distribution systems have total trihalomethane (TTHM) concentrations near or over the regulatory limit, aeration systems may be appropriate because they simultaneously mix water while decreasing TTHM levels. The technical aspects of aeration technology are discussed in much more detail in the following references:

- Chapter 4 of Manual M64, *Gas Transfer Applications in Water* (AWWA 2021)
- Reducing Volatile Disinfection By-Products in Treated Drinking Water Using Aeration Technologies (AWWA 2015)

An example of surface aeration units is shown in Figure 9-7. Spray aeration is another option.

Source: IXOM Watercare

Figure 9-6 A solar-powered active mixer installation

NEW TANK DESIGN CONSIDERATIONS

Additional storage may be necessary to meet water demands or provide fire flow. Following are some options that can be considered to provide additional supply while also limiting nitrification.

Tank Alternatives

Sometimes tanks are used to provide pressure and fire flow in a hydraulically limited part of the distribution system. If this part of the distribution system experiences low demand, then the tank can experience nitrification. When designing a new area of the distribution system, installing a booster system from a lower pressure zone to provide the required fire flow can be considered. Similarly, regulating down from a higher zone can be considered. The tank in the low-demand pressure zone can then be sized more appropriately to meet demand.

Number and Size of Tanks

One design option when constructing new tanks is to install multiple smaller tanks rather than one large tank. This allows for operational flexibility during low demand seasons. For example, on a site with three tanks, one can be emptied and taken out of service to reduce

Source: Corona Environmental Consulting

Figure 9-7 A surface aerator for mixing and total trihalomethane (TTHM) reduction.

water age during low-demand periods. It is important to determine if the required fire flow can still be met with reduced storage.

Tank Location

Selecting a tank location depends on many factors like land availability, public acceptance, water supply needs, etc. It is important to analyze the distribution system and identify the potential locations where water storage tanks would not exacerbate the nitrification issues.

Piping and Mixing of Tanks

It is important that the new tanks are designed with separate inlets and outlets that are located on opposite ends of the tank to prevent short circuiting, limit dead zones, and promote tank turnover. Tanks should also be unbaffled when used in the distribution system because baffles inhibit mixing, which is beneficial in a disinfection basin but not when trying to homogenize water. The tanks can be designed with either a passive or an active mixing system to prevent stratification within the water storage tank.

Future Provisions

If it is known or suspected that a new tank will be located in an area that is prone to nitrification or known to have nitrification issues, the tank can be designed with provisions to mitigate nitrification within the storage tank. For example, chloramine boosting system connections, as well as power and communication signal capability, can be implemented during the design to ease the installation of a system in the future.

WATER QUALITY MONITORING STATIONS

Water quality monitoring stations are useful for obtaining critical nitrification data in real time. There are several models of water quality monitoring equipment that measure different water quality parameters. Analyzers can be installed either individually or in combination to monitor the water quality in the distribution system. For example, analyzers for pH and chlorine residual can be used together to monitor for the presence of nitrification. The locations and parameters that should be monitored are discussed further in Chapter 6. Many of the analyzers on the market can be integrated into a water utility's supervisory control and data acquisition (SCADA) system. SCADA can be programmed to generate an alarm when a parameter is out of the desired range, which could potentially save operator time in going out to the distribution system to take samples. Moreover, there are reagentless analyzer alternatives available for some of the water quality parameters that simplify installation and operation, and some can run on solar power and batteries.

The monitoring stations can be located in areas of the distribution system that are susceptible to nitrification, especially areas where water age is expected to be high. Selection of a site where the water quality monitoring equipment can be secured to prevent vandalism is critical. For example, fenced areas at booster pump stations or water storage tanks are ideal. Note that analyzers should not be supplied from the outlet of a booster pump station; doing so will greatly impact water quality as fresh (when pumping) and stagnant (when not pumping) water will be analyzed. Other considerations for locating the monitors at a site are sample piping, power connection, network connectivity, and wastewater disposal. Furthermore, if applicable, freeze protection of the monitoring equipment should be evaluated.

Overall, multiparameter water quality monitoring stations can have several advantages. These stations can be simply used for monitoring the water quality in the distribution system or to identify when there is a need for operational change in the distribution system, such as system flushing. While there are several advantages of these stations, it is important to note the operations and maintenance (O&M) costs associated with each analyzer. Depending on the make and model of the analyzer, the number of replaceable components changes, and as such, all instruments require periodic maintenance. Operators need to periodically visit the water quality monitoring stations to replenish the reagents, as well as to calibrate and maintain each analyzer according to manufacturer recommendations.

CHLORAMINE BOOSTING SYSTEMS

A decrease in total chlorine residual within storage facilities can lead to nitrification, as discussed in Chapter 4. Limiting the amount of free ammonia and maintaining an adequate chloramine residual within a distribution system are key in preventing nitrification. One way to improve water quality involves adding chlorine to recombine with the free ammonia already present in the water to a target an optimal chlorine-to-ammonia ratio. The measured chlorine-to-ammonia ratio target is typically in the range of 4.5:1–5:1 on a mass basis (i.e., units of mg-N:mg-Cl_2). If the chloramine residual is too low or there is not adequate free ammonia, then the concentration of chloramine that is formed may not be sufficient to restore the chloramine residual to desired levels. To restore the chloramine residual concentrations, it may be necessary to add both chlorine and ammonia. Restoring chloramine chemistry and increasing the residual chloramine concentration in the distribution system can be achieved with either manual or automated boosting systems. Components of manual and automated chloramine boosting systems are discussed further below.

Manual Chloramine Boosting System

For the purposes of this section, a manual chloramine boosting system is defined as requiring an operator to be onsite in order for boosting to occur. The operator will physically add the chlorine and possibly ammonia either through an eductor, through a tank hatch, or into an injection port. On the other hand, a boosting system is considered automated if the boosting is programmed to occur even if an operator is not present. The automated system will require chemical storage facilities and feedback loops for the boosting to occur correctly.

Manual chloramine boosting systems are typically installed at storage tanks in the distribution system and are often incorporated with a water quality monitoring station using an online chlorine analyzer and active mixer, although a simpler setup is possible. Utilities that perform proactive nitrification monitoring on storage tanks or have a water quality monitoring station connected to their SCADA system can determine which storage tanks are showing signs of nitrification (parameters listed in Chapter 6). For the identified storage tanks, the amount of chlorine needed to recombine free ammonia to form chloramines can be calculated by multiplying the free ammonia concentration as N by 4.7±0.3 to determine the chlorine dose as Cl_2. For manual chloramine boosting systems, chlorine can be added in liquid (sodium hypochlorite), gaseous, or tablet forms. Hypochlorite generators can also be used. The addition of chlorine often helps stave off nitrification and increases the overall chloramine residual within the tank.

While it is possible to simply open the tank hatch and pour in the calculated amount of sodium hypochlorite, it is not preferred. One reason is that before chlorine has had time to disperse, the packets of fluid concentrated with chlorine can cause localized breakpoint chlorination reactions to occur, leading to a loss of chlorine and ammonia. It is therefore ideal to quickly distribute and homogenize chlorine levels in the water storage tank. One option is to slowly add chlorine with chemical feed pumps into a tank. Alternatively, a venturi system can be used to suck chemicals through an eductor out of a tote or gallon jug and into the tank. A fraction of the water from the storage tank passes through a venturi where chlorine is dosed and then is returned to the storage tank. Both systems require an active tank mixer. Such approaches are safer for operators because they can avoid climbing ladders carrying chemicals.

Another approach that does not involve active mixing is adding chlorine to the tank inlet and/or outlet. The turbulence at these locations disperses chlorine to combine with ammonia and efficiently form chloramines. As with the other types of injections, this must be done slowly to avoid localized breakpoint reactions.

Due to chloramine decay and nitrification, the manual chloramine boosting may have to be repeated periodically depending on the characteristics of the distribution system (water age, temperature, pH, etc.).

While different designs and models exist on the market for a manual boosting system, a utility can construct their own mixing/manual boosting system in house. The decision to buy an existing product or make one in house will depend on cost, any potential permitting requirements, in-house staff expertise and availability, electrical availability, and tank access.

In some instances, there is not enough available free ammonia to reach the target chloramine concentration. In that situation, a manual boosting system can be used to add both chlorine and ammonia. Either liquid ammonium sulfate or ammonium hydroxide is typically used depending on cost, availability, compatibility with equipment, and safety considerations. There are safety measures to consider to ensure that the chemicals do not mix at high concentrations. For this reason, it is crucial to provide adequate mixing in between the addition of each chemical. Manual boosting can sometimes be a labor-intensive process and would not be recommended for tanks that need a chloramine boost regularly.

Overall, there are quite a few benefits of manual boosting. The relatively low cost compared to other capital improvements (an automated chloramine boosting system, piping improvements, etc.) makes it a more affordable option, especially for small systems. If certain distribution system tanks only struggle to maintain an adequate chloramine residual during parts of the year, manual chloramine boosting may be a good solution. Since chemicals are only added periodically, operators may bring chlorine and ammonia totes with them when boosting is necessary, so permanent onsite chemical storage is not required, which can make permitting and construction much easier.

One of the challenges with manual boosting is that it can require more operator hours to boost or recombine a tank and raise the monochloramine residual. The process of adding chemicals can range anywhere from a few minutes to a full day of work, depending on the starting residual as well as the size of the tank and the capacity of the manual chloramine boosting system. Along with the hours, it requires operator calculation for the amount of chlorine and ammonia to be added. Any human error on this step could lead to a less-than-ideal chlorine-to-ammonia ratio. For example, if excess free chlorine is added, it may begin to destroy the monochloramine residual rather than increase it. Therefore, it is important for the operators to go through a thorough training and recurring follow-up training as needed. During the manual chloramine boosting, it is recommended that operators collect and analyze samples for free chlorine, combined chlorine, and free ammonia to ensure proper chloramine formation. If the storage tank has online analyzers, they can be used for assisting the operators during the manual chloramine boosting process.

Automated Chloramine Boosting

There are circumstances when a full chloramine boost needs to be achieved relatively frequently or the incoming water already has a low chloramine residual. In those cases, it may be more efficient to install an automated chloramine boosting system rather than a manual system. Tanks at the end of the system, larger reservoirs with longer retention time, tanks that feed multiple other tanks, and tanks that experience frequent nitrification events are locations best suited for placement of an automated boosting system.

An automated boosting system is designed to inject either chlorine or both chlorine and ammonia to increase the overall monochloramine concentration at a set ratio to a target residual. If the tank is generally at the target disinfectant residual with excess free ammonia, then it would be a good candidate for the chlorine-only boosting system. The chlorine and ammonia chemical feed systems are like the manual boosting system but are typically located in permanent housing. While mixing is usually a recommended feature in manual boosting systems, it is generally required for automated systems. In addition to these components, an automated boosting system has a water quality monitoring station and a system controller.

There are different configurations for the water quality monitoring station depending on the manufacturer, but there is at least an online total chlorine analyzer in all models. The other probes/sensors can measure oxidation reduction potential (ORP), ammonia, pH, and temperature, depending on the make and model of the system. Water quality measurements are taken by way of a sample line and conveyed to the water quality analyzer. The water quality analyzer consists of a set of sensors, flow cells, and controllers that continuously measure water chemistry inside the tank. Measured water quality data, like chlorine residual, temperature, and pH, are communicated to the system controller. The system controller compares received data to preset thresholds and alert levels established during setup. If disinfectant adjustments are needed, the system controller issues commands to the chemical feed pumps (either chlorine, ammonia, or both) to adjust dosing to meet setpoint levels.

The operation logic for the automated chloramine boosting systems varies depending on the manufacturer/model. One example utilizes the breakpoint curve and a total chlorine analyzer. In considering the breakpoint curve, the system will stop adding chlorine if the measured total chlorine residual decreases as more chlorine is added. If the total chlorine residual is less than the target, ammonia is added along with chlorine to get to the top of the curve and to the target concentration. On the other hand, some other systems have an integrated ORP analyzer. The control system uses the ORP levels to maintain the set chlorine-to-ammonia ratio. While the online analyzers can be tied into a utility's SCADA system, it is a good practice to periodically collect grab samples to confirm total chlorine, monochloramine, and free ammonia levels. Regardless of the control logic, these systems are generally equipped with an extensive alarm management system built into the controller that will alert the user of any irregularities within the system and produce an automated response from an alert on the screen to system shutdown to ensure safe operating conditions. Additional alarms that should be considered are leak detection, intrusion, and chemical inventory alarms.

Before moving forward with an automated boosting system at a specific tank, consider chemical storage requirements. Both the physical location of chemical storage as well as any local regulations associated with chemical storage should be carefully considered. Access for chemical deliveries and chemical transfer procedures should be considered. Some common challenges with automated boosting systems are the higher capital costs associated with construction, including the increased permitting requirements due to onsite chemical storage. Access to the tank is needed to run conduit to the mixer and the tubing for the chemical and online analyzer feed. Secondary containment is often required. Site constraints can often be an issue. There are trailer models, as well as the option of installing a boosting system within an existing or new building or housing (Seidel 2015).

There are many benefits to installing an automated boosting system. If placed at a tank that feeds other tanks, one system can increase the chloramine residual and prevent nitrification in a large portion of the distribution system, including multiple tanks (Seidel et al. 2018). In addition, operators do not need to make manual calculations when using the automated boosting system because all the calculations and logic are programmed in, so there is less room for human error. Along with preventing human error, this system can be safer because the operators do not directly handle the chemicals.

Disinfection Byproducts and Nitrification

Although chloramine boosting systems have been shown to reduce or eliminate nitrification (Seidel et al. 2018), the use of additional disinfectant is a concern, as it can increase the formation of disinfection byproducts (DBPs). A recent study investigated the effect of nitrification on the formation of DBPs (Zeng and Mitch 2016). This study found that nitrification events increased the concentrations of regulated DBPs (trihalomethanes and haloacetic acids) and emerging DBPs (nitrosamines, haloketones, halo aldehydes, haloacetonitriles, and haloamides). In distribution systems without the signs of nitrification, increases in regulated or emerging DBPs were relatively low and negligible in some cases. Based on the findings of this study, DBP formation may be reduced by the prevention of nitrification events.

Table 9-2 Operations and maintenance schedule for chloramine-boosting system

Time Period	Description
Weekly and as needed	• Routine checking of the system (leaks, chemical levels, grab samples for cross-referencing, inspection of chemical feed unit, etc.) • Calibration of sensors • Restocking of chemicals
Monthly or as needed	• Calibration of feed pumps
Yearly and as needed	• Winterize/dewinterize the system (if applicable) • Chemical feed pump maintenance • Inspect/replace chemical tubing • Inspect/replace sensors • Inspect/replace water mixer

Operational Consideration for Chloramine Boosting

One of the main drawbacks associated with chlorine boosting is the substantial O&M costs associated with these systems. For example, a chloramine boosting system would have power consumption, chemical use, equipment maintenance, sample monitoring and testing, and staff labor. It is worth noting that some systems experience a reduction in labor when using an automated booster system when compared to the time spent responding to nitrification events (see Table 9-2). In some cases, automated chloramine boosting systems are installed at water storage tanks located in residential neighborhoods. Residents may have concerns about chemical deliveries to the station.

PIPING IMPROVEMENTS

Piping improvements in the distribution system can reduce nitrification. These improvements typically include relining or replacing corroded/old water mains, eliminating dead ends, and sizing/resizing the water mains. Following, these improvements are discussed in further detail.

Line or Replace Corroded/Old Water Mains

Nitrification can occur in any common distribution system pipeline material; however, deteriorated pipelines, especially unlined cast-iron lines, can create water quality problems such as rusty or red water, reduce secondary disinfectant levels, and support excessive biofilm growth, including nitrifiers (Zhang and Edwards 2010; Zhang et al. 2010; Liu et al. 2005). Therefore, deteriorated pipelines can promote nitrification, and their replacement could improve overall water quality. However, pipeline replacement or rehabilitation programs are costly and should be carefully planned if they are to be an efficient and cost-effective part of a capital improvement program (Kirmeyer et al. 2000).

Eliminate Dead Ends

Water age in a pipeline depends on the volume of the pipe and the water demand. In locations where the water demand is low and the pipe diameter is sufficiently large, water age is generally high, especially if those locations are at the extremities of the distribution system. For that reason, water age tends to be high in a dead end, so dead ends can be prone to nitrification if the main suffers from long water age. Looping the distribution system to eliminate dead ends can provide better circulation in the area and limit nitrification. Note that if sufficient demand does not exist within the loop, a loop can also become stagnant. As such, it is important to evaluate the impacts of looping using a hydraulic model before implementation.

Sizing/Resizing Water Mains

Water demand is one of the main factors that governs the sizing of water mains. Designing a piping system with a smaller pipe diameter would result in lower water age and thus can help with nitrification. However, it is important to consider fire-flow conditions as smaller pipe diameters would restrict the flow. Furthermore, smaller mains result in greater system pressure losses and increased pumping costs, possibly even requiring replacement of high-service pumps. Therefore, it is important to size pipelines properly (neither oversize nor undersize) to balance the water demands and water quality.

Flushing Systems

To create an artificial water demand and reduce the water age in select locations of the distribution system, automated flush devices can be installed. Automated flush devices have the most beneficial impact where water age is highest, usually at dead ends of the distribution system. Figure 9-8 provides an example of an automated flushing device, which can be operated manually or automatically. Automated flush devices can be set for timed flushes or to flush when the chlorine residual is low. There are also automated flush valves that can be operated remotely using cloud-based software.

Flushing devices typically discharge water within the range of 10–30 gal per min (gpm) for limited time frames. This discharge rate can correspond to a considerable quantity of potable water discharged to either a storm drain or sanitary sewer. Therefore, automated flush devices are typically installed in areas with no drought concerns. It is important to identify the local discharge requirements before installation. In some areas, it is necessary to dechlorinate discharge water. Automated flush devices that are equipped with a dechlorination chamber are available.

It is likewise important to note that flushing devices are different from unidirectional flushing as discussed in Chapters 7 and 8. Typical flow rates of flushing valves (10–30 gpm) are usually not high enough to create scouring in the pipelines and are mainly used to reduce water age.

Flushing devices do not pose any unintended consequence with regards to nitrification or the water quality. However, other implications must be evaluated. One of the main considerations is the fact that treated potable water is being discharged. Depending on the number of flushing devices, duration of flushing, and flushing flow rate, there can be a significant quantity of water discharged. There are three main considerations associated with this discharge. Potable water that is not consumed by the customers is lost revenue for the utility. Secondly, this area was not a good location for water discharge. Furthermore, there can be concerns with the public perception; especially in locations where water resources are limited, flushing of potable water may not be received well.

To overcome the problems associated with discharged water generated during flushing, new flushing strategies are being developed. There are flushing trucks for distribution systems that are designed to minimize water waste. The truck connects in between two hydrants within a close vicinity. Water is taken into the truck from one hydrant, pushed through a series of filters on the truck, and then pushed back into the system through the second hydrant. Chlorine can also be injected to combine with free ammonia to increase the chloramine residual, or more chlorine can be added to achieve breakpoint chlorination and produce a free chlorine residual. Loops can be isolated by closing valves within the system to focus the scouring flow on specific pipes. This circulation is used to create scouring velocities that clean the deposits on the pipes, which are filtered through cartridge filters present on the truck. Scouring the pipes reduces growth of nitrifying bacteria within the distribution system and helps maintain a chloramine residual.

Source: Mueller

Figure 9-8 A hydrant flushing unit

OTHER TECHNOLOGIES

Based on ongoing research, new products are being developed that can be used as a strategy to reduce nitrification. One of these strategies is use of the disinfection properties of ultraviolet (UV) light. A UV disinfection system can be used to inactivate nitrifying bacteria (White and Adams 2010). These systems are typically installed in water storage tanks. The use of UV or near-UV disinfection at the storage tank may inhibit nitrification. However, the benefits of inhibition of nitrification at the water storage tank in the distribution system are not well understood. It should be noted that further research is necessary to evaluate the efficiency of UV or near-UV disinfection for controlling nitrification.

COMBINATION OF CAPITAL IMPROVEMENTS

Each of the engineering changes and capital improvements listed in this section can provide some benefits with nitrification. However, depending on the extent of nitrification, a single improvement by itself may not be sufficient. It may be necessary to implement several of the options or use a combination of approaches to achieve target goals with respect to nitrification. For example, pressure zone circulation can be implemented together with eliminating pipeline dead ends to reduce the water age.

OPERATIONS AND MAINTENANCE COSTS

Other than a few engineering changes and capital improvements (e.g., piping improvements), all the other modification alternatives are likely to require some O&M. Therefore, O&M costs are likely to be associated with each modification. Some alternatives can have O&M costs as simple as power consumption: for example, costs associated with active tank mixing or pressure zone circulation. Some of the other alternatives can require more extensive O&M and thus have a higher O&M cost. For example, a chloramine-boosting system would not only have power consumption but also chemical use, equipment maintenance, sample monitoring and testing, etc. While most of the alternatives have some O&M costs, these alternatives also have the potential to reduce the O&M cost associated with nitrification events.

KEY TAKEAWAYS

Engineering changes and capital improvements should have a goal of reduction of water age and/or improving chloramine residual. Engineering changes and capital improvements have been shown to resolve long-term or recurring nitrification episodes. These modifications may require significant upfront investment and can potentially increase the O&M costs. For that reason, operational changes should be considered before making any engineering changes or capital improvements. Although some cost is associated with engineering changes and capital improvements, there are also savings of time for the operations staff. However, engineering changes and capital improvements can have unintended consequences. Any improvement should be evaluated thoroughly before implementation. Frequent and continued monitoring should always be done for nitrification before and after the implementation of improvement(s).

REFERENCES

American Water Works Association (AWWA). 2021. Manual M64, *Gas Transfer Applications in Water: Addition and Removal*. Denver: AWWA.

Ghosh, A., C. Seidel, E. Townsend, R. Pacheco, and C. Corwin. 2015. *Reducing Volatile Disinfection By-Products in Treated Drinking Water Using Aeration Technologies (No. 4441)*. Denver: WRF.

Kirmeyer, G.J., M. Friedman, J. Clement, A. Sandvig, P.F. Noran, K. Martel, D. Smith, M. LeChevallier, C. Volk, E. Antoun, D. Hiltebrand, D., J. Dyksen, and R. Cushing. 2000. *Guidance Manual for Maintaining Distribution System Water Quality*. Denver: AWWA Research Foundation.

Liu, S., J.S. Taylor, and D. Webb. 2005. "Water Quality Profiles During Nitrification in a Pilot Distribution System Study." *Journal of Water Supply: Research and Technology-Aqua*, 54:3:133–145.

Seidel, A. 2015. "Utility Finds Way to Manage a Tank's Chloramine Residual." *Opflow*, 41:12:16–18.

Seidel, A., M. Miller, and P. Jorgensen. 2018. "Fine-tune Chloramine Disinfectant Residual Levels in Distribution Networks." *Opflow*, 44:12:22-25.

White, B., and M. Adams. 2010. "Battling Nitrification With Blacklights." *Opflow*, 36:2:10–13.

Zeng, T., and W. A. Mitch. 2016. "Impact of Nitrification on the Formation of N-Nitrosamines and Halogenated Disinfection Byproducts within Distribution System Storage Facilities." *Environmental Science & Technology*, 50:6:2964–2973.

Zhang, Y., and M. Edwards. 2010. "Nutrients and Metals Effects on Nitrification in Drinking Water Systems." *Journal AWWA*, 102:7:56–66.

Zhang, Y., A. Griffin, and M. Edwards. 2010. "Effect of Nitrification on Corrosion of Galvanized Iron, Copper, and Concrete." *Journal AWWA*, 102:4:83–93.

This page intentionally blank.

Appendix **A**

Cell Culture Media and Molecular Methods for the Enumeration of Nitrifying Microorganisms

(Connects to Chapter 5, "Microbiology of Nitrifying Microorganisms")

Table A-1 Media composition for enrichment and enumeration of ammonia-oxidizing bacteria per liter of solution

Chemicals	Soriano and Walker (1968)	Watson (1971)	Watson et al. (1971)	Matulewich et al. (1975)	Lipponen et al. (2002)	Oldenburg et al. (2002)	Zhang et al. (2008)
$(NH_4)_2SO_4$	500 mg	130 mg	2,000 mg	500 mg	330 mg	1,700 mg	500 mg
$MgSO_4 \bullet 7H_2O$	40 mg	200 mg	200 mg	50 mg	40 mg	200 mg	40 mg
KH_2PO_4	200 mg				100 mg	1,010 mg	200 mg
K_2HPO_4		87 mg	15.9 mg	500 mg		20,100 mg	
$CaCl_2 \bullet 2H_2O$	40 mg	20 mg	20 mg	20 mg	15 mg	20 mg	40 mg

(continued)

Table A-1 Media composition for enrichment and enumeration of ammonia-oxidizing bacteria per liter of solution (*Continued*)

Chemicals	Soriano and Walker (1968)	Watson (1971)	Watson et al. (1971)	Matulewich et al. (1975)	Lipponen et al. (2002)	Oldenburg et al. (2002)	Zhang et al. (2008)
Chelated Fe (sequestrene 138 Fe)	0.1 mg						0.1 mg
Chelated Fe (13% Geigy Chemical)		1 mg	1 mg				
FeEDTA solution						1 mL	
Na_2EDTA					4,292 μg		
$Na_2MoO_4 \bullet 2H_2O$		100 μg	100 μg	2.4 μg	24 μg	100 μg	100 μg
$MnCl_2 \bullet 4H_2O$		200 μg	200 μg		99 μg	200 μg	
$MnSO_4 \bullet 2H_2O$							172 μg
$CoCl_2 \bullet 6H_2O$		2 μg	2 μg		24 μg	2 μg	4 μg
$CuSO_4 \bullet 5H_2O$		20 μg	20 μg			20 μg	
$CuCl_2 \bullet 2H_2O$					17 μg		13.3 μg
$ZnSO_4 \bullet 7H_2O$		100 μg	100 μg			100 μg	100 μg
$FeCl_2 \bullet 4H_2O$					1,988 μg		
$NiCl_2 \bullet 6H_2O$					24 μg		
$ZnCl_2$					68 μg		
H_3BO_3					62 μg		
Trace metal mix No.44				1 mL			
0.5% phenol red	1 drop					1 drop	
NaCl				500 mg			
$KHCO_3$				20 mg			
$NaHCO_3$							84 mg

Table A-2 Medium composition for enrichment and enumeration of nitrite-oxidizing bacteria

Chemicals	Lipponen et al. (2002)
$NaNO_2$	34.5 mg
$MgSO_4 \bullet 7H_2O$	40 mg
KH_2PO_4	100 mg
$CaCl_2 \bullet 2H_2O$	15 mg
Na_2EDTA	4,292 μg
$Na_2MoO_4 \bullet 2H_2O$	24 μg
$MnCl_2 \bullet 4H_2O$	99 μg
$CoCl_2 \bullet 6H_2O$	24 μg
$CuCl_2 \bullet 2H_2O$	17 μg
$ZnCl_2$	68 μg
$FeCl_2 \bullet 4H_2O$	1,988 μg
$NiCl_2 \bullet 6H_2O$	24 μg
H_3BO_3	62 μg
Distilled water	1,000 mL

Table A-3 Medium composition for enrichment and cultivation of complete ammonia-oxidizers

	Chemicals	Per Liter
	NH_4Cl	2.7 mg
Modified AOM medium:	KH_2PO_4	50 mg
	KCl	75 mg
	$MgSO_4{\bullet}7H_2O$	50 mg
	NaCl	584 mg
	$CaCO_3$	4 g*
Trace element solution†:	$MnSO_4{\bullet}1H_2O$	34.4 mg
	H_3BO_3	50 mg
	$ZnCl_2$	70 mg
	$Na_2MoO_4{\bullet}2H_2O$	72.6 mg
	$CuCl_2{\bullet}6H_2O$	20 mg
	$NiCl_2{\bullet}6H_2O$	24 mg
	$CoCl_2{\bullet}6H_2O$	80 mg
	$FeSO_4{\bullet}7H_2O$	1 g‡
Selenium-wolfram solution†:	NaOH	0.5 g
	$Na_2SeO_3{\bullet}5H_2O$	3 mg
	$Na_2WO_4{\bullet}2H_2O$	4 mg

* Serves as a solid buffering system and growth surface by remaining mostly undissolved
† Prepared as a separate solution with 1 mL added to the NH_4Cl containing modified AOM medium
‡ Except for $FeSO_4{\cdot}7H_2O$, all salts for the trace element solution were dissolved in 997.5 mL Milli-Q water with 2.5 mL of 37% HCl added before addition of $FeSO_4{\cdot}7H_2O$.
AOM = ammonia-oxidizing microorganisms

Table A-4 Examples of polymerase chain reaction primers developed to identify nitrifying microorganisms

Organism	Gene Target	Primer Name	Reference
Ammonia-oxidizing archaea	*amoA*	Arch-amoAF and Arch-amoAR	Francis et al. (2005)
	amoA	GenAOAF and GenAOAR	Meinhardt et al. (2015)
AOB of the *Betaproteobacteria*	16S rRNA	NIT-A and NIT-B	Voytek and Ward (1995)
AOB of the *Betaproteobacteria* (V-2 and V-3 variable domains)	16S rRNA	CTO189f-GC and CTO654r	Kowalchuk et al. (1997)
AOB	*amoA*	amoA-1F and amoA-2R	Rotthauwe et al. (1997)
AOB	*amoA*	amoA-3F and amoA-4R	McTavish et al. (1993) and Alzerreca et al. (1999)
AOB of the *Betaproteobacteria*	16S rRNA	AOB620 and AOB770rc	Regan et al. (2006)
AOB of the *Betaproteobacteria*	16S rRNA	AOB1150 and Nso1225	Regan et al. (2006)
AOB of the *Betaproteobacteria*	16S rRNA	βAMOf and βAMOr	McCaig et al. (1994)
AOB	*amoA*	GenAOBF and GenAOBR	Meinhardt et al. (2015)
Nitrosomonas oligotropha-like	*amoA*	amoNo550D2f, amoNo754r, amoNoTaq729	Harms et al. (2003)
Nitrobacter spp.	16S to 23S rRNA gene intergenic spacer region	NITISRf, NITISRr, NITISRp	Hawkins et al. (2008)

(continued)

Table A-4 Examples of polymerase chain reaction primers developed to identify nitrifying microorganisms (*Continued*)

Organism	Gene Target	Primer Name	Reference
Nitrobacter spp.	16S rRNA	Nitro-1198F, Nitro-1423R, Nitro-1374Taq	Graham et al. (2007)
Nitrospira spp.	16S rRNA	Eub338f and Ntspa0685	Hovanec et al. (1998)
Nitrospira inopinata	*amoA*	Nino_amoA19F and Nino_amoA 252R	Daims et al. (2015)
	nxr		
Clade A comammox	*amoA*	COMAF and COMAR	Yu et al. (2018)
	amoA	comaA-244F and comaA-659R	Pjevac et al. (2017)
Clade B comammox	*amoA*	comaB-244F and comaB-659R	Pjevac et al. (2017)

AOB = ammonia-oxidizing bacteria; comammox = complete ammonia-oxidizers

Table A-5 Probes for hybridization of nitrifying bacteria

Organisms Targeted	Probe Name	Reference
Ammonia-oxidizing bacteria (AOB):		
AOB of the *Betaproteobacteria*	β-AO233	Stephen et al. (1998)
AOB of the *Betaproteobacteria*	NIT-C	Voytek and Ward (1995)
AOB of the *Betaproteobacteria*	AOB770	Regan et al. (2006)
AOB of the *Betaproteobacteria*	Nso190	Mobarry et al. (1996)
AOB of the *Betaproteobacteria*, except *Nitrosococcus* mobilis	Nso1225	Mobarry et al. (1996)
Nitrosomonas Cluster	Nsm156	Mobarry et al. (1996)
Nitrosospira Cluster	Nsv443	Mobarry et al. (1996)
Nistrosococcus mobilis lineage	NmV	Pommerening-Röser et al. (1996)
Most halophilic and halotolerant AOB (*Nitrosomonas europaea, Nitrosomonas eutropha, Nitrosomonas cryotolerans, Nitrosomonas aestuarii, Nitrosomonas marina, Nitrosomonas halophila,* and *Nitrosococcus mobilis*)	NEU	Wagner et al. (1995)
Nitrosomonas, except the marine Nitrosomonas	amoA-Nm3	Regan et al. (2006)
Nitrosomonas oligotropha	amoA-Nm4	Regan et al. (2006)
Nitrosospira	amoA-Ns1	Regan et al. (2006)
All *Nitrosomonas*	Nmo254 and Nmo254a	Stephen et al. (1998)
Nitrosomonas Cluster 6a	NmoCL6a_205	Stephen et al. (1998)
Nitrosomonas Cluster 6b	NmoCL6b_376	Stephen et al. (1998)
Nitrosomonas Cluster 7	NmoCL7_439	Stephen et al. (1998)
All *Nitrosospira*	Nsp436	Stephen et al. (1998)
Nitrosospira Cluster 1	NspCL1_249	Stephen et al. (1998)
Nitrosospira Cluster 2	NspCL2_458	Stephen et al. (1998)
Nitrosospira Cluster 3	NspCL3_454	Stephen et al. (1998)
Nitrosospira Cluster 4	NspCL4_446	Stephen et al. (1998)

(continued)

Table A-5 Probes for hybridization of nitrifying bacteria (*Continued*)

Organisms Targeted	Probe Name	Reference
Nitrite-oxidizing bacteria:		
Nitrobacter spp.	Nb1000	Mobarry et al. (1996)
Nitrobacter	NIT2 and NIT3	Wagner et al. (1996)
Nitrobacter spp.	NITISRp	Hawkins et al. (2008)
Nitrospira	Ntspa0685 M	Hovanec et al. (1998)
Freshwater *Nitrospira* spp.	Nsr826	Schramm et al. (1998)
Freshwater *Nitrospira* spp.	Nsr1156	Schramm et al. (1998)
Nitrospira moscoviensis	Ntspa1026	Kane et al. (1993)

NOTE: Additional probes for *Nitrosomonas europaea* and a number of *Nitrosospira* have also been proposed by Okano and colleagues (2004).

This page intentionally blank.

Appendix B

Example of Nitrification Action/Control Plan

Many water systems have developed a nitrification action/control plan (NAP/NCP) to help them prevent and/or address nitrification episodes. NAP/NCPs are system specific. They can be detailed written documents, or they can be limited to a simple table that explains the monitoring parameters, sampling frequencies, baseline water quality, trigger levels for corrective actions, and follow-up actions. An NAP/NCP is a living document and should be updated as needed based on changes to water quality, system infrastructure, or operations.

NAP/NCPs are widely available in the literature, as well as from individual organizations. For example, the Texas Commission on Environmental Quality offers fact sheets, guidance documents, and templates that can be found at www.tceq.texas.gov/drinkingwater/disinfection/nitrification.html. The following presents an example of a simple NAP/NCP. The sampling frequencies, baseline and trigger levels, and response actions shown are not a prescription for all utilities, but rather, they are intended to demonstrate the minimum level of detail to be included in an NAP/NCP.

Table B-1 Examples of monitoring parameters and sampling frequency

Sampling Site	Distribution System Entry Point(s)	Distribution System Sites	Storage Tanks
Total chlorine or monochloramine	Daily	Weekly	Weekly
Free ammonia	Daily	Weekly	Weekly
Nitrite	Monthly	Weekly	Weekly
Water temperature	Weekly	Weekly	Weekly
pH	Weekly	Weekly	Weekly

Table B-2 Examples of baseline and trigger levels

Parameter	Unit	Baseline Levels	Trigger Levels
Total chlorine or monochloramine	mg/L Cl_2	1.5–2.5	<1.5
Free ammonia	mg/L N	ND–0.05	>0.1
Nitrite	mg/L N	ND–0.01	>0.01
Water temperature	°C (°F)	20–25 (68–75)	N/A
pH		7.5–8.5	N/A

N/A: not applicable; ND: nondetectable

Table B-3 Examples of response actions

Responses	Distribution System Entry Point(s)	Distribution System Sites	Storage Tanks
Short-term actions	If the trigger for free ammonia is reached, • check chlorine-to-ammonia-N ratio and • adjust to 4.5 to 1, if needed. If the trigger for total chlorine or monochloramine is reached, • increase total chlorine and ammonia concentrations leaving treatment facilities and • maintain chlorine-to-ammonia-N ratio of 4.5:1.	If total chlorine, monochloramine, or nitrite trigger is reached, • flush as needed and • monitor chlorine and nitrite in the storage tank(s) that supply the area.	If nitrite trigger is reached, • cycle the tank; • measure total chlorine or monochloramine, nitrite, and free ammonia one day after tank turnover; no further action is needed if values are back within baseline levels; and • breakpoint-chlorinate the tank if trigger levels are still exceeded.
Long-term actions		Conduct unidirectional flushing.	If nitrite trigger level is routinely exceeded in the same tank, • assess whether the tank is stratified or if dead zones exist due to the inlet/outlet configuration and • conduct weekly monitoring of nitrite on water entering the tank. If water entering the tank is not nitrified, • consider installing tank mixing equipment and/or booster chloramination station and • breakpoint-chlorinate if trigger levels are still exceeded. Clean each tank every three to five years.

1 PORTLAND WATER BUREAU NITRIFICATION MONITORING AND ACTION PLAN

Portland Water Bureau (PWB) had historically experienced nitrification in their distribution system. PWB developed and implemented a nitrification action plan to identify early indicators of nitrification, which would then lead to adoption of mitigation measures. These approaches helped them successfully manage nitrification in their system, and as such, this example is included here for illustration purposes.

Sites were chosen based on known problem areas, historic data, and recommendations from stakeholders.

1.1 Distribution System Sampling Plan

Plan development. The procedure for selecting the distribution system sites was as follows:

- storage tank sites were selected based on previous year's nitrification data, TCR data, tank monitoring data, review of any other nonregulatory data collected in the system, and stakeholder feedback;
- wholesaler connection sites were chosen based on location in the system and stakeholder feedback; and
- TCR monitoring sites were chosen based on previous year's nitrification data, TCR data, and any other nonregulatory data collected in the system.

Sampling locations and frequency. Samples will be collected at the following locations:

- Total Coliform Rule (TCR) Sites: 28 select sites will be monitored every two weeks. Samples will be collected as part of the routine TCR when possible.
- Tanks: 27 tank samples will be collected weekly as part of a special nitrification route.
- Wholesaler connections. All wholesale connection of the distribution system will be monitored every two weeks.

Water quality parameters to be analyzed. Table B-1 lists the parameters to be monitored. The Water Bureau Laboratory will be conducting all the laboratory analysis.

Timeline. The tentative start date for monitoring is early summer and will extend into the late fall as water quality parameters dictate.

Table B-1 Parameters to be analyzed

Parameter	Field	Laboratory
Total chlorine	x	
pH	x	
Temperature	x	
Oxidation-reduction potential	x	
Turbidity	x	
Free ammonia	x (select samples)	x
Nitrate		x
Nitrite		x
Heterotrophic plate count-R2A		x

1.2 Information Dissemination

The ability for decision makers to have the data in a timely manner is key. Exceptions-like reports will be autogenerated to a very small, select group of stakeholders. This will allow key personnel access to the data as they become available. Short meetings will also be held regularly to discuss these results. For wholesaler customers, dissemination of information in biweekly reports is possible.

2 ACTION PLAN

2.1 Preventing and Mitigating Nitrification Episodes in the System

There are certain water quality parameters that impact nitrification over which we have little control, such as temperature and total organic carbon; however, other parameters, such as water age, chlorine residual, and free ammonia, can be affected by changes in operations. Particularly effective measures include taking storage out of service seasonally in problematic areas, increased flushing (both spot and unidirectional), deep cycling tanks/decreasing tank levels (where possible), modified pumping/regulator operations, and increased tank cleaning (relative to their schedule).

Nitrification triggers and response levels. Draft trigger and response levels are listed in Table B-2. Also, action level responses in some cases will be a function of multiple parameters being exceeded.

Action level 1: Evaluate increased sampling
Ensure optimization at chlorine and ammonia injection points (both at the plant and booster chlorination sites)
Evaluate pumping operations to see if operations could be altered/synchronized up the cascade to bring fresher water to the area
Evaluate whether tanks can be cycled more effectively
Perform flushing in the area

Action level 2: Continue performing action level 1 responses
Drain and refill tanks with fresh water (disinfect if necessary)
Clean tank

Table B-2 Nitrification triggers and potential response levels

Parameter	Target	Action Level 1	Action Level 2	Action Level 3
Total chlorine (mg/L)	>1.0 (tanks) >0.5 (pipes)	<1.0 (tanks) <0.5 (pipes)	<0.5 (tanks) <0.5 (pipes)	<0.1 (DS)
Nitrite-N (mg/L)	<0.010 or background level	Increase relative to background	>0.3	
Nitrate-N (mg/L)	Background (Based on 10 years of data, the average nitrate entering the system is 0.025 mg/L.)	Increase relative to background		
Heterotrophic plate count-R2A (cfu/mL)	<500	>500 or a significant increase from the previous sampling date	>1000	
Free ammonia-N (mg/L)	<0.05 (pipes) 0.05–0.15 (tanks)	>0.10 (DS) >0.2 (tanks)	>0.35	
pH	As close to 8 as possible	Less than target		

DS = distribution system

Evaluate whether additional storage can be taken offline (may not be possible based on hydraulics)
If at a wholesaler connection, evaluate recent wholesaler consumption data
Notify affected wholesalers

Action level 3: Perform unidirectional flushing of the area (and in the process look for erroneously closed valves)
Evaluate breakpoint chlorination

Proactive measures. Proactive measures to decrease nitrification and water age in the system will be sent out in a separate document. These should be completed in early June before the onset of nitrification season.

This page intentionally blank.

Appendix C

Example of Storage Tank Monitoring Plan

(Connected to Chapter 6, "Monitoring for Nitrification Prevention and Control")

As described in this appendix, two types of sampling can be considered to assess the extent of nitrification and/or stratification in a storage facility: operational sampling and depth-dependent sampling. Ideally, an operational sampling would first be conducted during a fill cycle, a depth-dependent sampling would be performed when the tank is full, and another operational sampling would be conducted during a drain cycle. This sampling strategy can be performed before implementing tank improvements to assess the severity of the problem(s) and after improvements to evaluate the benefits of work done. Based on the goals of the assessment, sampling can be conducted when booster chlorination stations (if present) are in operation, when they are offline, or both.

Operational sampling in a storage tank. Operational sampling is performed by collecting water entering a tank during a fill cycle and water exiting the tank during a drain cycle. This sampling can determine the extent of nitrification in a tank. Operational sampling involves the following.

- Repeat sampling during a tank fill cycle, starting shortly after the tank starts filling (e.g., 15 min after a cycle begins), sampling throughout the fill cycle, and doing the last sampling near the end of the fill cycle (e.g., 15 min before filling ends).
- Repeat sampling during a tank drain cycle, starting shortly after the tank starts draining (e.g., 15 min after draining begins), sampling throughout the drain cycle, and doing the last sampling near the end of the drain cycle (e.g., 15 min before draining ends).

Parameters to measure should include water temperature, chlorine residual (total chlorine or monochloramine), pH, free ammonia, and nitrite. Additional parameters can be

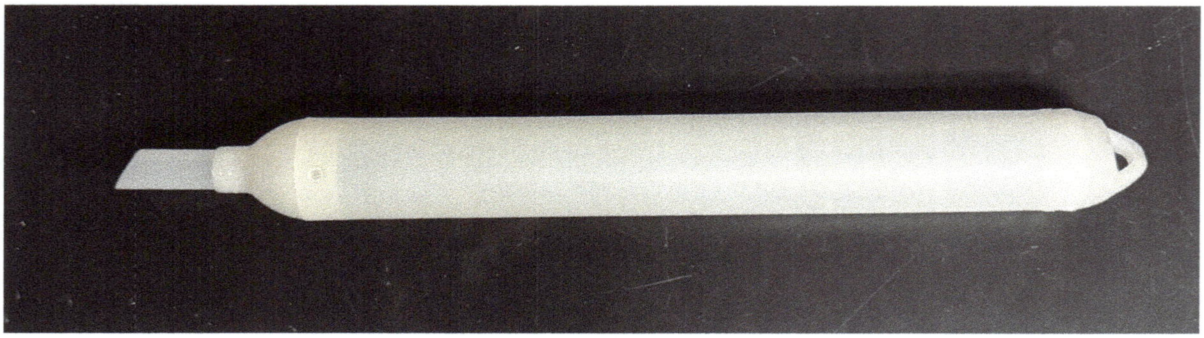

Figure C-1 Disposable polyethylene bailer used for depth-dependent sampling in a storage tank

included based on system conditions and water-quality challenges. The sampling date(s), times when the fill cycle begins and ends, sampling times, water-quality parameters measured, and other storage tank conditions (e.g., whether a booster chlorination station is present and if it is used during sampling) should be recorded.

Depth-dependent sampling in a storage tank. Depth-dependent sampling can assess whether tank stratification and water quality variations occur throughout a tank because of poor mixing or infrequent cycling. Depth-dependent sampling can be conducted by lowering a sampling device (shown in Figure C-1) to various depths through an access hatch. To account for water level variation in a storage tank, samples should be collected near the water surface (e.g., 1 ft below the water level), midwater level, and close to the bottom of the tank. To assess worst-case conditions, depth-dependent sampling should be performed before a tank starts draining, i.e., when water inside the tank is oldest. Comparison of depth-specific samples with those collected from the tank outlet will indicate whether stratification exists within the tank.

Parameters to measure at each depth should include water temperature, chlorine residual (total chlorine or monochloramine), pH, free ammonia, and nitrite. Additional parameters (e.g., disinfection byproducts, turbidity, total organic carbon, ultraviolet absorbance) can be included based on system conditions and water quality challenges beyond nitrification. The sampling date, time, whether the tank is filling or draining, water level in the tank, sample depths, water quality parameters measured, and other storage tank conditions (e.g., whether a booster chlorination station is present and if it is used during sampling) should be recorded.

Appendix D

Units of Measure with Metric Conversions

Linear Measurement

inch (in.)	×	25.4	=	millimeters (mm)
inch (in.)	×	2.54	=	centimeters (cm)
foot (ft)	×	304.8	=	millimeters (mm)
foot (ft)	×	30.48	=	centimeters (cm)
foot (ft)	×	0.3048	=	meters (m)
yard (yd)	×	0.9144	=	meters (m)
mile (mi)	×	1,609.3	=	meters (m)
mile (mi)	×	1.6093	=	kilometers (km)
millimeter (mm)	×	0.03937	=	inches (in.)
centimeter (cm)	×	0.3937	=	inches (in.)
meter (m)	×	39.3701	=	inches (in.)
meter (m)	×	3.2808	=	feet (ft)
meter (m)	×	1.0936	=	yards (yd)
kilometer (km)	×	0.6214	=	miles (mi)

Area Measurement

square meter (m^2)	×	10,000	=	square centimeters (cm^2)
hectare (ha)	×	10,000	=	square meters (m^2)
square inch ($in.^2$)	×	6.4516	=	square centimeters (cm^2)
square foot (ft^2)	×	0.092903	=	square meters (m^2)
square yard (yd^2)	×	0.8361	=	square meters (m^2)

acre	×	0.004047	=	square kilometers (km^2)
acre	×	0.4047	=	hectares (ha)
square mile (mi^2)	×	2.59	=	square kilometers (km^2)
square centimeter (cm^2)	×	0.16	=	square inches ($in.^2$)
square meters (m^2)	×	10.7639	=	square feet (ft^2)
square meters (m^2)	×	1.1960	=	square yards (yd^2)
hectare (ha)	×	2.471	=	acres
square kilometer (km^2)	×	247.1054	=	acres
square kilometer (km^2)	×	0.3861	=	square miles (mi^2)
Volume Measurement				
cubic inch ($in.^3$)	×	16.3871	=	cubic centimeters (cm^3)
cubic foot (ft^3)	×	28,317	=	cubic centimeters (cm^3)
cubic foot (ft^3)	×	0.028317	=	cubic meters (m^3)
cubic foot (ft^3)	×	28.317	=	liters (L)
cubic yard (yd^3)	×	0.7646	=	cubic meters (m^3)
acre foot (acre-ft)	×	123.34	=	cubic meters (m^3)
ounce (US fluid) (oz)	×	0.029573	=	liters (L)
quart (liquid) (qt)	×	946.9	=	milliliters (mL)
quart (liquid) (qt)	×	0.9463	=	liters (L)
gallon (gal)	×	3.7854	=	liters (L)
gallon (gal)	×	0.0037854	=	cubic meters (m^3)
peck (pk)	×	0.881	=	decaliters (dL)
bushel (bu)	×	0.3524	=	hectoliters (hL)
cubic centimeters (cm^3)	×	0.061	=	cubic inches ($in.^3$)
cubic meter (m^3)	×	35.3183	=	cubic feet (ft^3)
cubic meter (m^3)	×	1.3079	=	cubic yards (yd^3)
cubic meter (m^3)	×	264.2	=	gallons (gal)
cubic meter (m^3)	×	0.000811	=	acre-feet (acre-ft)
liter (L)	×	1.0567	=	quart (liquid) (qt)
liter (L)	×	0.264	=	gallons (gal)
liter (L)	×	0.0353	=	cubic feet (ft^3)
decaliter (dL)	×	2.6417	=	gallons (gal)
decaliter (dL)	×	1.135	=	pecks (pk)
hectoliter (hL)	×	3.531	=	cubic feet (ft^3)
hectoliter (hL)	×	2.84	=	bushels (bu)
hectoliter (hL)	×	0.131	=	cubic yards (yd^3)
hectoliter (hL)	×	26.42	=	gallons (gal)
Pressure Measurement				
pound/square inch (psi)	×	6.8948	=	kilopascals (kPa)
pound/square inch (psi)	×	0.00689	=	pascals (Pa)
pound/square inch (psi)	×	0.070307	=	kilograms/square centimeter (kg/cm^2)
pound/square foot (lb/ft^2)	×	47.8803	=	pascals (Pa)
pound/square foot (lb/ft^2)	×	0.000488	=	kilograms/square centimeter (kg/cm^2)

pound/square foot (lb/ft^2)	×	4.8824	=	kilograms/square meter (kg/m^2)
inches of mercury	×	3,376.8	=	pascals (Pa)
inches of water	×	248.84	=	pascals (Pa)
bar	×	100,000	=	newtons per square meter
pascals (Pa)	×	1	=	newtons per square meter
pascals (Pa)	×	0.000145	=	pounds/square inch (psi)
kilopascals (kPa)	×	0.145	=	pounds/square inch (psi)
pascals (Pa)	×	0.000296	=	inches of mercury (at 60°F)
kilogram/square centimeter (kg/cm^2)	×	14.22	=	pounds/square inch (psi)
kilogram/square centimeter (kg/cm^2)	×	28.959	=	inches of mercury (at 60°F)
kilogram/square meter (kg/m^2)	×	0.2048	=	pounds per square foot (lb/ft^2)
centimeters of mercury	×	0.4461	=	feet of water

Weight Measurement

ounce (oz)	×	28.3495	=	grams (g)
pound (lb)	×	0.045359	=	grams (g)
pound (lb)	×	0.4536	=	kilograms (kg)
ton (short)	×	0.9072	=	megagrams (metric ton)
pounds/cubic foot (lb/ft^3)	×	16.02	=	grams per liter (g/L)
pounds/million gallons (lb/mil gal)	×	0.1198	=	grams per cubic meter (g/m^3)
gram (g)	×	15.4324	=	grains (gr)
gram (g)	×	0.0353	=	ounces (oz)
gram (g)	×	0.0022	=	pounds (lb)
kilograms (kg)	×	2.2046	=	pounds (lb)
kilograms (kg)	×	0.0011	=	tons (short)
megagram (metric ton)	×	1.1023	=	tons (short)
grams/liter (g/L)	×	0.0624	=	pounds per cubic foot (lb/ft^3)
grams/cubic meter (g/m^3)	×	8.3454	=	pounds/million gallons (lb/mil gal)

Flow Rates

gallons/second (gps)	×	3.785	=	liters per second (L/s)
gallons/minute (gpm)	×	0.00006308	=	cubic meters per second (m^3/s)
gallons/minute (gpm)	×	0.06308	=	liters per second (L/s)
gallons/hour (gph)	×	0.003785	=	cubic meters per hour (m^3/h)
gallons/day (gpd)	×	0.000003785	=	million liters per day (ML/day)
gallons/day (gpd)	×	0.003785	=	cubic meters per day (m^3/day)
cubic feet/second (ft^3/s)	×	0.028317	=	cubic meters per second (m^3/s)
cubic feet/second (ft^3/s)	×	1,699	=	liters per minute (L/min)
cubic feet/minute (ft^3/min)	×	472	=	cubic centimeters/second (cm^3/s)
cubic feet/minute (ft^3/min)	×	0.472	=	liters per second (L/s)
cubic feet/minute (ft^3/min)	×	1.6990	=	cubic meters per hour (m^3/h)
million gallons/day (mgd)	×	43.8126	=	liters per second (L/s)
million gallons/day (mgd)	×	0.003785	=	cubic meters per day (m^3/day)
million gallons/day (mgd)	×	0.043813	=	cubic meters per second (m^3/s)
gallons/square foot (gal/ft^2)	×	40.74	=	liters per square meter (L/m^2)

gallons/acre/day (gal/acre/day)	×	0.0094	=	cubic meters/hectare/day (m^3/ha/day)
gallons/square foot/day (gal/ft^2/day)	×	0.0407	=	cubic meters/square meter/day (m^3/m^2/day)
gallons/square foot/day (gal/ft^2/day)	×	0.0283	=	liters/square meter/day (L/m^2/day)
gallons/square foot/minute (gal/ft^2/min)	×	2.444	=	cubic meters/square meter/hour (m^3/m^2/h) = m/h
gallons/square foot/minute (gal/ft^2/min)	×	0.679	=	liters/square meter/second (L/m^2/s)
gallons/square foot/minute (gal/ft^2/min)	×	40.7458	=	liters/square meter/minute (L/m^2/min)
gallons/capita/day (gpcd)	×	3.785	=	liters/day/capita (L/d per capita)
liters/second (L/s)	×	22,824.5	=	gallons per day (gpd)
liters/second (L/s)	×	0.0228	=	million gallons per day (mgd)
liters/second (L/s)	×	15.8508	=	gallons per minute (gpm)
liters/second (L/s)	×	2.119	=	cubic feet per minute (ft^3/min)
liters/minute (L/min)	×	0.0005886	=	cubic feet per second (ft^3/s)
cubic centimeters/second (cm^3/s)	×	0.0021	=	cubic feet per minute (ft^3/min)
cubic meters/second (m^3/s)	×	35.3147	=	cubic feet per second (ft^3/s)
cubic meters/second (m^3/s)	×	22.8245	=	million gallons per day (mgd)
cubic meters/second (m^3/s)	×	15,850.3	=	gallons per minute (gpm)
cubic meters/hour (m^3/h)	×	0.5886	=	cubic feet per minute (ft^3/min)
cubic meters/hour (m^3/h)	×	4.403	=	gallons per minute (gpm)
cubic meters/day (m^3/day)	×	264.1720	=	gallons per day (gpd)
cubic meters/day (m^3/day)	×	0.00026417	=	million gallons per day (mgd)
cubic meters/hectare/day (m^3/ha/day)	×	106.9064	=	gallons per acre per day (gal/acre/day)
cubic meters/square meter/day (m^3/m^2/day)	×	24.5424	=	gallons/square foot/day (gal/ft^2/day)
liters/square meter/minute (L/m^2/min)	×	0.0245	=	gallons/square foot/minute (gal/ft^2/min)
liters/square meter/minute (L/m^2/min)	×	35.3420	=	gallons/square foot/day (gal/ft^2/day)

Work, Heat, and Energy

British thermal units (Btu)	×	1.0551	=	kilojoules (kJ)
British thermal units (Btu)	×	0.2520	=	kilogram-calories (kg-cal)
foot-pound (force) (ft-lb)	×	1.3558	=	joules (J)
horsepower-hour (hp·h)	×	2.6845	=	megajoules (MJ)
watt-second (W-s)	×	1.000	=	joules (J)
watt-hour (W·h)	×	3.600	=	kilojoules (kJ)
kilowatt-hour (kW·h)	×	3,600	=	kilojoules (kJ)
kilowatt-hour (kW·h)	×	3,600,000	=	joules (J)
British thermal units per pound (Btu/lb)	×	0.5555	=	kilogram-calories per kilogram (kg/cal/kg)
British thermal units per cubic foot (Btu/ft^3)	×	8.8987	=	kilogram-calories/cubic meter (kg-cal/m^3)
kilojoule (kJ)	×	0.9478	=	British thermal units (Btu)
kilojoule (kJ)	×	0.00027778	=	kilowatt-hours (kW·h)
kilojoule (kJ)	×	0.2778	=	watt-hours (W·h)
joule (J)	×	0.7376	=	foot-pounds (ft-lb)
joule (J)	×	1.0000	=	watt-seconds (W-s)
joule (J)	×	0.2399	=	calories (cal)
megajoule (MJ)	×	0.3725	=	horsepower-hour (hp·h)
kilogram-calories (kg-cal)	×	3.9685	=	British thermal units (Btu)
kilogram-calories per kilogram (kg-cal/kg)	×	1.8000	=	British thermal units per pound (Btu/lb)

kilogram-calories per liter (kg-cal/L)	×	112.37	=	British thermal units per cubic foot (Btu/ft^3)
kilogram-calories/cubic meter (kg-cal/m^3)	×	0.1124	=	British thermal units per cubic foot (Btu/ft^3)
Velocity, Acceleration, and Force				
feet per minute (ft/min)	×	18.2880	=	meters per hour (m/h)
feet per hour (ft/h)	×	0.3048	=	meters per hour (m/h)
miles per hour (mph)	×	44.7	=	centimeters per second (cm/s)
miles per hour (mph)	×	26.82	=	meters per minute (m/min)
miles per hour (mph)	×	1.609	=	kilometers per hour (km/h)
feet/second/second (ft/s^2)	×	0.3048	=	meters/second/second (m/s^2)
inches/second/second (in./s^2)	×	0.0254	=	meters/second/second (m/s^2)
pounds force (lbf)	×	4.44482	=	newtons (N)
centimeters/second (cm/s)	×	0.0224	=	miles per hour (mph)
meters/second (m/s)	×	3.2808	=	feet per second (ft/s)
meters/minute (m/min)	×	0.0373	=	miles per hour (mph)
meters per hour (m/h)	×	0.0547	=	feet per minute (ft/min)
meters per hour (m/h)	×	3.2808	=	feet per hour (ft/h)
kilometers/second (km/s)	×	2.2369	=	miles per hour (mph)
kilometers/hour (km/h)	×	0.0103	=	miles per hour (mph)
meters/second/second (m/s^2)	×	3.2808	=	feet/second/second (ft/s^2)
meters/second/second (m/s^2)	×	39.3701	=	inches/second/second (in./s^2)
newtons (N)	×	0.2248	=	pounds force (lbf)

This page intentionally blank.

Index

Note: *f* indicates figure; *t* indicates table

C

D

E

F

G

H

I

M

N

O

P

R

W

This page intentionally blank.

AWWA Manuals

M1, *Principles of Water Rates, Fees, and Charges*, #30001
M2, *Instrumentation and Control*, #30002
M3, *Safety Management for Water Utilities*, #30003
M4, *Water Fluoridation Principles and Practices*, #30004
M5, *Water Utility Management*, #30005
M6, *Water Meters—Selection, Installation, Testing, and Maintenance*, #30006
M7, *Problem Organisms in Water: Identification and Treatment*, #30007
M9, *Concrete Pressure Pipe*, #30009
M11, *Steel Pipe—A Guide for Design and Installation*, #30011
M12, *Simplified Procedures for Water Examination*, #30012
M14, *Backflow Prevention and Cross-Connection Control: Recommended Practices*, #30014
M17, *Fire Hydrants: Installation, Field Testing, and Maintenance*, #30017
M19, *Emergency Planning for Water and Wastewater Utilities*, #30019
M20, *Water Chlorination/Chloramination Practices and Principles*, #30020
M21, *Groundwater*, #30021
M22, *Sizing Water Service Lines and Meters*, #30022
M23, *PVC Pipe—Design and Installation*, #30023
M24, *Planning for the Distribution of Reclaimed Water*, #30024
M25, *Flexible-Membrane Covers and Linings for Potable-Water Reservoirs*, #30025
M27, *External Corrosion Control for Infrastructure Sustainability*, #30027
M28, *Rehabilitation of Water Mains*, #30028
M29, *Water Utility Capital Financing*, #30029
M30, *Precoat Filtration*, #30030
M31, *Distribution System Requirements for Fire Protection*, #30031
M32, *Computer Modeling of Water Distribution Systems*, #30032
M33, *Flowmeters in Water Supply*, #30033
M36, *Water Audits and Loss Control Programs*, #30036
M37, *Operational Control of Coagulation and Filtration Processes*, #30037
M38, *Electrodialysis and Electrodialysis Reversal*, #30038
M41, *Ductile-Iron Pipe and Fittings*, #30041
M42, *Steel Water-Storage Tanks*, #30042
M44, *Distribution Valves: Selection, Installation, Field Testing, and Maintenance*, #30044
M41, *Fiberglass Pipe Design*, #30045
M46, *Reverse Osmosis and Nanofiltration*, #30046
M47, *Capital Project Delivery*, #30047
M48, *Waterborne Pathogens*, #30048
M49, *Quarter-Turn Valves: Head Loss, Torque, and Cavitation Analysis*, #30049
M50, *Water Resources Planning*, #30050
M51, *Air Valves: Air-Release, Air/Vacuum and Combination*, #30051
M52, *Water Conservation Programs—A Planning Manual*, #30052
M53, *Microfiltration and Ultrafiltration Membranes for Drinking Water*, #30053
M54, *Developing Rates for Small Systems*, #30054
M55, *PE Pipe—Design and Installation*, #30055
M56, *Nitrification Prevention and Control in Drinking Water*, #30056
M57, *Algae: Source to Treatment*, #30057
M58, *Internal Corrosion Control in Water Distribution Systems*, #30058
M60, *Drought Preparedness and Response*, #30060
M61, *Desalination of Seawater*, #30061
M62, *Membrane Applications for Water Reuse*, #30062
M63, *Aquifer Storage and Recovery*, #30063
M65, *On-Site Generation of Hypochlorite*, #30065

M66, *Cylinder and Vane Actuators and Controls—Design and Installation,* #30066

M68, *Water Quality in Distribution Systems,* #30068

M69, *Inland Desalination and Concentrate Management,* #30069

M77, *Condition Assessment of Water Mains,* #30077

M80, *Biological Drinking Water Treatment* #30080

www.ingramcontent.com/pod-product-compliance
Ingram Content Group UK Ltd.
Pitfield, Milton Keynes, MK11 3LW, UK
UKHW050728071026
15590UKWH00012B/2256

9 781647 172190